Organic Crop Breeding

Organic Crop Breeding

Editors
EDITH T. LAMMERTS VAN BUEREN
Louis Bolk Institute, Driebergen
and
Wageningen University, Wageningen
The Netherlands

JAMES R. MYERS
Oregon State University
Corvallis, Oregon, United States

A John Wiley & Sons, Ltd., Publication

This edition first published 2012 © 2012 by John Wiley & Sons, Inc.

Wiley-Blackwell is an imprint of John Wiley & Sons, formed by the merger of Wiley's global Scientific, Technical, and Medical business with Blackwell Publishing.

Registered office: John Wiley & Sons Ltd, The Atrium, Southern Gate, Chichester, West Sussex, PO19 8SQ, UK

Editorial offices: 2121 State Avenue, Ames, Iowa 50014-8300, US
The Atrium, Southern Gate, Chichester, West Sussex, PO19 8SQ, UK
9600 Garsington Road, Oxford, OX4 2DQ, UK

For details of our global editorial offices, for customer services, and for information about how to apply for permission to reuse the copyright material in this book please see our website at www.wiley.com/wiley-blackwell.

Authorization to photocopy items for internal or personal use, or the internal or personal use of specific clients, is granted by Blackwell Publishing, provided that the base fee is paid directly to the Copyright Clearance Center, 222 Rosewood Drive, Danvers, MA 01923. For those organizations that have been granted a photocopy license by CCC, a separate system of payments has been arranged. The fee codes for users of the Transactional Reporting Service are ISBN-13: 978-0-4709-5858-2/2012.

Designations used by companies to distinguish their products are often claimed as trademarks. All brand names and product names used in this book are trade names, service marks, trademarks, or registered trademarks of their respective owners. The publisher is not associated with any product or vendor mentioned in this book. This publication is designed to provide accurate and authoritative information in regard to the subject matter covered. It is sold on the understanding that the publisher is not engaged in rendering professional services. If professional advice or other expert assistance is required, the services of a competent professional should be sought.

Library of Congress Cataloging-in-Publication Data

Organic crop breeding / editors, Edith T. Lammerts van Bueren, James R. Myers.
  p. cm.
  Includes bibliographical references and index.
  ISBN 978-0-470-95858-2 (hard cover : alk. paper)   1. Plant breeding.   2. Organic farming.   I. Lammerts van Bueren, E. T. (Edith T.)   II. Myers, James Robert.
  SB123.O74 2012
  631.5′2–dc23
                                                                                  2011036447

A catalog record for this book is available from the British Library.

Wiley also publishes its books in a variety of electronic formats. Some content that appears in print may not be available in electronic books.

Set in 10.5/12 pt Times by Aptara® Inc., New Delhi, India
Printed and bound in Malaysia by Vivar Printing Sdn Bhd

1   2012

Dedication

We dedicate this book to all organic growers whose knowledge, vision, and wisdom has helped us to see the marvellous complexities of organic plant breeding through farmers' eyes.

Contents

Contributors xiii
Foreword xix
William F. Tracy
Preface xxi
Edith T. Lammerts van Bueren and James R. Myers
Acknowledgments xxiii

Section 1   General Topics Related to Organic Plant Breeding   1

Chapter 1   Organic Crop Breeding: Integrating Organic Agricultural Approaches and Traditional and Modern Plant Breeding Methods   3
Edith T. Lammerts van Bueren and James R. Myers

Introduction   3
How Different Are Organic Farming Systems?   4
Consequences for Cultivar Requirements   5
From Cultivar Evaluation to Organic Seed Production and Plant Breeding Programs   6
The History of Organic Crop Breeding in Europe and the United States   8
Perspectives and Challenges for Breeding for Organic Agriculture   11
Conclusion   12
References   12

Chapter 2   Nutrient Management in Organic Farming and Consequences for Direct and Indirect Selection Strategies   15
Monika Messmer, Isabell Hildermann, Kristian Thorup-Kristensen, and Zed Rengel

Introduction   15
Availability of Nutrients in Organic Farming   16
Roots: The Hidden Potential   17
Even Greater Complexity: Plant-Microbe-Soil Interactions   21
Importance of Selection Environments   27
Breeding Strategies   30
References   32

Chapter 3	**Pest and Disease Management in Organic Farming: Implications and Inspirations for Plant Breeding**	39
	Thomas F. Döring, Marco Pautasso, Martin S. Wolfe, and Maria R. Finckh	
	Introduction	39
	Plant Protection in Organic Farming	42
	Key Target Areas of Plant Breeding for Organic Plant Protection	46
	Breeding Goals for Ecological Plant Protection	49
	Plant Breeding Approaches Directly Targeting Pests or Diseases	50
	Plant Breeding Approaches with Indirect Effects on Plant Health	53
	Discussion and Conclusions	54
	References	55
Chapter 4	**Approaches to Breed for Improved Weed Suppression in Organically Grown Cereals**	61
	Steve P. Hoad, Nils-Øve Bertholdsson, Daniel Neuhoff, and Ulrich Köpke	
	Background	61
	Crop Competitiveness Against Weeds	62
	Crop Traits Involved in Weed Suppression	63
	Selection of Traits and Their Evaluation in Plant Breeding Programs	64
	Selection Strategies	68
	Understanding Crop-Weed Interactions to Assist Plant Breeding	70
	Concluding Remarks and Wider Perspectives	71
	References	72
Chapter 5	**Breeding for Genetically Diverse Populations: Variety Mixtures and Evolutionary Populations**	77
	Julie C. Dawson and Isabelle Goldringer	
	Introduction	77
	Benefits of Genetic Diversity for Organic Agriculture	79
	On-Farm Conservation of Useful Genetic Diversity	80
	Breeding Strategies	81
	Conclusion	94
	References	94
Chapter 6	**Centralized or Decentralized Breeding: The Potentials of Participatory Approaches for Low-Input and Organic Agriculture**	99
	Dominique Desclaux, Salvatore Ceccarelli, John Navazio, Micaela Coley, Gilles Trouche, Silvio Aguirre, Eva Weltzien, and Jacques Lançon	
	Introduction	99
	Centralized and Decentralized Breeding: Definitions	100
	What Can Be Decentralized in Breeding and Why?	100
	Participatory Approaches	102
	PPB: A Single Term Yielding Different Approaches	102

	Some Examples of PPB for Organic and Low Input Agriculture in Southern Countries	106
	Some Examples of PPB for Organic and Low Input Agriculture in Northern Countries	113
	General Conclusions and Limits of PPB Approaches in Organic Farming	119
	References	120
Chapter 7	**Values and Principles in Organic Farming and Consequences for Breeding Approaches and Techniques**	**125**
	Klaus P. Wilbois, Brian Baker, Maaike Raaijmakers, and Edith T. Lammerts van Bueren	
	Introduction	125
	Arguments Against Genetic Engineering	126
	Organic Basic Principles	127
	Toward Organic Breeding	130
	From Values to Criteria: Evaluation of Breeding Techniques	131
	How to Deal with Varieties Bred with Non-compliant Techniques?	132
	Toward Appropriate Standards to Promote Organic Plant Breeding	135
	Discussion and Challenges for Organic Plant Breeding	136
	References	136
Chapter 8	**Plant Breeding, Variety Release, and Seed Commercialization: Laws and Policies Applied to the Organic Sector**	**139**
	Véronique Chable, Niels Louwaars, Kristina Hubbard, Brian Baker, and Riccardo Bocci	
	Introduction	139
	The Developments of Plant Breeding and the Emergence of Seed Laws	139
	Variety Registration	142
	Seed Quality Control and Certification	144
	Special Needs for Organic Agriculture	146
	A Recent Development in Europe: Conservation Varieties	148
	Intellectual Property Rights and Plant Breeding	151
	Discussion	154
	Conclusions	156
	Notes	156
	References	157
Section 2	**Organic Plant Breeding in Specific Crops**	**161**
Chapter 9	**Wheat: Breeding for Organic Farming Systems**	**163**
	Matt Arterburn, Kevin Murphy, and Steve S. Jones	
	Introduction	163
	Methods	163
	Traits for Selection in Organic Breeding Programs	168
	A Case Study for EPB: Lexi's Project	170

A Case Study for Breeding within a Supply Chain Approach:
Peter Kunz and Sativa ... 171
Conclusion ... 171
References ... 172

Chapter 10 Maize: Breeding and Field Testing for Organic Farmers ... 175
Walter A. Goldstein, Walter Schmidt, Henriette Burger, Monika Messmer, Linda M. Pollak, Margaret E. Smith, Major M. Goodman, Frank J. Kutka, and Richard C. Pratt

Introduction ... 175
What Kind of Maize do Organic Farmers Want? ... 175
Are There Viable Alternatives to Single Cross Hybrids? ... 176
Testing and Using Alternative Hybrids ... 178
Are There Benefits for Breeding under Organic Conditions? ... 178
For Which Traits Is It Necessary to Test under Organic Conditions? ... 179
Choice of Parents for Breeding Programs ... 181
Breeding Programs ... 182
Future Directions ... 186
Notes ... 186
References ... 188

Chapter 11 Rice: Crop Breeding Using Farmer-Led Participatory Plant Breeding ... 191
Charito P. Medina

Introduction ... 191
MASIPAG and Participatory Rice Breeding ... 192
Beyond PPB: Farmer-Led Rice Breeding ... 193
The Breeding Process ... 194
Outcomes of the MASIPAG Program ... 198
Outlook ... 200
References ... 201

Chapter 12 Soybean: Breeding for Organic Farming Systems ... 203
Johann Vollmann and Michelle Menken

Introduction ... 203
Agronomic Characters ... 204
Seed Quality Features ... 208
Considerations on Breeding Methods ... 211
References ... 212

Chapter 13 Faba Bean: Breeding for Organic Farming Systems ... 215
Wolfgang Link and Lamiae Ghaouti

Purposes of Breeding and Growing Faba Bean ... 215
Genetic and Botanical Basics of Breeding Faba Bean ... 216
Methodological Considerations ... 218
Traits To Be Improved in Faba Bean Breeding ... 221

	Open Questions, Need for Action	223
	References	224
Chapter 14	**Potato: Perspectives to Breed for an Organic Crop Ideotype**	**227**
	Marjolein Tiemens-Hulscher, Edith T. Lammerts van Bueren, and Ronald C.B. Hutten	
	Introduction	227
	Required Cultivar Characteristics	228
	Introgression Breeding and Applied Techniques	232
	Participatory Approach: An Example from the Netherlands	233
	Outlook	234
	References	234
Chapter 15	**Tomato: Breeding for Improved Disease Resistance in Fresh Market and Home Garden Varieties**	**239**
	Bernd Horneburg and James R. Myers	
	Introduction	239
	Botanical and Genetic Characteristics of Tomato	240
	Rationale for Breeding Tomatoes within Organic Systems	240
	Breeding Needs with Focus on Organic Production	243
	Case Studies: Breeding for Late Blight Resistance in Europe and North America	245
	Outlook	247
	References	248
Chapter 16	**Brassicas: Breeding Cole Crops for Organic Agriculture**	**251**
	James R. Myers, Laurie McKenzie, and Roeland E. Voorrips	
	Introduction	251
	Rationale for Breeding within Organic Systems	251
	Plant Biology	252
	Traits Needed for Adaptation to Organic Production	253
	Consideration of Breeding Methods	257
	A Farmer Participatory Broccoli Breeding Program	258
	Outlook	260
	References	261
Chapter 17	**Onions: Breeding Onions for Low-Input and Organic Agriculture**	**263**
	Olga E. Scholten and Thomas W. Kuyper	
	Introduction	263
	Robust Onion Cultivars	264
	Breeding for Improved Nutrient Acquisition	265
	Mycorrhizal Symbiosis and Product Quality	269
	Conclusion	270
	References	271
Index		**273**

Contributors

Silvio Aguirre
CIPRES
Pueblo Nuevo, Nicaragua

Matt Arterburn
Department of Biology
Washburn University
Topeka, KS, USA

Brian Baker
The Organic Center, P.O. Box 20513, Boulder, CO 80308, USA; formely with Alfred State College Sustainability Institute
Alfred, NY, USA

Nils-Øve Bertholdsson
Swedish University of Agricultural Sciences
Department of Plant Breeding and Biotechnology
Alnarp, Sweden

Riccardo Bocci
Italian Association for Organic Agriculture (AIAB)
Roma, Italy

Henriette Burger
KWS SAAT AG
Einbeck, Germany

Salvatore Ceccarelli
ICARDA
Aleppo, Syria

Véronique Chable
Institut National de la Recherche Agronomique (INRA)
Sciences for Action and Development (SAD-Paysage)
Rennes, France

Micaela Colley
Organic Seed Alliance
Port Townsend, WA, USA

Julie C. Dawson
Department of Plant Breeding and Genetics, Cornell University, Ithaca, NY 14853, USA; formerly with Institut National de la Recherche Agronomique (INRA)
Gif sur Yvette, France

Dominique Desclaux
INRA
Mauguio, France

Thomas F. Döring	The Organic Research Centre, Elm Farm Hamstead Marshall, UK
Maria R. Finckh	Ecological Plant Protection Group Faculty of Organic Agricultural Sciences University of Kassel Witzenhausen, Germany
Lamiae Ghaouti	Department of protection, production and vegetal biotechnology, Institut Agronomique et Vétérinaire Hassan II, B.P. 6202 Rabat-Instituts Madinat Al Irfan C.P 10101, Morocco; formerly with Georg-August-Universität Crop Science Department Göttingen, Germany
Isabelle Goldringer	Institut National de la Recherche Agronomique (INRA) Gif sur Yvette, France
Walter A. Goldstein	Mandaamin Institute, Elkhorn, Wisconsin; formerly with Crops and Soils Research Michael Fields Agricultural Institute East Troy, WI, USA
Major M. Goodman	Department of Crop Science North Carolina State University Raleigh, North Carolina, USA
Isabell Hildermann	Research Institute of Organic Agriculture (FiBL) Ackerstrasse Frick, Switzerland
Steve P. Hoad	Scottish Agricultural College West Mains Road Edinburgh, UK
Bernd Horneburg	Department of Crop Sciences Georg-August-University of Göttingen Göttingen, Germany
Kristina Hubbard	Organic Seed Alliance Port Townsend, WA, USA
Ronald C.B. Hutten	Wageningen UR Plant Breeding Wageningen University and Research Centre Wageningen, The Netherlands
Steve S. Jones	Washington State University Research and Extension Center Mt. Vernon, WA, USA
Ulrich Köpke	Institute of Organic Agriculture University of Bonn Bonn, Germany

Frank J. Kutka	Seed We Need Project Dickinson, ND, USA
Thomas W. Kuyper	Department of Soil Quality Wageningen University Wageningen, The Netherlands
Edith T. Lammerts van Bueren	Louis Bolk Institute Driebergen, The Netherlands; and Wageningen UR Plant Breeding Wageningen University and Research Centre Wageningen, The Netherlands
Jacques Lançon	Cirad c/o ICRAF Nairobi, Kenya
Wolfgang Link	Georg-August-Universität Göttingen Crop Science Department 37075, Göttingen, Germany
Niels Louwaars	Plantum, P.O. Box 462, 2800 AL Gouda, The Netherlands; formerly with Centre for Genetic Resources The Netherlands (CGN) Wageningen University and Research Centre Wageningen, The Netherlands
Laurie McKenzie	Department of Horticulture Oregon State University Corvallis, OR, USA
Charito P. Medina	MASIPAG Laguna, Philippines
Michelle Menken	Minnesota Crop Improvement Association St. Paul, MN, USA Department of Agronomy and Plant Genetics University of Minnesota St. Paul, MN, USA
Monika Messmer	Research Institute of Organic Agriculture (FiBL) Ackerstrasse Frick, Switzerland
Kevin Murphy	Department of Crop and Soil Sciences Washington State University Pullman, WA, USA
James R. Myers	Department of Horticulture Oregon State University Corvallis, OR, USA
John Navazio	Organic Seed Alliance Port Townsend, WA, USA

Daniel Neuhoff	Institute of Organic Agriculture University of Bonn Bonn, Germany
Marco Pautasso	Center for Synthesis and Analysis on Biodiversity, French Foundation for Research on Biodiversity. Montpellier Cedex, France; formerly with Conservation Science Group Department of Zoology Cambridge University Cambridge, UK
Linda M. Pollak	Guthrie Center, IA, USA Formerly with USDA-ARS Corn and Soybean Unit Iowa State University Ames, Iowa, USA
Richard C. Pratt	Department of Horticulture and Crop Science Ohio Agricultural Research and Development Center The Ohio State University Wooster, OH, USA
Maaike Raaijmakers	Bionext, Laan van Vollenhove 3221, 3706 AR Zeist, The Netherlands
Zed Rengel	Soil Science and Plant Nutrition School of Earth and Environment The University of Western Australia Crawley, WA, Australia
Walter Schmidt	KWS SAAT AG Einbeck, Germany
Olga E. Scholten	Wageningen UR Plant Breeding Wageningen University and Research Centre Wageningen, The Netherlands
Margaret E. Smith	Department of Plant Breeding and Genetics Cornell University Ithaca, NY, USA
Kristian Thorup-Kristensen	Department of Agriculture and Ecology, University of Copenhagen Taastrup, Denmark
Marjolein Tiemens-Hulscher	Louis Bolk Institute Driebergen, The Netherlands
Gilles Trouche	CIRAD, UMR AGAP Avenue Agropolis Montpellier, France

Johann Vollmann	BOKU – University of Natural Resources and Life Sciences Vienna Department of Crop Sciences Vienna, Konrad Lorenz Str. 24, 3430 Tulln, Austria
Roeland E. Voorrips	Wageningen UR Plant Breeding Wageningen University and Research Centre Wageningen, The Netherlands
Eva Weltzien	International Crops Research Institute for the Semi-Arid Tropics (ICRISAT) Bamako, Mali
Klaus P. Wilbois	FiBL Deutschland e.V. Research Institute for Organic Agriculture Frankfurt am Main, Germany
Martin S. Wolfe	The Organic Research Centre, Elm Farm Hamstead Marshall, UK

Foreword

This book is a long overdue addition to the plant breeding literature. Until now, what has been written on the subject of breeding for organic production systems has been scattered among many journals and conference proceedings. These other sources seldom, if ever, combine both philosophy and practical experience, and often the older sources have focused on one species or a group of allied species. This volume overcomes both weaknesses. It is rich in both practical knowledge and philosophy of breeding plants for organic systems. Field and vegetable crops and self-pollinated and cross-pollinated species are thoroughly discussed.

While many will seek out this volume in their desire to become practitioners of breeding for organic agriculture, all those interested in plant breeding should read it to understand the theory and philosophy as well as the legal and structural aspects that underlie breeding for organic systems. Undoubtedly some readers will question some ideas, but many will be exposed to new perspectives. And this is precisely why this book is so important.

Increasingly, breeding for organic agriculture is gaining attention in the public and private sectors in both Europe and in the United States. This has come about through the realization that organic production systems usually represent a different environment from conventional systems, and due to potential genotype $\times$ system interactions, varieties not adapted to a specific production system may not be the best performers in that system. Many of the methods used in organic plant breeding have been developed in poorer countries under low-input conditions without a seed production infrastructure. The sharing of innovative methods among these different areas has opened an exciting time in the breeding for organic and sustainable systems.

The editors have organized a thoughtful review on the topic of organic plant breeding and tapped as authors the leaders in the field. The book is split into two sections: Section 1, *General topics related to organic plant breeding* and Section 2 *Organic plant breeding in specific crops*. The first section will be of interest to all those interested in plant breeding. After an introductory chapter, the next three chapters cover the unique environmental challenges presented by organic systems and various ways plant breeders have risen to meet those challenges. The next four chapters are perhaps the most provocative of the book and are sure to generate discussion among students of plant breeding. Chapters 5 and 6 deal with technical challenges that should be of interest to all breeders and those are the development of genetically diverse cultivars that can respond to evolutionary pressures and the dichotomy of centralized versus decentralized research and extension systems. Chapter 7 tackles values and principles in organic breeding and will be eye-opening to many. Chapter 8 discusses intellectual property (IP), germplasm ownership, and commercialization issues that constrain organic plant breeding. Given the much smaller and decentralized nature of organic plant breeding efforts the IP and distribution mechanisms that have been developed for large commercial ventures don't fit, often constrain, and sometimes eliminate the ability of organic plant breeding to be successful.

Section 2 with nine chapters on organic breeding of specific crops will be a valuable resource to breeders of these and other allied crops.

The editors hailing from Europe and the United States represent the two regions where organic plant breeding has shown the greatest advances. They have assembled a team of authors that reflects this diversity, and in the process of writing these chapters have fostered cross pollination among the regions. The result is a cosmopolitan approach to the subject.

The field is rather new, and we are seeing an exponential increase in the literature as the first round of studies is completed. This book establishes the baseline for a growing discipline within plant breeding and is designed to contribute to furthering knowledge and innovation in the field of organic plant breeding.

William F. Tracy
Madison, Wisconsin

Preface

Organic crop breeding is a relatively young and growing discipline described as either plant breeding for organic agriculture or plant breeding under organic conditions, acknowledging the organic principles that underlie this field. Both approaches are captured in this book as "organic crop breeding."

Research into organic plant breeding has, to a large extent, been focused on cereal crops, but for many other crops this field is still in its infancy. This book brings some balance, in that it includes several crop-specific book chapters, of which cereals are part, but other groups representing legumes, cole crops and various vegetable crops are also covered.

While it is well established that organic production systems represent different growing environments compared to conventional systems, and that genotype x production system interaction is an important consideration, we do not have adequate knowledge of what components of an organic system are most important and what varietal traits allow optimal adaptation to such a system. The first section of the book provides some illumination on this subject in its coverage of general topics of organic plant breeding.

This book will be able to be read independently by those with some background in plant breeding; however, in general, this book will not discuss basic breeding principles, but will focus on the specific issues that are of importance when considering breeding for the organic sector. The book is divided into two sections covering (1) general topics related to organic crop breeding and (2) crop-specific topics. The general topics section discusses the basic organic principles and their consequences for plant breeding, and reviews the state-of-the-art of current breeding research.

Chapter 1 is an introductory chapter showing how organic breeding is a cross between organic agricultural approaches and traditional and modern plant breeding techniques and discusses the basic differences between organic, conventional high- and low-input agriculture, and the history of organic plant breeding in the United States and Europe.

Chapter 2 discusses the consequences of nutrient management in organic farming systems for crop improvement and includes complex issues such as breeding for nutrient efficiency and root system improvement.

Chapter 3 focuses on the fact that organic farming refrains from chemical pest management. Although resistance breeding is a familiar topic to most breeders, this chapter describes where and in which way organic management can reduce pressure of pest and diseases, where crop breeding needs to contribute, and in which ways innovative approaches should be further explored.

Chapter 4 deals with a relatively new topic for breeders on approaches to breed for improved weed suppression – departing from the experience in organically grown cereals, but with application to other crops. Chapter 5 addresses the central issue of biodiversity in organic agriculture. Biodiversity is considered a vital tool in creating higher levels of resilience in farming systems. The chapter describes strategies to exploit genetic diversity using cultivar mixtures and evolutionary breeding approaches.

Chapter 6 discusses the potentials of participatory approaches for low-input and organic agriculture in Western contexts, and arises from the fact that organic plant breeding is an economic niche for the commercial breeding industry that can profit from the experience of organic growers.

Chapter 7 describes the "Values and principles in organic farming and consequences for breeding approaches and techniques." It underscores the importance of understanding the origins of organic agriculture, its world-view, and the rationale for the rejection of certain modern plant breeding techniques. Overall, it provides a framework in which organic plant breeding can develop. Chapter 8 discusses the consequences of the current "Laws and policies that govern plant breeding and seed supply" and recommends modifications that support the emerging organic breeding sector.

The second half of the book provides applied examples of the general approaches discussed in chapters 1 through 8 in specific crops. Research groups and breeders were invited to describe the most relevant traits for their crop species, including their experience in breeding for organic farming systems and perspectives on traits required for better adapted cultivars. For this book, crops were selected based on whether experience in breeding for organic systems was available. The crops included: Wheat, maize, rice, soybean, faba bean, potato, tomato, crucifers, and onion. They each give either a general review of the state-of-the-art and breeding perspectives, and/or emphasize a specific breeding approach, and included some examples of current organic breeding programs.

With this compilation, we seek to provide subject matter of interest to students, researchers, and other professionals from universities and institutes related to breeding research and those in plant breeding and seed companies. It is our hope and expectation that this book will be relevant to organic and conventional agriculture alike, and that it will facilitate the search for more sustainable farming systems for the twenty-first century and beyond.

Edith T. Lammerts van Bueren
James R. Myers

Acknowledgments

The idea for this book came from Justin Jeffryes, executive editor at Wiley-Blackwell. The editors of this book gratefully took up his invitation and are thankful for his valuable and expert assistance during preparation of the book.

This book was an opportunity to bring the European and U.S. research and breeding networks together, as well as include some authors from the southern latitudes. Most of the chapters have mixed U.S. and European authorship, and the authors have worked with great inspiration, together sharing knowledge and experience. The editors are most grateful to the various author groups, because without their commitment, this book would not have achieved a high level of quality!

We would like to individually thank those who patiently assisted us in the finalization of the book. Laurie McKenzie read and provided feedback on a number of chapters as they came in, and Berend-Jan Dobma assisted with the exacting process of finally formatting and assembly of the chapters.

Finally, we send thanks to our families for their patience and support as we burned the midnight oil in bringing this task to fruition.

Section 1
General Topics Related to Organic Plant Breeding

1 Organic Crop Breeding: Integrating Organic Agricultural Approaches and Traditional and Modern Plant Breeding Methods

Edith T. Lammerts van Bueren and James R. Myers

Introduction

Organic agriculture is continuously growing worldwide on land and farms in more than 160 countries as well as in the global marketplace (Willer and Kilcher, 2011). Globally, there are 37.2 million hectares of organic agricultural land (including in-conversion or transition hectarage), which is about 0.9% of all arable lands. Of the total organic area in 2009, most (24.9%) is in Europe, followed by Latin America (23.0%), Asia (9.6%), North America (7.1%), and Africa (2.8%). Some individual countries (mainly those in Europe) had higher percentages due to support by national policies, e.g., Austria (18.5%), Sweden (12.6%), and Italy (9.0%; Willer and Kilcher, 2011).

Organic agriculture has its origins in the early 1900s with individuals advocating that "living soil" was a fundamental value of sound agriculture (Balfour, 1943; Howard, 1940; Pfeiffer, 1947; Steiner, 1958; Rodale, 1961). It was not until the 1970s that the organic movement grew substantially, however. Growth of the movement coincided with consumers' and farmers' reactions against the unsustainable environmental impact of the agriculture of that time. In the 1990s, organic agriculture became large enough to attract the interest of major food suppliers. In 2008–2009 organic products occupied about 5% of the market and were worth 55 billion US dollars, or 40 billion euros (Willer and Kilcher, 2011). To date, increasing development in the organic sector is influenced by three main drivers: *Values* (see four basic principles of the International Federation for Organic Agricultural Movements in Chapter 7), *protest* (promoting organic agriculture as an alternative strategy) and *market* (an economic interesting niche market). Alrøe and Noe, 2008.

Regulations translating the values and principles of the organic sector into rules and standards (IFOAM, 2005; Luttikholt, 2007) have been harmonized to promote global trade. The four basic principles of the organic movements as described by the world umbrella organization IFOAM, include (a) *the principle of health:* Expressing the concept of wholeness and integrity of living systems and supporting their immunity, resilience, and sustainability; (b) *the principle of ecology:* Promoting diversity in site-specific ecological production systems; (c) *the principle of fairness:* Serving equity, respect, justice, and stewardship of the shared world; and (d) *the principle*

Organic Crop Breeding, First Edition. Edited by Edith T. Lammerts van Bueren and James R. Myers.
© 2012 John Wiley & Sons, Inc. Published 2012 by John Wiley & Sons, Inc.

of care: Enhancing efficiency and productivity in a precautionary and responsible way (IFOAM, 2005; Luttikholt, 2007).

These principles have been codified in governmental regulations such as the National Organic Program (NOP) in the United States (USDA, 2002) and in Europe by the European Commission (EC, 2007).

It was only in the early 1990s that crop breeding and seed production came to the fore as an issue for organic growers and consumers in response to the emerging field of genetic engineering (GE) and strengthening of intellectual property rights. The organic sector began to discuss ways to actively stimulate crop improvement to meet organic principles.

In this chapter we will describe how organic management differs from conventional agricultural management, what plant traits are required for optimal adaptation to organic farming systems, and ways to acquire such adaptation via cultivar selection, seed production, and breeding. We also summarize the history and future perspectives for organic crop breeding in the United States and Europe.

How Different Are Organic Farming Systems?

When the U.S. National Organic Standards Board convened to advise the USDA on developing organic regulations, they described organic agriculture as:

> "... an ecological production management system that promotes and enhances biodiversity, biological cycles, and soil biological activity. It is based on minimal use of off-farm inputs and on management practices that restore, maintain, and enhance ecological harmony" (USDA, 2002).

Organic farming is more than merely replacing chemical pesticides and fertilizers with organic ones. Emanating from the principles of health and ecology, the aim has been to move away from curative measures and to amplify agro-ecological system resilience by developing preventative strategies at the system level (e.g., Kristiansen et al., 2006; see table 1.1). The goal is to stimulate a high level of internal system self-regulation through functional diversity in and above the soil, as opposed to depending on external inputs for regulation (Østergård et al., 2009).

In considering differences among current farming systems in Western societies (e.g., conventional with high-external inputs systems, conventional systems reducing external inputs to become more sustainable, and organic farming systems) organic farming systems are the most extreme of the three types in refraining from chemical-synthetic inputs and in using preventative rather than curative measures. Although conventional low-external input farming seeking sustainability can be considered an intermediate between high-external input farming and organic farming, there is still a critical difference. It aims to reduce the input levels through precision farming methods and integrated pest management but still relies on chemical inputs to quickly correct during crop growth. In contrast, organic farming systems that cannot (easily) "escape" by applying curative methods rely on indirect, long-term strategies of fostering systems resilience. Organic farming systems focus on soil building through increasing organic matter, which increases water holding capacity and buffers against perturbations to the system. Such systems generally lack short-term controls (e.g., by applying mineral fertilizers with ready water-soluble nutrients or pesticides) to modify the growing environment during the season. Because organic farmers have fewer means to mitigate environmental variation, the varieties grown in organic agriculture will exhibit larger genotype by

Table 1.1 Overview of the main difference in crop management between conventional agriculture, sustainable low-external input farming systems and organic farming systems

Category	Conventional Farming System	Sustainable, Low-External Input Farming Systems	Organic Farming Systems
Biodiversity	Not a specific issue	Attention given to natural predators	Biodiversity is both product of and tool for maintaining a resilient farming system. Including diversity in (beneficial) soil organisms, crop species, and varietal diversity in space and time
Fertilization	Application of high inputs of mineral nutrients, including hydroponics systems; aiming at maximum crop growth with readily soluble nutrients	Application of reduced levels of mineral fertilizers, precision fertilization, including use of green manures; aiming at optimal crop growth with reduced loss of nutrients by leaching	Application of reduced levels of organic fertilizers (animal manure, compost and green manures); slow release of nutrients; aiming at long-term soil fertility and high biological soil activity
Rotation	1:1 to 1:3	1:1 to 1:4	1:6 to 1:10
Crop protection	Highly dependent on synthetic-chemical crop protectants	IPM approach to crop protection. Scouting, and trapping; reduced application of synthetic-chemicals added with biological agents	Access to only organically approved inputs
Weed management	Herbicides	Reduced herbicide use and mechanical weeding	Mechanical weeding combined with crop rotation design, or no tillage systems including mulching; stale seedbed, and crop competition
Seed treatment and sprouting inhibitors	Chemical	Chemical and physical	Physical (hot water or steam) and organic additives, e.g., mustard powder
Tillage	Increasing use of no-till in major field crops	Application of minimum or no-till	Reliance on tillage but seeking to apply minimum- or no-till systems to field crops

environment interactions, with greater emphasis placed on cultivar traits that allow adaptation to variable growing conditions (Lammerts van Bueren et al., 2007).

Another important difference among the aforementioned farming systems is that the main source of nitrogen (N) in organic farming systems is mineralization of organic matter, making N availability less controllable (Mäder et al., 2002). Under low temperatures in spring, soil microbiota that mineralize organic matter are not active enough to provide sufficient N, causing crop growth to lag and allowing weeds to compete. This requires cultivars that can cope with early season low fertility and produce vigorous growth to cover the soil as early as possible.

Consequences for Cultivar Requirements

A conclusion drawn from the description in the previous section is that conventional agriculture has more external means to adapt the environment to optimal crop growth, whereas organic farming systems need cultivars to adapt to the given environment. Crops bred for conventional production may be adapted to a narrower range of environmental conditions, especially those controlled by the external inputs of the grower. Therefore, cultivar selection is more critical for organic than for conventional farmers. The emphasis is on choosing flexible, robust cultivars that are adapted to such farming systems and that possess yield stability and can compensate for unfavorable conditions.

Organic growers have largely depended on cultivars bred for conventional systems, but not all are optimal for organic farming systems because traits associated with independence from external inputs have not received high priority in current breeding programs.

Traits

A focus on breeding for organic agriculture would require a shift from emphasis on maximizing the yield level in combination with the use of "crop protectants" to an emphasis on optimal yield stability. One of the main characteristics of organic farming is a multilevel approach to increasing system stability to reduce risk of failures. A similar approach could apply to cultivar development to adapt to less controllable and unfavorable growing conditions (see table 1.2). The aim would then not only be adaptation to low nutrient levels supported by improved interaction with beneficial soil mycorrhizae, but also morphological and phytochemical traits that reduce disease susceptibility (wax layers in *Brassica* species, open plant architecture), enhance weed competition (early vigor and planophile growth habit), and increase in flavonoids and glucosinolates (pest feeding deterrents; Stamp, 2003; Züst et al., 2011).

From Cultivar Evaluation to Organic Seed Production and Plant Breeding Programs

Just as conventional colleagues do, organic farmers are always looking for the best cultivars to meet their needs. As described above, cultivar choice is a valuable tool of organic farmers to increase system and yield stability. Many research projects have emphasized farmer participatory trials to evaluate current cultivars to select the best performing cultivars under organic growing conditions. The next step in evolution to an organically based breeding program has been to produce organic seed of the most suitable conventionally bred cultivars. The subsequent step has been to identify "ecological" traits that should be included in current breeding programs. Often breeders interested in breeding for low-input or organic farming have also found that the protocols for cultivar testing need to be adapted to allow appropriate cultivars to enter the market (e.g., as in Europe; Löschenberger et al., 2008; Rey et al., 2008; see Chapter 8).

The final step in program evolution has been to develop appropriate cultivars through breeding programs that are conducted under organic conditions. Table 1.3 shows an overview of such steps that currently coexist in the market. These steps represent a continuum that, depending on the goals of the breeding program, may fall somewhere in between. For example, rather than maintaining two distinct programs for conventional and organic, some private companies do their early breeding in conventionally managed environments, then test later generations in both conventional and organic trials (Löschenberger et al., 2008).

Organic Seed Production

Crucial to engaging breeding companies to in breeding better adapted cultivars is stimulating organic seed production of the best performing cultivars. Even before organic breeding became an issue,

Table 1.2 Differences in plant ideotype between high input conventional and low-input organic cropping systems

Conventional	Organic
Above-ground traits	
Performs well at high population density	Optimal performance at lower densities
Increased harvest index	Increased harvest index, but not as dramatic as for conventional production
Erect architecture and leaves, shortened plant stature	Taller plants, spreading canopy to be productive in low input situations
Weeds controlled by herbicides	Weeds limited by competition (plant height, spreading architecture), plants tolerate cultivation
Yield is maximized with high level of inputs	Maximized sustainable yield achievable with input of nutrients from organic sources
Pest and disease resistance to specific complex of organisms; need for resistance to diseases of monoculture systems	Pests and pathogens of monoculture potentially less severe, pathogen and pest complex differ; induced resistance relatively important; secondary plant compounds important for pathogen and pest defense
Rhizosphere traits	
Root architecture unknown	Exploratory root architecture; able to penetrate to lower soil horizons
Adapted to nutrients in readily available form	Adapted to nutrients from mineralization – not readily available; need for nutrient use efficiency; responsive to mycorrhizae
Legume-specific traits	
Nitrogen production by rhizobia of lesser importance	Rhizobia more important; discrimination against infective rhizobia important for N acquisition
Harvest and marketing traits	
Improved labor efficiency	Incorporate traits that improve working conditions
Improved processing, packing, and shipping efficiency	Improved nutrition, taste, aroma, and texture
Crop shaped by mechanical harvest constraints	Traits priorities set jointly by researcher and farmer

Table 1.3 Time schedule to develop organic seed production and plant breeding

Time	Activity	Product
Current	Selection of best-performing conventionally bred cultivars; No use of GM cultivars	Conventionally bred cultivars
	No chemical post-harvest seed treatments	Conventionally produced, untreated seeds
Short-term	Organic seed production of the best suitable conventionally bred cultivars	Conventionally bred cultivars
	Organic seed treatments	Organic seed production
Mid-term	Including "ecological" plant traits in conventional breeding programs	Low-input cultivars
	Adapted protocols for organic variety trialing (e.g., VCU) to allow adapted cultivars to pass testing thresholds	Organic seed production
Long-term	Whole breeding cycle under organic conditions	Organically bred cultivars
	Including the concept of integrity of plants	Organic seed production

organic seed production (free of pre- and post-harvest chemicals) had already started to develop on a small scale in the 1970s. This was mainly driven by small enterprises concerned about lost genetic diversity that had once preserved older heirloom and regional varieties. With more stringent rules on the use of organic seed incorporated into organic regulations in the United States and the Europe Union, the conventional seed industry became interested in serving this market. As a result, the availability and use of certified organic seed has increased and seed businesses have matured. Both small organic enterprises and larger commercial companies (who have traditionally only serviced conventional markets) are dealing with organic seed production of both horticultural and field crops.

At present, one can generally distinguish four types of organic seed production businesses:

- Fully organic and independent seed companies;
- More or less independent daughter seed companies linked through formal partnerships to conventional seed companies;
- An integrated part of a conventional seed companies;
- Conventional seed companies that chooses not to produce organic seed.

Confronted with a limited assortment of suitable cultivars, organic growers have become more aware of the need for greater cultivar choice with a greater diversity in cultivar types (e.g., open-pollinated, F_1 hybrids, or variety mixtures). In this context, organic crop breeding has become an emergent sector in business and science.

The History of Organic Crop Breeding in Europe and the United States

Breeding activities for organic agriculture in Europe and the United States have had distinctive historical trajectories that are products of different laws and policies concerning seeds and plant breeding.

Europe

In Europe, plant breeding within the organic sector started in the 1950s on a small scale conducted mostly by biodynamic farmers considering selection as part of their farming system. They felt that it was imperative to allow cultivars to co-evolve over time into more resilient farm organisms (e.g., Kunz and Karutz, 1991). In that context Martin and Georg Schmidt developed ("regenerated") winter rye through the ear-bed method by sowing the kernels according the position in the ears and developing a procedure for selection resulting in a very tall (2 m) winter rye cultivar named 'Schmidt Roggen' (Wistinghausen, 1967). In the 1980s a biodynamic working group of farmer breeders and specialized breeders/researchers met in Dornach, Switzerland, to discuss several research methodologies. In 1985 a group was formed in Germany ("Initiative Kreiz") that led to the founding of Kultursaat in 1994, which is an association for biodynamic breeding research and maintenance of cultivated species (see www.kultursaat.org; Fleck, 2009). This group now consists of approximately 40 breeders and farmer-breeders, each improving one or more crop species in the context of a biodynamic farm. They currently have registered about 40 new vegetable cultivars. Kultursaat considers breeding to be a public activity and the group is supported financially by donors and seed funds. The seed production of these new cultivars and older open-pollinated and heirloom varieties is organized by the company Bingenheimer Saatgut AG in Germany. In addition to vegetable breeders, several biodynamic cereal breeders began programs in the mid-1980s in Germany and Switzerland (e.g., Kunz and Karutz, 1991; Müller et al., 2000).

Toward the end of the 1990s, broader support emerged to address gaps in improving cultivars for organic farming systems, and breeding research was initiated by organic research institutes and universities in Europe, which were funded on a project basis by national governments and as European cross-country consortia: For example, a large-scale collaborative breeding project SOLIBAM from 2010–2015 for several crops (www.solibam.eu).

To stimulate knowledge exchange on breeding for organic agriculture, a European Consortium for Organic Plant Breeding (ECO-PB) was founded in 2001. Among other functions, it has organized several conferences (see www.eco-pb.org) to provide a venue for information exchange. ECO-PB acts as an umbrella organization that supports harmonization of the national organic seed regimes by organizing roundtable and workshop meetings on the issue (e.g., Lammerts van Bueren and Wilbois, 2008).

Increasing numbers of conventional breeders became interested in breeding for the organic sector. Not only was there a need to serve this growing market, but they also saw it as an investment in breeding for the future, as conventional agriculture moves toward increasing sustainability. During the Eucarpia Conference on Breeding for Organic and Low-input Agriculture in 2007 in Wageningen, the European Association of Plant Breeders and Researchers (Eucarpia) founded the Section "Organic and Low-input Agriculture" (www.eucarpia.org). To support publishing of peer-reviewed results, a special issue of Euphytica was published in 2008 (Lammerts van Bueren et al., 2008). Also the Proceedings of the Second Conference of the Eucarpia Section Organic and Low-input Agriculture revealed many research projects aimed at improved selection methods or strategies for organic plant breeding (Goldringer et al., 2010). In addition to the fact that several European universities have various research programs set up to develop methods to obtain adapted varieties, Wageningen University in the Netherlands initiated an endowed chair specialized in Organic Plant Breeding in 2005. Kassel University in Germany established a full-time chair for Organic Plant Breeding and Biodiversity in 2011.

United States

Origins of organic plant breeding in the United States are not well documented. The emergence of organic organizations came about in the 1970s, and around the same time, seed companies and nonprofit organizations (NGOs) with an interest in seeds arose (Dillon and Hubbard, 2011).

While little is known about the varieties in use by the early practitioners of organic agriculture in the United States, most were almost certainly non-hybrid, open-pollinated (OP) heirloom varieties. The early seed companies were selling predominantly OP varieties to organic growers, and NGOs such as Abundant Life and Seed Saver's Exchange were focused primarily on preserving heirlooms as a counter to the loss of biodiversity that was beginning in the formal seed sector. The use of OPs and heirlooms was tied to environmental, economic, and social sustainability values and was a reaction against what organic practitioners saw happening in the conventional seed industry. The ability to save OP seed fit well with the "back to the land" movement of the 1970s, which embraced organic agriculture and had a strong belief in self-reliance. With the exception of field corn, where most varieties offered commercially are F_1 hybrid, catalogs selling organic seed carry proportionally more OPs than F_1 hybrids (Dillon and Hubbard, 2011).

The Organic Food Production Act passed in the 1990 farm bill gave the federal government the authority to craft a national organic standard. This resulted in the publication of the National Organic Program (NOP) in 2000, and after receiving feedback from stakeholders, the program was implemented by USDA in 2002. It was apparent from the beginning that access to certified organic seed was a limitation, as apparent in the NOP advisory on certified organic seed which states "The producer must use organically grown seeds...except...non-organically produced, untreated seeds and planting stock may be used to produce an organic crop when an equivalent organically produced variety is not commercially available" (USDA, 2002). This exception was instituted because certifying agencies and regulators recognized that the organic seed market was not large enough to supply the needs of the organic sector, and that many organic growers relied on conventional sources for their seed needs. Over time, this loophole shrunk. In 2010, certification inspectors were requiring that a grower check at least three seed sources to determine if their desired variety or equivalent was available as certified organic seed. If not, then the grower was allowed to use untreated conventionally grown seed. Much of the recent surge in plant breeding and trialing activities have sought to increase the portfolio of varieties available to growers in organic form.

Organic plant breeding activities by the private sector are not well documented, but probably began in the 1970s or 1980s on the part of companies that were selling organic seed. Organic plant breeding in the public sector was formalized in the mid-1990s with funding from federal grant programs such as the USDA Sustainable Agriculture Research and Education program (SARE), federal Risk Management Agency (RMA), USDA Value Added Producer Grants program (VAPG), and USDA Integrated Organic Program (IOP), which later became the Organic Research and Education Initiative (OREI). NGO organizations funding organic research included the Organic Farming Research Foundation (OFRF) and the Farmers Advocating for Organics fund (FAFO). At least 57 projects were funded through these venues from the mid-1990s to 2010 (Dillon and Hubbard, 2011). Funding has been distributed to both field and vegetable crops with the majority going to wheat and vegetables. Over this time period approximately $9.1 million (€6.3 million) has been invested with most of it administered recently through USDA OREI grants. A large part of these projects has been a farmer participatory component. One of the most difficult aspects of this funding is that it has typically been for a year or a few years at a time, preventing continuity in programs that generally require a decade to develop. It is only recently that varieties from the first programs have been released and the number is expected to grow.

Organization of meetings for information exchange has been relatively recent in the United States. Among the first sponsored by the American Society of Agronomy, Crop Science Society of America, Soil Science Society of America (ASA-CSSA-SSSA) was an Organic Symposium in 2003 (Podoll, 2009). Another Organic Symposium was conducted as part of the 2005 Annual Meeting was about "Organic Seed Production and Breeding for Organic Production Systems." In 2007, the American Society of Horticultural Sciences Annual Meeting sponsored a colloquium on "Breeding Horticultural Crops for Sustainable and Organic Production." Regional meetings, such as Organicology in the Pacific Northwest, Ecofarm in California, MOSES in Wisconsin, and the NOFA conferences in the North east have all had plant breeding and seed components.

Comparison of European and U.S. Experiences

One of the obstacles to European organic breeding efforts has been the registration requirement for any variety in commercial trade (see also Chapter 8). The same obstacle has not been present in the United States, where growers have had freer access to older traditional varieties. This has perhaps caused the U.S. breeding effort to lag behind that of Europe because there has not been the regulatory-driven need to breed new varieties. In general, the European organic breeding effort has been ahead of the U.S. effort, when the first breeding efforts took place, in the organization of conferences, and in establishing chairs of organic breeding at public institutions.

Funding for organic breeding is roughly similar on both sides of the Atlantic. Given the limited information available, it is difficult to compare European and U.S. private sector efforts, but European companies have invested more heavily than U.S. companies, and in fact, some of the largest organic seed companies in the U.S. are European based.

Since the late nineteenth century, the U.S. has strongly supported public plant breeding through the land-grant university system. However, in the last two decades, the ranks of public plant breeders have declined. Some of this has come about by reduced federal funding to support these positions, and some have been converted into biotechnology positions. Another source of pressure on public plant breeding has been strengthening the private sector through the advent of stronger intellectual property rights. As private companies have taken over plant breeding efforts in many crops, there has been less of a need for public plant breeders in those crops, with a subsequent increase in the difficulty for public plant breeders to obtain operating funds for their research. One consequence of the loss of public plant breeders has been fewer graduate students trained in plant breeding, which has alarmed private seed companies because they do not see where their future cadre of plant breeders will come from (Ransom et al., 2006).

With the increase in funding for organic research in general, and plant breeding in particular, a niche has been created where public plant breeders can operate. Because private seed companies are reluctant to invest in organic plant breeding, public breeders can conduct research programs on crops that would otherwise be the domain of the private sector for traits of importance to organic production (and ultimately sustainable agriculture). At the same time, these types of projects provide a venue for training the next generation of plant breeders.

Perspectives and Challenges for Breeding for Organic Agriculture

Although organic seed production is increasing annually, specific organic breeding programs are few, and many are focused on cereals.

Currently three types of breeding programs are operating (Wolfe et al., 2008):

a. Conventional breeding programs resulting in cultivars (by chance) also suitable for organic farming systems;
b. Conventional breeding programs aimed at cultivars adapted to low-input and organic agriculture;
c. Organic breeding programs fully conducted under organic growing conditions.

The organic sector is too small to financially support enough programs to improve a wide range of crops, so the reality is that these three types of programs will run in parallel for at least the next two decades (Osman et al., 2007). Another model calls for cooperation of organic breeding programs with conventional breeding companies and institutes that recognize the need for sustainability – and who anticipate developing societal recognition from the contribution that breeding for low-input and organic agriculture can make.

The challenges for research to support the development of breeding for organic farming systems focus on the following three categories:

a. Defining which selection criteria are relevant per crop;
b. Defining selection methods;
c. Developing appropriate socio-economic and legal conditions to stimulate organic breeding programs.

In many regards, organic production is still a black box, in terms of knowing what traits are important for adapting a cultivar to an organic system. We know in general terms what conditions are limiting, but what the optimal plant traits might be is open to investigation. An example would be phosphorus use efficiency, which can be achieved through mychorrizal symbioses or through developing vigorous root systems that are better at exploring the soil. Both approaches have advantages and tradeoffs, and it is not yet possible to know which is better for a particular situation.

There is a need to design more efficient breeding methods for organically adapted crops. Farmer participatory methods have been used in organic plant breeding because when working in a new system, breeders do not always know what traits are important to growers. Farmer participatory methods can work quite well, especially in situations that empower farmers and lead them to take over breeding activities, but they have drawbacks in terms of resources required to visit the sites or bring farmers to a central site. There is also a need for breeding methods that can work around techniques or traits that are considered not compatible with organic principles. Examples include cytoplasmic male sterility derived through somatic hybridization and disease resistances developed with the aid of embryo culture.

The socio-economic, policy, and legal frameworks to facilitate organic production are currently lacking. The increasing use of utility patents to protect crop varieties (in the United States) and methods (both in Europe and the United States) are limiting germplasm exchange and thereby reducing the rate of genetic progress. The variety registration system in Europe has difficulties in coping with the heterogeneous materials developed by organic plant breeders who seek diversity in their varieties.

Conclusion

Organic crop breeding is rising from its infancy, has been maturing as a business, and is becoming a scientific discipline. It is contributing not only to the needs of organic farmers who require cultivars better adapted to their farming systems, but also to the development of sustainable agriculture aiming to reduce external inputs. Organic crop breeding is an essential strategy in arriving at such sustainable farming systems.

References

Alrøe, H.F., and E. Noe. 2008. What makes organic agriculture move – protest, meaning, or market? A polyocular approach to the dynamics and governance of organic agriculture. *International Journal of Agricultural Resources, Governance and Ecology* 7:5–22.

Balfour, E.B. 1943. The living soil. Faber and Faber LTD, London. Retreived from http://www.soilandhealth.org/.

Dillon, M., and K. Hubbard 2011. State of Seed 2011. Port Townsend, WA: Organic Seed Alliance. Retrieved from www.seedalliance.org.

European Commision (EC). 2007. Council Regulation No 834/2007 of 28 June 2007 on organic production and labelling of organic products and repealing Regulation (EEC) No 2092/91. Retrieved from http://ec.europa.eu/agriculture/organic/eu-policy/legislation_en.

Fleck, M. 2009. Approaches and achievements of biodynamic vegetable breeding by Kultursaat e.V. (Germany) using the example of RODELIKA one of the first certified biodynamic varieties. *In*: Proceedings of the 1st International IFOAM-Conference on Animal and Plant Breeding, Santa Fe, New Mexico, USA, 25.-28.08.2009. Retrieved from http://orgprints.org/16513/1/santafe09.pdf.

Goldringer, I., J. Dawson, F. Rey, and A. Vettoretti. 2010. Breeding for Resilience: A strategy for organic and low-input farming systems? Proceedings of Eucarpia conference Section Organic and Low-input Agriculture, December 1–3, 2010, Paris, France: INRA and ITAB.

Howard, A. 1940. *An Agricultural Testament. Nature's methods of soil management*. Oxford: Oxford University Press.

IFOAM. 2005. Principles of organic agriculture. Retrieved from http://www.ifoam.org/about_ifoam/principles/.

Kristiansen, P., A. Taji, and J. Reganold (eds.). 2006. *Advances in Organic Agriculture*. Melbourne, Australia: CSIRO Publishing.

Kunz, P., and C. Karutz. 1991. Pflanzenzüchtung dynamisch – die Züchtung standortangepasster Weizen und Dinkelsorten. Dornach, Switzerland: Forschungslabor an Goetheanum.

Lammerts van Bueren E.T., H. Østergård, I. Goldringer, and O. Scholten. 2007. Abstract book of Eucarpia Symposium Plant breeding for organic and sustainable, low-input agriculture: Dealing with genotype-environment interactions. November 7–9 2007. Wageningen, The Netherlands: Wageningen University. Accessed April 26, 2011. www.eucarpia.org.

Lammerts van Bueren, E.T., and K.P. Wilbois. 2008. Report on the ECO-PB meeting on international attuning of the assortment, supply and demand of organic seed in vegetable production in north-west Europe. September 25–26, 2008. Driebergen, The Netherlands: ECO-PB.

Lammerts van Bueren E.T., H. Østergård, I. Goldringer, and O. Scholten. 2008. Preface to the special issue: Plant breeding for organic and sustainable, low-input agriculture: Dealing with genotype-environment interactions. *Euphytica* 163:321–322.

Löschenberger, F., A. Fleck, H. Grausgruber, H. Hetzendorfer, G. Hof, J. Lafferty, M. Marn, A. Neumayer, G. Pfaffinger, and J. Birschitzky. 2008. Breeding for organic agriculture: The example of winter wheat in Austria. *Euphytica* 163:469–480.

Lukkiholt, L.W.M. 2007. Principles of organic agriculture as formulated by the International Federation of Organic Agriculture Movements. *NJAS Wageningen Journal of Life Sciences* 54:347–360.

Mäder, P., A. Fliessbach, D. Dubois, L. Gunst, P. Fried, and U. Niggli. 2002. Soil fertility and biodiversity in organic farming. *Science* 296:1694–1697.

Müller, K.J., P. Kunz, H.H. Spiess, B. Heyden, E. Irion, and C. Karutz. 2000. An overnational cereal circuit for developing locally adapted organic seeds of wheat. In: Alföldi, T., W. Lockeretz, and U. Niggli (eds.). IFOAM 2000 – The World Grows Organic, Proceedings 13th International IFOAM Scientific Conference, Basel, August 28–31, 2000. Hochschulverlag Zürich: IFOAM.

Osman, A.M., K.-J. Müller, and K.-P. Wilbois (eds.). 2007. Different models to finance plant breeding. Proceedings of the ECO-PB International workshop on different models to finance plant breeding, February 27, Frankfurt, Germany. European Consortium for Organic Plant breeding, Driebergen/Frankfurt. Accessed April 26, 2011. www.eco-pb.org.

Østergård, H., M.R. Finckh, L. Fontaine, I. Goldringer, S.P. Hoad, K. Kristensen, E.T. Lammerts van Bueren, F. Mascher, L. Munk, and M.S. Wolfe. 2009. Time for a shift in crop production: Embracing complexity through diversity at all levels. *J Sci Food Agric* 89:1439–1445.

Pfeiffer, E. 1947, repr. 1983. Soil fertility, renewal & preservation, bio-dynamic farming and gardening. East Grinstead, UK: Lanthorn Press.

Podoll, T.M. 2009. Participatory plant breeding's contributions to resilience and the triple bottom line of sustainability – healthy ecosystem, vital economy, and social inclusion. M.S. thesis. Ames, IA: Iowa State University.

Ransom, C., C. Patricka, K. Ando, and J. Olmstead. 2006. Report breakout group 1. What kind of training do plant breeders need, and how can we most effectively provide that training? *Hort Science* 41:53–54.

Rey, F., L. Fontaine, A. Osman, and J. Van Waes. 2008. Value for cultivation and use testing of organic cereal varieties. What are the key issues? Proceedings of the COST ACTION 860 – SUSVAR and ECO-PB Workshop on February 28–29, 2008, Brussels, Belgium. France: ITAB.

Rodale, J.I. 1961. How to grow vegetables and fruits by the organic method. Emmaus, PA: Rodale Books.

Stamp, N. 2003. Out of the quagmire of plant defense hypotheses. *The Quarterly Review of Biology* 78:23–55.

Steiner, R. 1958. Agriculture – A course in eight lectures given at Koberwitz, Silesia in 1924. English translation 1958. London: Bio-Dynamic Agricultural Association.

U.S. Department of Agriculture. 2002. USDA Agricultural Marketing Service, National Organic Program, National Organic Program Regulations (Standards). Retrieved from http://www.ams.usda.gov/nop/standards.

U.S. Department of Agriculture. 2010. Data sets of organic production. Retrieved from http://www.ers.usda.gov/data/organic/.

Willer, H., and L. Kilcher (eds.). 2011. The world of organic agriculture: Statistics and emerging trends 2011. Frick, Switzerland: IFOAM, Bonn, and FiBL.

Wistinghausen, E. 1967. Gesetzmaesigkeiten beim Roggen (*Secale cereale*): Die Ahrenbeetmethode von Martin Schmidt. *Elemente der Naturwissenschaft* 6:24–34.

Wolfe, M.S., J.P. Baresel, D. Desclaux, I. Goldringer, S. Hoad, G. Kovacs, F. Löschenberger, T. Miedaner, H. Østergård, and E.T. Lammerts van Bueren. 2008. Developments in breeding cereals for organic agriculture. *Euphytica* 163:323–346.

Züst, T., B. Joseph K.K. Shimizu, D.J. Kliebenstein, and L.A. Turnbull. 2011. Using knockout mutants to reveal the growth costs of defensive traits. Proceedings of the Royal Society B. Biological Sciences. doi: 10.1098/rspb.2010.2475.

2 Nutrient Management in Organic Farming and Consequences for Direct and Indirect Selection Strategies

Monika Messmer, Isabell Hildermann, Kristian Thorup-Kristensen, and Zed Rengel

Introduction

The world is faced with the need to increase food production to feed a rapidly increasing population. A crucial component in this endeavor is improvement and maintenance of soil fertility. Soil fertility is a measure of the ability of soil to sustain crop growth in the long term, and can be determined by physical, chemical, and biological processes intrinsically linked to soil organic matter content, and quality (Bhupinderpal-Singh and Rengel, 2007; Diacono and Montemurro, 2010). Organic agriculture relies on the use of organic fertilizers – as such, the nutrient cycling provided by the decomposition of organic matter is an essential aspect of food production.

Crop residues are an important source of organic matter that can be returned to soil for nutrient recycling and improving soil physical, chemical, and biological properties. Globally, the total crop residue production is estimated at 3.8 billion Mg Year^{-1} (74% cereals, 8% legumes, 3% oil crops, 10% sugar crops and 5% tubers; Bhupinderpal-Singh and Rengel, 2007). These residues should be returned to soil, uniformly spread over an entire field to prevent depleting the soil in nutrients and organic carbon (C) (Brennan et al., 2000). The nature of crop residues and their management can significantly affect the amount of nutrients available for subsequent crops and the content and quality of soil organic matter (Bhupinderpal-Singh and Rengel, 2007).

Organic agriculture strives for closed nutrient cycles; therefore, inputs should be limited to the farm level. Nutrient input depends on organic fertilizers like green manure, compost or animal manure for building soil fertility. Thus, crops should be adapted to the slow and irregular release of nutrients that might temporarily be in short supply. Root morphology and the capacity to establish beneficial plant microbial interactions both play an important role in nutrient uptake. This aspect has been widely neglected in most conventional breeding programs but might be of special importance in organic farming. In addition, organic farming systems are more heterogeneous compared to conventional farms with respect to crop rotation, type and quantity of organic fertilizer, weed control, or tillage system. Breeders are therefore confronted with developing cultivars that perform well in very different environments. Not only genotypic effects but also genotype × environment (G×E) interactions need to be considered in order to define the most promising breeding strategies.

Organic Crop Breeding, First Edition. Edited by Edith T. Lammerts van Bueren and James R. Myers.
© 2012 John Wiley & Sons, Inc. Published 2012 by John Wiley & Sons, Inc.

Availability of Nutrients in Organic Farming

When we want to breed crop cultivars that are better adapted to organic farming practices than current cultivars bred under the conditions of modern conventional farming, it is important to understand how conditions differ between organic and conventional farming. Nutrient availability is sometimes lower in organic systems as demonstrated by several studies comparing organic and conventional rotations (Marinari et al., 2010; Mäder et al., 2007; Olesen et al., 2009). This is mainly due to reduced nutrient input into organic systems. But at times, nutrient availability is not always lower in organic farming. Nutrient balances on organic farms are difficult to determine, as farmers may introduce substantial amounts of nutrients through the import of feed and bedding materials for animals (Gustafson et al., 2003). In most countries, various nutrient sources accepted for organic farming may include some animal manure obtained directly from conventional farms. Large amounts of nutrients are usually applied for vegetable production.

Nutrient availability is not easy to control in organic farming when using inorganic fertilizers with slow nutrient release, which may be variable and often unknown quality. Organic farmers have to rely on the nutrient sources available on the farm or acceptable nutrient sources available from the market. These sources depend not only on local regulations but also on many other conditions. The economic value of nutrient input to organic farms can be very high, such that farmers have experimented with growing grain legumes to use the seeds as fertilizer, or harvesting and drying legume biomass to be used as green manure (Almeida et al., 2008). Such fertilizers will be very expensive relative to nutrient density, but in some situations organic farmers are willing to pay high prices to obtain available nutrients. Unlike inorganic fertilizers available to conventional farmers, there are few possibilities for modifying the nutrient composition of the fertilizers applied in organic farming. In many situations farmers may inadvertently raise the level of too many other nutrients contained as part of an organic fertilizer when aiming at a certain level of, for example, nitrogen. Such problems may be found in all types of organic farming, but will be most pronounced for high value crops, where farmers are willing to invest substantially to secure sufficient nutrient supply. For example, organic broccoli is seldom grown at low nutrient supply, while organic cereal crops may often be. Soil organic matter content is, on average, higher under organic farming compared to conventional farming, due to higher input of organic fertilizers (Leifeld and Fuhrer, 2010). Thus, nutrient mineralization during the growing season is expected to be higher under organic farming. While, on average, the nutrient availability in organic farming is lower than in conventional farming, the absolute amount of plant available nutrients will differ greatly across organic farms and across crops.

There seem to be two basic differences that ideal organic crop cultivars should be adapted to:

- Limited nutrient availability during early growth, where mineralization from organic nutrient sources is restricted. Under the conventional farming conditions where our current crop cultivars have been bred, fertilizers with easily available nutrients are added to the crops from the start, creating very different conditions for early growth and nutrient uptake than in organic farming.
- Less readily available nutrient sources. These may include nutrients gradually mineralizing during the growth period, mineralizing late when they are not easy for the crop to use, and nutrients found deeper in the soil profile or nutrients bound in more recalcitrant forms in soil minerals or organic matter. These sources of nutrients may be just as available in conventional farming, but the crops will be much less dependent on them.

Roots: The Hidden Potential

Genetic Variation of Root Morphology

Variation in rooting depth among genotypes is more difficult to demonstrate, as the intra-species variation is much smaller than among species, and because root studies are costly and the experimental error is substantial. A few examples of genotype variation in rooting depth under field conditions have been published (Acuna et al., 2010; Thorup-Kristensen, 1998), which indicate that such genotypic differences may be large enough to significantly affect nitrogen (N) uptake from deeper soil layers. In experiments with ryegrass it was possible to breed for root systems with different depth distribution, and thereby for higher root density in the deeper soil layers (Crush et al., 2010; Thorup-Kristensen and Van den Boogaard, 1999). Root architecture is critically important in determining soil exploration and therefore nutrient acquisition. Architectural traits under genetic control include basal-root gravitropism, adventitious-root formation and lateral branching (Lynch, 2007).

Few investigators have conducted root studies in the field and tested/validated the occurrence and effects of differences in root morphology. In a minirhizotron study of cauliflower root growth, two cultivars differed in their ability to grow roots from the crop row into the inter-row soil (Thorup-Kristensen and Van den Boogaard, 1998). One cultivar (Plana) distributed its roots very evenly within the soil volume whereas the other (Siria) had more roots beneath the crop row than in the inter-row soil. This was reflected in their depletion of available soil N, where Siria left more N in the inter-row soil than Plana. Johnson et al. (2000) were able to demonstrate the association between taproot length and the ability to extract water from deep in the soil profile in *Lactuca* species.

What Root Traits Are Ideal for Nutrient Uptake/Efficiency under Organic Farming Conditions?

It is not completely clear what root characteristics would be desirable under nutrient-limited conditions as found under most organic farming conditions. As shown by Costigan (1984, 1988) young plants can be nutrient limited even when soil nutrient availability is high, due to a combination of limited root contact with the soil medium and high growth rates and therefore high nutrient demand compared to plant size. In organic production this problem is more intense due to the lack of easily available nutrients from inorganic fertilizers, e.g., under cold spring conditions. This sometimes leads to strategies with delayed sowing in organic farming, to wait until soil temperatures are higher and more favorable for nutrient release as well as for root growth. If cultivars with more extensive early root growth and branching can be identified, this could improve nutrient uptake during the early growth and promote crop establishment and development of yield potential. However, before this is adopted as a universal strategy, it should be kept in mind that in some situations the same characteristics may lead to increased early nutrient uptake and soil depletion at the expense of later nutrient uptake, which could be more directly important to yield and quality formation. For example, the root morphologies of 21 modern maize inbred lines were studied under different temperature regimes. While the lateral root length was most closely related to plant dry weight at cool temperatures (15 to 13°C), the axial root length was closely related to plant dry weight at 24 to 20°C (Hund et al., 2008, 2009). The authors concluded that selection for long laterals on the primary roots holds promise for improving early season vigor in environments and cropping systems with reduced soil warming in spring. On the other hand, long laterals on primary roots might be disadvantageous under warmer conditions.

Improved ability to take up nutrients such as soil phosphorus (P) from recalcitrant sources by increasing length and improving root distribution should also be investigated. Plants have many different mechanisms for the uptake of such nutrients (Gahoonia et al., 2007; Marschner, 1998), and genetic variation for this trait exists among crop plants (Gahoonia and Nielsen, 2004b; Gahoonia et al., 2005). One effect that varies significantly among cultivars of several crops is variation in root hair density along roots as well as root hair length (Gahoonia and Nielsen, 2004a). P-efficiency of maize inbred lines was related to the larger root system, greater ability to acidify the rhizosphere, and positive response of APase production and excretion under low P conditions in calcareous soils (Liu et al., 2004).

Another approach to increasing nutrient use efficiency is to increase the extent of the root zone, allowing the crop to exploit nutrients from a larger soil volume. The ability of crops to develop deep root systems varies strongly (Kristensen and Thorup-Kristensen, 2004a; Smit and Groenwold, 2005; Thorup-Kristensen, 2006a). This especially affects the uptake of N from deeper soil layers (Acuna et al., 2010; Kristensen and Thorup-Kristensen, 2004b; Thorup-Kristensen, 2001). It appears advantageous and possible to breed for deeper root systems at least in some crops (Bonos et al., 2004; Crush et al., 2010). Deep root system development depends on root growth rates and duration. In most crops, root growth slows as maturity advances when plant resources are prioritized for seed production. Genotypes with increased late root growth capacity may develop an overall deeper rooting system and may better exploit nutrients from deep soil layers during the grain-filling period, which could clearly be advantageous.

Our understanding of the presence of available plant nutrients in deeper soil layers is very limited. The results of Thorup-Kristensen (2006a) and Thorup-Kristensen et al. (2009) show that high potassium (K) levels may be available in deep soil layers, and that even in an unfertilized organic crop rotation the amount of available N in soil layers between 1 and 2.5 m deep may sometimes become high. Furthermore, deep rooting is relevant to drought stress when water is only available in deep soil layers.

Increasing the soil volume exploited by roots may also involve an increased exploitation of inter-row soils by crops grown in rows, or increased exploitation of the uppermost few centimeters of soil layer, which in some situations is rich in nutrients compared to the rest of the soil volume. This layer is often not well exploited due to root architectural constraints or unfavorable growth conditions, in terms of temperature and humidity.

There is large variation among row crops in their ability to develop high root density in the inter-row soil (Thorup-Kristensen, 2006b), and among species where this ability is limited, increased horizontal root growth may increase nutrient uptake from the inter-row soil. As an example, Thorup-Kristensen and Van den Boogaard (1999) found that carrots left higher amounts of N in the inter-row soil than beneath the crop row. In contrast, higher maize yields were related to higher plant density combined with genotypes with less horizontal but more efficient vertical root growth. This indicates that more efficient exploitation of inter-row soil in row crops may be solved by improved planting systems with reduced row distance, rather than by breeding new genotypes adapted to a suboptimal planting pattern.

Increased exploitation of the uppermost few centimeters of soil has been a main focus of the work by Lynch and Brown (2001), which has examined developing beans and other legumes for low fertility tropical soils. The authors demonstrated that genotypes with a different root angling leading to higher root density in the topsoil could increase P uptake and growth of bean genotypes in depleted soils where much of the P availability is concentrated in the topsoil. Such a situation is probably not common in most organic farming systems where soil tillage by ploughing is commonly performed for various reasons, including weed management. In such a situation, the enriched topsoil will be more like 15 to 25 cm deep. However, with the adoption of conservation tillage in organic farming, the root distribution in different soil layers will remain an important issue.

Influence of Dwarfing Genes on Roots and Nutrient Uptake

It has often been theorized that the development of shorter cultivars using dwarfing genes would not only reduce shoot length but also root length. However, experimental data are rare and conflicting (McCaig and Morgan, 1993; Miralles et al., 1997). While dwarfing genes in isogenic wheat lines increased root length in the gel chambers, the total root length (measured after four weeks) was reduced in the field (Wojciechowski et al., 2009). Dwarfing genes in isogenic durum lines did not show any effect on root penetration ability (Kubo et al., 2005). Chloupek et al. (2006) observed larger "root system size" of barley lines containing dwarfing genes among 12 barley genotypes by an electrical capacitance method related to fresh matter root weight.

Semi dwarfing traits may have other effects on nutrient efficiency as showed by Manske et al. (2002). In their study, semi-dwarf lines of wheat produced higher yields, but their ability to store P in the straw and then remobilize it into the seed during grain filling was reduced, leading to a reduced P harvest index and a more critical dependence on P uptake after anthesis for incorporation of P into the wheat grain. Blum et al. (1997) showed a similar effect with limited storage of carbohydrates in the straw of semi-dwarf sorghum genotypes, which thereby became more sensitive to terminal drought as there was less carbohydrate to remobilize for grain filling.

Methodologies to Assess Root Characteristics in the Field

The classical methods for studying root growth under field conditions are soil sampling and washing out roots from the soil, trench wall (Achat et al., 2008), and core break techniques (Stone et al., 2001). The last two are very simplified and based on creating a new surface of soil by digging a trench wall, or by breaking a soil core, and then observing the roots at this surface. Currently the mini-rhizotron technique (see fig. 2.1) is used for many root studies in the field (Thorup-Kristensen, 2001). These methods have all been used for years, and several books and reviews exist (Neumann et al., 2009).

For breeding purposes washing roots out of soil samples is too time consuming, and trench wall techniques are very destructive to the experimental fields and cannot be used routinely. Depending on the equipment available the core break method and the mini-rhizotron method may be applied for studying root growth variation among parent lines in breeding programs but not for large-scale screening. Therefore, less time consuming methods have been applied, e.g., root-pulling resistance where the force needed to pull the plant from the soil is measured as an indication of root system size (Kamara et al., 2006). Chloupek (1972, 1977) suggested electrical capacitance measurement in order to evaluate root size of single plants. Field studies in maize demonstrated that electrical capacitance and root fresh mass were correlated and may facilitate a non-destructive identification of genotypes with large roots (van Beem et al., 1998). Other indirect methods are the measurement of root function by uptake of isotope labeled nutrients (Hauggaard-Nielsen and Jensen, 2005; Kristensen and Thorup-Kristensen, 2004b) or the assessment of root depth by herbicide placement in deep soil layers (Meyers et al., 1996). Methods with placement of isotope labeled nutrients may be the most promising for studying genotype variation in root function under field conditions, especially since isotope discrimination measurements have become rather cheap routine

Figure 2.1 Installing a minirhizotron to study root systems in the field by drilling a hole for a minirhizotron (top). The drilling is done twice, first with a spiral auger with a smaller diameter than the minirhizotron, and then with a piston auger with a diameter matching the diameter of the minirhizotron (with counting grid painted on it) to leave a smooth soil surface to ensure maximum contact between soil and glass surface (bottom left). The top of a 3-m long minirhizotron installed and ready for use (bottom right). (Photos: K. Thorup-Kristensen, 2009).

methods. However, these methods are far from being realistic for large-scale routine screening among breeding lines.

Is Large-Scale Screening for Root Growth and Architecture Possible?

As there are no methods for field screening of large numbers of breeding lines that are realistic in the near future, a main focus is on methods for root screening under lab conditions. Very simple methods such as germinating seeds in "root pouches" (Kaspar et al., 1984; Liao et al., 2006) or in a petri dish and measuring tap root length after a few days have been tested and were positively correlated with root depth of the same genotype in the field (Kaspar et al., 1984). Currently there is a focus on high throughput phenotyping of root growth of plants grown in transparent tubes (Hund et al., 2009; Yazdanbakhsh and Fisahn, 2009). If methods as simple and far removed from real field conditions such as germinating seeds in root pouches can give relevant information, study of root growth in transparent tubes filled with growth medium must be expected to give even better results. A previous study has also shown a strong correlation between root growth of peas in soil-filled boxes and in the field (Thorup-Kristensen, 1998).

Before such methods can be applied in large scale screening, they should be validated with different genetic materials. There is a need to determine optimal parameters of root growth to select for, as many different parameters can be measured and their relevance will differ greatly.

Seedling test in root pouches or transparent tubes facilitates assessing seminal root number (Hund et al., 2009), angle, root hair density and length (Yazdanbakhsh and Fisahn, 2009), and hypocotyl stem length. Simple systems may also be developed where uptake of isotope labeled nutrients can be measured directly.

Some researchers have attempted to select in hydroponic systems (Frantz et al., 1998), where factors such as root/shoot ratio or nutrient uptake rates per unit root mass may be measured with relatively low effort (Lainé et al., 1993). However, while hydroponic methods may be appealing because variation in nutrient uptake rates can be easily measured, they may be too far removed from the field situation.

As methods for studying root growth and activity are difficult and demanding, we are continuously looking for simpler methods that can give relevant information about the root system. It is crucial that all these methods are thoroughly validated in field trials under different pedo-climatic conditions and with different genetic material before they can be applied in breeding programs. Identifying genetic loci to facilitate marker-assisted selection for root traits will promote breeding for more efficient capture of water and nutrients (De Dorlodot et al., 2007). Indeed, marker-assisted selection for root-related genetic loci has been used to alter root architecture and at the same time increase grain yield in field-grown rice (Steele et al., 2007) and also to increase root size and simultaneously grain yield in maize (Landi et al., 2005). However, wide-scale use of root-related genetic information in breeding is hampered by relatively small mapping populations and inaccurate phenotyping (De Dorlodot et al., 2007).

Even Greater Complexity: Plant-Microbe-Soil Interactions

Plant growth is strongly influenced by interactions in the plant-soil-microbe continuum. These interactions are particularly obvious in the rhizosphere (a thin layer of soil surrounding roots), whereby plant-microbe interactions may alter chemical, physical, and biological properties of the rhizosphere soil (Crowley and Rengel, 1999; Hartmann et al., 2009; Jones et al., 2009; Kohler et al., 2009; Lambers et al., 2009; Marschner and Rengel, 2003, 2010; Rengel, 2002, 2005; Rengel and Marschner, 2005; Solaiman et al., 2007; Richardson et al., 2009).

Despite the importance of rhizosphere processes in influencing nutrient availability, until recently these processes have not received a major consideration in modern agriculture, where the practice has been to provide N, P, and K in luxurious quantities as synthetic fertilizers. Hence, selection and breeding have produced cultivars that are responsive to highly soluble fertilizers, but they are lacking traits necessary for growth under low fertility or with slow-release sources of nutrients as found in organic agriculture (Guppy and McLaughlin, 2009; Lynch, 1998; Marschner and Rengel, 2003). To develop improved cultivars for sustainable, low-input agriculture, a better knowledge of the rhizosphere processes, particularly plant-soil-microbe interactions that contribute to nutrient uptake efficiency is essential (Crowley and Rengel, 1999; Hartmann et al., 2009; Jones et al., 2009; Lambers et al., 2009; Rengel and Marschner, 2005).

The management of crop residues in agriculture can significantly affect biomass, activity, and composition of soil micro- and macro-organisms and their functions by altering the supply of carbon and other nutrients and the physico-chemical characteristics of the soil environment (Bhupinderpal-Singh and Rengel, 2007). Microbial biomass represents only a small proportion (1–5%) of the total C, N, S, and P pool in arable soils (Stockdale and Brookes, 2006; Werth and Kuzyakov, 2010). Microorganisms can act as a direct source and sink for nutrients, with gross turnover times of microbial biomass C and N estimated at 1 to 2 years (Bhupinderpal-Singh and Rengel, 2007; Diacono and Montemurro, 2010; Werth and Kuzyakov, 2010).

Nutrients with limited mobility in soils (P, K, Fe, Zn, Mn, and Cu) are usually present in relatively large total amounts, but the plant-available fraction and the concentration in the soil solution may be insufficient to satisfy plant requirements. In such cases, a desirable characteristic of efficient genotypes would be a capacity to increase the nutrient fraction available to them so that they may take up relatively larger amounts of that nutrient (uptake efficiency), even though nutrient utilization efficiency (e.g., increased amounts of dry matter produced per unit of nutrient taken up) may also be important (Rengel, 2001; Rengel and Marschner, 2005; Sadeghzadeh and Rengel, 2011).

Plants may form symbiotic associations that aid the uptake of nutrients from the atmosphere or soil, such as with *Rhizobia* or *mycorrhizae* (Lambers et al., 2009; Mattoo and Teasdale, 2010). Plants may also modify the rhizosphere (secretion of chelating agents, reducing compounds, etc.) to increase the availability of nutrients (Jones et al., 2009).

Even though plant-microbe-soil interactions may influence plant tolerance to a range of abiotic and biotic stresses, these interactions are arguably most important in nutrient cycling. We will concentrate particularly on N, P, and Zn.

Nitrogen-Use Efficiency

N-use efficiency measures the capacity of a given genotype to take up, and to convert to yield, N applied as fertilizer (units of dry matter produced per units of N taken up). Worldwide, N-use efficiency in cereal production (wheat, maize, rice, barley, sorghum, pearl millet, oats, and rye) is approximately 33%. The unaccounted 67% represents a U.S. $15.9 billion annual loss of N fertilizer (assuming fertilizer-soil equilibrium; Raun and Johnson, 1999).

N-use efficiency is related to N-uptake efficiency of roots and metabolic efficiency of absorbed N in roots and shoots (Jiang et al., 2000). Genotypic variation in N-use efficiency has been reported for a number of crops (Rengel, 2005; Svecnjak and Rengel, 2006), thus setting the stage for selection and breeding for increased N-use efficiency (Dawson et al., 2008a). For example, a breeding strategy for improved N efficiency in canola is based on selection for low N content in dropped leaves and for high N harvest index by reducing N yield of straw (Svecnjak and Rengel, 2006).

Wild crop plant relatives may serve as a source of genes for improved and more effective N_2 fixation (de Faria et al., 2010). Enzymes involved in the N assimilation pathway are important contributors to N-use efficiency (Andrews et al., 2004). Genotypic differences in activities of glutamine synthetase and glutamate synthase in nodules were related to different symbiotic N_2 fixation efficiency in faba bean (Caba et al., 1993).

Symbiotic and Associative N_2 Fixation

Dinitrogen fixation provides more N to the agricultural ecosystems worldwide than the total amount of fertilizer N applied (Rengel, 2002; Salvagiotti et al., 2008). However, a significant potential exists to further improve symbiotic and associative N_2 fixation by breeding genotypes with a greater capacity to sustain these interactions with bacteria.

Various rhizobia strains differ in N_2 fixation capacity; it is therefore possible to select for more efficient strains (Rengel, 2002; Salvagiotti et al., 2008). Recent work showed that competitive strains respond more readily to flavonoids (signal molecules promoting the formation of nodules by symbiotic bacteria) than non-competitive strains, suggesting that improvement of legume inoculants can occur through selection of flavonoid-responsive strains (Maj et al., 2010).

Different legume genotypes differ in efficacy of symbiotic N_2 fixation they can support (for examples see Rengel, 2002; Salvagiotti et al., 2008). Genetic factors underlying differential efficacy of the symbiotic N_2 fixation are being elucidated. Quantitative Trait Loci (QTL) regulating nodulation and N_2 fixation have been identified in several crops (e.g., in pea crops; Bourion et al., 2010). A continuous and coordinated selection of the most effective combinations of host and microbial symbionts is a prerequisite for profitable and sustainable agricultural systems.

The contribution of N_2 fixation via associative bacteria is dependent on rainfall, temperature, and C availability in soil; such contribution is estimated to be relatively minor (<10 kg N ha^{-1} per year) under dryland conditions (e.g., in Australia), but the contribution can be considerably greater in tropical systems where high temperatures and high rainfall coincide (Unkovich and Baldock, 2008). For example, N_2 fixation via associative bacteria may contribute up to 30% of N needs of wheat and up to 150 kg N ha^{-1} per year in sugar cane fields in Brazil. Moreover, it has been suggested that Brazilian cereal and sugar cane genotypes have inadvertently been selected for efficient utilization of N fixed by associative diazotrophs (Baldani et al., 2002; Hungria et al., 2010), thus significantly decreasing or completely eliminating the need for N fertilization.

P Efficiency

Although the total amount of P in the soil may be high, it is mainly ($>80\%$) present in forms unavailable to plants because of adsorption, precipitation, or conversion to organic forms. P solubility can be increased by excretion of organic acid anions into the rhizosphere and/or changing the rhizosphere pH. Phosphatases released by plant roots or soil microorganisms can mineralize organic P and specialized root structures such as cluster or dauciform roots are particularly effective in mobilizing P (Soleiman et al., 2007). Comparing wheat genotypes with contrasting P efficiency revealed differences in their microbial community composition in the rhizosphere. Interestingly, there were no differences in mycorrhizal colonization between the genotypes. A positive correlation among microbial P in the rhizosphere and shoot dry weight, P uptake per plant, and available P in the rhizosphere suggested the importance of this source for subsequent host P uptake (Marschner et al., 2006). In wheat, P-efficient genotypes appear to have rhizosphere microbial communities that mobilize this element but also have high turnover rates, releasing it for subsequent plant uptake. By contrast, in the rhizosphere of P-inefficient genotypes, rhizosphere microbial communities also mobilize this element, but due to the low turnover rate, P remains locked-up in the microbial biomass (Marschner and Rengel, 2010).

Symbiosis between Plants and Arbuscular Mycorrhizal Fungi (AMF)

Various types of mycorrhizal associations (arbuscular, ectomycorrhizal, and ericoid mycorrhizas) have been found in 336 plant families, representing 99% of flowering plants, with more than 10,000 plants having literature records of mycorrhizal associations (Brundrett, 2009). However, all members of the *Brassicaceae* and *Polygonaceae* families and many *Chenopodiaceae* species are non-mycorrhizal plants, including agriculturally important crops like canola, broccoli, buckwheat, and sugar beet. The AMF symbiosis is the most widespread plant symbiosis, with different combinations of host plant and AM fungi resulting in differential effects on gene expression, cellular functions, root morphology, and nutrient status (Feddermann et al., 2010; Manjarrez et al., 2009; Mattoo and Teasdale, 2010). There is genetic variability in the AMF colonization capacity of various genotypes of host species (Brundrett, 2009; Manjarrez et al., 2009; Rengel, 2002; see also Chapter 17 in this book). Wide variability also exists in populations of mycorrhizal fungi in their hyphal growth and thus competitive ability (Drew et al., 2006; Feddermann et al., 2010).

The knowledge of genetics and physiology of the colonization process will be instrumental in designing screening procedures and molecular markers to breed genotypes for more efficient AMF symbiosis (Manjarrez et al., 2009). There is substantial ongoing research effort to identify plant genes required for mycorrhizal development (Manjarrez et al., 2009; Morandi et al., 2009).

Arbuscular mycorrhizal fungi are less host-specific than rhizobia, but their rate of colonization and other benefits to the host are genotype-dependent (Marschner and Rengel, 2010). Mycorrhizal responsiveness in terms of P uptake was lower in modern wheat genotypes than in older cultivars (Zhu et al., 2001), suggesting that current plant breeding practices have selected for genotypes that are less dependent on mycorrhiza for their P uptake.

Zn Efficiency

Plant genotypes are widely different in their tolerance to Zn deficiency, both in Zn uptake and utilization. A wide range of wheat, barley, rice, bean, chickpea, and maize germplasm has been studied, indicating there is enough genotypic variation to allow breeding for nutritional improvement (Rengel, 2001; Rengel and Marschner, 2005; Sadeghzadeh et al., 2010; Sadeghzadeh and Rengel, 2011). Nutritional traits are generally stable across environments, despite some reported GxE interactions (Sadeghzadeh and Rengel, 2011); it is therefore possible to combine high micronutrient traits with high yield.

When grown in soils of low Zn availability, Zn-efficient wheat genotypes had greater fertilizer efficiency (grain yield produced per unit of fertilizer applied) and a greater harvest index compared to the Zn-inefficient genotypes (Rengel and Graham, 1995). Zn-efficient wheat genotypes offer some promise in managing soils low in available Zn. Given that Zn deficiency is typically patchy in a single field, growing Zn-efficient genotypes can overcome spatial variability in the horizontal and vertical direction as well as temporal variation in Zn availability to plants (Rengel, 2001).

Plant Growth Promoting Rhizosphere Organisms (PGPR)

PGPRs are a heterogeneous group of microorganisms that stimulate plant growth (Dutta and Podile, 2010; Lugtenberg and Kamilova, 2009). They have been isolated from a range of host plants and do not appear to be highly plant-species specific. Several mechanisms of plant growth promotion have been proposed, and a given PGPR often exhibits several positive traits (Richardson et al., 2009). Plant growth promotion may be due to:

- P solubilization,
- N_2 fixation,
- nitrogenase activity,
- phytohormone production,
- pathogen suppression (antibiotics, siderophores), and
- stimulation of other beneficial microorganisms such as N_2 fixers or mycorrhizal fungi.

A large number of microorganisms show the capacity to solubilize P in vitro (Whitelaw, 2000; Whitelaw et al., 1999). In several pot or field experiments, increased P uptake and plant growth after inoculation with P-solubilizing microorganisms have been reported (Whitelaw, 2000). For example, dual inoculation of wheat with AMF and PGPR increased grain yield by 41% and P efficiency by 95% as compared to the un-inoculated control (Mäder et al., 2011). In addition, soil enzyme activity (alkaline and acid phosphatase, urease, and dehydrogenase) was improved.

Molecular genetic studies modifying auxin production of PGPR (Baudoin et al., 2010) or altering transporters involved in root exudation (Badri et al., 2009) differentially influenced rhizosphere microbial communities. These studies clearly showed the complexity of interactions happening in the rhizosphere, whereby changing a single property of a single organism had measurable effects on other organisms in the rhizosphere.

Root Exudates as Trigger for Plant-Microbe-Soil Interactions

Plants exude a variety of organic compounds (carboxylate anions, phenolics, carbohydrates, amino acids, enzymes, other proteins, etc.) and inorganic ions (protons, phosphate and other nutrients, etc.) into the rhizosphere to change chemistry and biology of the rhizosphere and enhance adaptation to a particular environment (Crowley and Rengel, 1999; Jones et al., 2009). Common conditions under which plants increase exudation of organic compounds into the rhizosphere encompass deficiency of various nutrients: E.g., P, Fe, Zn, or Mn (Mattoo and Teasdale, 2010; Rengel, 2005).

Understanding the role that root exudates play in increasing plant adaptation to the given set of environmental conditions is increasing rapidly (Badri et al., 2009; Baudoin et al., 2010; Eilers et al., 2010; Jones et al., 2009). The two most popular strategies in increasing root exudation are (1) overexpressing genes coding for critical enzymes in the biosynthesis of carboxylate anions (Delhaize et al., 2007; Deng et al., 2009) and (2) engineering plants to increase exudation of phosphatase and phytase (George et al., 2004, 2005, 2009; Richardson, 2009), which are not acceptable in the organic agriculture (see Chapter 7 in this book). However, the emphasis may be on utilizing genotypic differences between existing crop germplasm, including wild relatives (Rengel, 2005) and, particularly for cereals, rye as the cereal most tolerant to a wide range of nutrient deficiency, and ion toxicity stresses (Ma et al., 2002; Osborne and Rengel, 2002). Root exudation is highest at the root tip where the microbial density is still low; with increasing distance from the root tip, exudation generally decreases while microbial density increases (Marschner, 1995). Thus, the region of highest release of root exudate and the region of highest microbial population density are spatially separated.

Compared to the soil organic matter, root exudates represent an easily degradable nutrient source for microorganisms, and some microbial species proliferate rapidly in the rhizosphere (Hartmann et al., 2009; Marschner and Rengel, 2003). The microbial biomass may be up to 36% of root dry weight. These are usually species with high growth rates and relatively high nutrient requirements such as pseudomonads (Hartmann et al., 2009). Microorganisms such as N_2 fixers are also attracted by signaling substances excreted by roots (Rengel, 2002). Microorganisms can also enhance the release of root exudates and produce growth factors that influence root growth (Hartmann et al., 2009; Marschner and Rengel, 2003).

Plant species differ in composition and amount of root exudates (Jones et al., 2004; Nuruzzaman et al., 2006; Roelofs et al., 2001) and in the composition of the rhizosphere microflora (Lambers et al., 2009; Marschner et al., 2004, 2005; Marschner and Timonen, 2005). Even genotypes within a species may have a specific rhizosphere microflora (Hartmann et al., 2009; Marschner et al., 2006, 2007). Additionally, the microbial community composition depends on growth stage, P and N fertilization, and other soil factors (Marschner et al., 2006, 2007).

Exudation of phosphatases occurs from the apical root region and increases under P-deficient conditions in a number of crop species (Yadav and Tarafdar, 2003), with variation noted among genotypes of a single species (George et al., 2008). The activity of phosphatases decreased with the distance from the root surface (Tarafdar and Jungk, 1987). Plant roots do not exude phytases (Richardson et al., 2009) – it is only through microorganisms (e.g., *Aspergillus niger*) excreting phytases that phytin may become a source of P to plants (Richardson et al., 2000, 2001).

Is Large-Scale Screening for Plant-Microbe-Soil Interactions Possible?

Screening techniques need to be developed by targeting specific processes that cause (or prevent) nutrient deficiency, rather than those that appear as a consequence of it. Selection based on increased root exudation might be one possibility for large-scale screening of segregating genotypes considering the progress in the metabolomics. Since increasing the root exudation capacity may adversely affect crop yield because of increased partitioning of assimilates toward roots, a better (and more economic) solution would obviously be to increase exudation of only targeted compounds that perform a specific function in the rhizosphere – for example, a function that will result in improved tolerance to nutrient deficiencies or in the more effective nodulation process (Rengel, 2005; Rengel and Marschner, 2005). For such a selection pressure to be included in the breeding program, breeders would need to have well-defined screening procedures that rely heavily on understanding the physiological processes involved. Such knowledge appears to be inadequate at present but might evolve in the near future.

For Zn efficiency, different screening methods have been used to evaluate genotypes: Nutrient solution culture, field evaluations, and greenhouse soil bioassays (Sadeghzadeh and Rengel 2011). Soil-free systems such as chelator-buffered hydroponic systems have been used commonly to study plant mineral nutrition at realistically low Zn concentrations (Parker and Norvell, 1999; Rengel, 2001). In a study on differential Zn efficiency of two barley genotypes, the responses to Zn fertilization in chelator-buffered nutrient solution were consistent with their response in soils in terms of visual Zn deficiency symptoms and shoot and root Zn concentration and content (Sadeghzadeh et al., 2009, 2010).

Field-based techniques are more laborious than pot or solution cultured ones. The results can be variable because the severity of the nutrient deficiency varies among sites and years due to the effects of other growth limiting factors (such as drought and disease) as well as the irregular spatial distribution of Zn in soil (Sadeghzadeh and Rengel, 2011).

Screening under glasshouse condition in pots is generally easier than under field conditions because it is fast, cost-effective and can overcome problems of soil heterogeneity. The growth containers should be big enough to eliminate root restriction, reduce sink strength at the whole plant level, and reduce photosynthetic enhancement (Arp, 1991). However, pot screening is less realistic than field conditions, especially for grain-yield production. Nevertheless, using 150 barley doubled-haploid lines grown to maturity, a strong correlation between Zn efficiency in field and glasshouse conditions was shown (Sadeghzadeh et al., 2010).

It is expected that in the near future molecular methods will dominate selection for nutrient efficiency. Development of molecular markers for regions of the genome segregating with the trait of a particular nutrient efficiency will be crucial in that respect (Sadeghzadeh et al., 2010). However, the correct identification of such molecular markers depends strongly on the reliability of the phenotypic assessment in the field. While some molecular markers may be population-dependent (e.g., markers for Fe efficiency in soybeans; Lin et al., 2000), making them unsuitable for use in breeding programs, the advent of association mapping will allow for the identification of generally valid genetic regions for the trait of interest with high precision.

Importance of Selection Environments

Most cultivars presently used in organic farming are derived from breeding programs for mainstream agriculture (Lammerts van Bueren et al., 2002, 2011). Selection is generally performed under uniform and highly controlled conditions applying seed treatment, herbicides, and, in most cases, high level of fast-releasing fertilizer and pesticides. These conditions do not mirror the situation of organic farming, which aims for closed nutrient cycles with minimal external input. Instead, nutrient release is dependent on the temperature and biological activity of the soil and its interaction with the declining residues from a previous leguminous fertility-building crop and/or farmyard manure. As a consequence, it is not always possible to match nutrient release with crop demand over time (Dawson et al., 2008a; Jones et al., 2010) resulting in considerable yield reduction compared to high input systems (e.g., 5–47% reduction for cereals, Przystalski et al., 2008; 44% reduction for lentil, Vlachostergios and Roupakias, 2008). Therefore, it is necessary to identify cultivars that show high nutrient-use efficiency (i.e., genotypes that produce higher yields per unit of nutrient applied or absorbed than others) under organic growing conditions (Dawson et al., 2008a; Fageria et al., 2008).

Are Conventionally Bred Cultivars Suitable for Organic Farming?

Under the hypothesis that old cultivars released before the agricultural intensification of the "green revolution" might be better adapted to the nutrient-limited conditions of organic environments, several studies compare the performance of modern conventionally bred cultivars and old cultivars and landraces. Carr et al. (2006) compared 17 old and modern spring wheat cultivars under organic farming conditions and concluded that modern cultivars out-yielded old cultivars also under organic conditions. In contrast, no genetic variation was found between 10 old and modern conventionally and organically bred winter wheat varieties under organic conditions on fertile soil (Hildermann et al., 2009b, 2010), while modern organically bred varieties were superior under low yielding organic conditions (Hildermann et al., 2009a). Similarly, no major yield improvement of modern compared to historical winter wheat cultivars was found under organic growing conditions in the United States (Murphy and Jones, 2007). Investigation of 17 to 18 old and modern cultivars of oat and wheat under different N regimes in Finland revealed that modern oat and wheat cultivars had improved N-use efficiency (especially uptake efficiency) compared to old cultivars, whereas no difference was found for barley (Muurinen et al., 2006). Improvement of N-uptake efficiency in modern compared to old wheat cultivars was also detected in Italy (Guarda et al., 2003) and France (Brancourt-Hulmel et al., 2003). Wissuwa and Ae (2001) reported that traditional rice varieties were more efficient in P uptake than modern cultivars under low P conditions. The responsiveness of modern wheat to AMF cultivars in pot trials was reduced compared to old cultivars (Hetrick et al., 1992, 1993; Zhu et al., 2001), but this could not be confirmed in field trials (Friedel et al., 2008; Hildermann et al., 2010).

Explanations for these contradictory results from comparisons in organic versus conventional farming might be that (1) the registered cultivars represent only a small proportion of the available genetic diversity, (2) the experimental conditions of organic farming can vary from high to low soil fertility, and (3) the intensity of conventional farming varies strongly among countries. These studies also highlight the fact that organic, low external input farming is not identical to low N input, but represents a range of much more complex situations (Dawson et al., 2008b).

Comparison of Direct and Indirect Selection

Since yield improvement of conventionally bred cultivars under organic farming conditions is very limited, it can be presumed that valuable genotypes that might be profitable for organic farming are eliminated during the selection process under conventional farming. On the other hand, selection under very controlled growing conditions allows a more precise estimation of the genotypic value of breeding lines, due to minimized confounding environmental effects compared to organic farming conditions. Therefore, experimental data are needed to determine whether indirect (correlated) selection for adaptation to organic farming is as effective as direct selection. A quantitative genetic criterion for the efficiency of indirect selection is the ratio of the expected indirect (CR) to direct (R) gain from selection (Falconer and Mackay, 1996):

Direct selection response under organic farming (org):

$$R_{ORG} = i_{org} \times h_{org} \times \sqrt{V_{G(org)}}$$

Indirect selection gain or correlated response under conventional farming (con) for the improvement under organic farming (org): (2.1)

$$CR_{ORG} = i_{con} \times h_{con} \times r_{G(con/org)} \times \sqrt{V_G(org)}$$

Efficiency of indirect selection:

$$CR/R = i_{con} \times h_{con} \times r_{G(con/org)} / i_{org} \times h_{org}$$

In this formula, i is the selection intensity (i.e., the percentage of selected lines under conventional (con) and organic farming (org), respectively, h is the square root of the heritability coefficient, $\sqrt{V_{G(org)}}$ is the square root of the genotypic variance under organic farming, and r_G is the genotypic correlation coefficient. If the same selection intensity is applied, the efficiency of indirect selection for a given trait depends on the heritability under the different farming systems and the genotypic correlation between the two farming systems.

The genotypic correlation (r_G) between two traits y and z can be calculated by dividing the genotypic covariance of y and z [COV_G (yz)] by the square root of the genotypic variance of trait y [V_G (y)] and the square root of the genotypic variance of trait z [V_G (z)]. Accordingly, the phenotypic correlation is based on the phenotypic covariance and phenotypic variances. Applied to the assessment under different farming systems we obtain following formulas for the genotypic and phenotypic correlations for a given trait:

$$\text{Genotypic correlation: } r_G = COV_{G(org/con)} / \sqrt{V_{G(org)}} \sqrt{V_{G(con)}}$$
$$\text{Phenotypic correlation: } r_P = COV_{P(org/con)} / \sqrt{V_{P(org)}} \sqrt{V_{P(con)}}$$

(2.2)

If the genotypic correlation is moderate (<0.7) and heritabilities are comparable under organic and conventional selection environments, direct selection in the target environment is more efficient. Thus, specific breeding programs should be carried out in the target environment to maximize selection gain.

While there is substantial evidence that breeding for low N input conditions is more efficient under severe N stress than under high-input conditions – as was demonstrated for wheat (Brancourt-Hulmel et al., 2005), barley (Ceccarelli, 1996; Sinebo et al., 2002; Ryan et al., 2008), oat (Atlin and Frey, 1990), maize (Presterl et al., 2003), and rice (Mandal et al., 2010) – only a few studies have been conducted to demonstrate the difference between direct and indirect selection for organic farming. Studies based on 79 F_6-derived recombinant inbred lines of spring wheat cultivated on conventionally and organically managed land for three years revealed that direct selection was more efficient than indirect selection, resulting in less than 50% of genotypes selected in common (Reid et al., 2009). This implies that breeding spring wheat specific to organic agriculture should be conducted on organically managed land. To verify these results the whole population was grown in replicated yield trials on six organic and six conventional sites in diverse locations. Conventionally managed sites exhibited positive and significant genetic correlations, whereas organically managed sites were more variable, resulting in lower realized breeding gain under organic farming. Although there were considerable rank changes, five of the eight top-10% under organic farming were also among the top-15% under conventional farming (Reid et al., 2011). This would support the strategy of Löschenberger et al. (2008) to apply indirect selection under low input conditions in early generations where high genotypic correlations were found for many traits, whereas in the later generations the replicated yield trials are grown in parallel under organic and low input conditions to separate breeding material for organic and conventional farming.

Similar results were reported for genetically broad populations of maize (Burger et al., 2008; Messmer et al., 2009), strongly supporting direct selection under organic farming for complex traits like grain yield, whereas indirect selection was very efficient for highly heritable traits like dry matter content. Although heritability values were assumed to be lower under organic farming due to higher experimental error rate, this was compensated in these studies by greater genotypic variance evoked under organic conditions. In contrast to these results, lower heritability under organic compared to conventional farming and high efficiency of indirect selection was found for grain yield of one maize population in the United States (Lorenzana and Bernardo, 2008), not justifying additional screening under organic farming (see also Chapter 10).

Low genotypic correlations were found for faba bean breeding material tested among different organic on-farm conditions (Ghaouti and Link, 2009), indicating that performance was controlled by substantially different sets of genes. The authors recommended direct- and site-specific selection in order to increase selection gain for typical organic locations (see also Chapter 13 in this book), whereas Horneburg and Becker (2008) advocated multi-location selection for organic outdoor tomato breeding in order to combine site-specific adaptation.

Breeding Strategies

Organic farming practices with low external inputs but very diverse crop rotations and organic fertilizer sources have a great impact on soil fertility and nutrient availability (Dawson et al., 2008a). Breeding crops adapted to specific farming practices will therefore be a major key for sustainable management of resources and eco-functional intensification of organic farming as proposed by Schmid et al. (2009).

Breeding for Improved Rhizosphere-Related Traits

Nutrient-efficient plants will play a major role in increasing crop yields in the next decades (Fageria et al., 2008), not only in organic agriculture but also on soils where the effectiveness of fertilizers may be limited by chemical and biological reactions, topsoil drying, subsoil constraints, and/or disease interactions (Rengel, 2005). Different breeding strategies like maintaining photosynthesis under nutrient stress, improved nutrient acquisition by enhanced root morphology, modified root exudates, and symbiotic interaction with soil micro-organisms have been proposed as well as the synchronization of nutrient demand and nutrient availability under organic farming (Dawson et al., 2008a; Fageria et al., 2008). Root growth and function (root architecture, root hair length, root metabolism) have gained increased attention for the development of high-yielding and eco-friendly crops (Garnett et al., 2009; den Herder et al., 2010; Lynch, 2007). Genetic variation in rhizo-deposition such as the efflux of protons, carbon compounds, organic acids, and more complex metabolites is important for the mobilization of nutrients such as P and transition metals and in facilitating association with soil microorganisms (Berg et al., 2009; Lynch, 2007). Genetic variation among cultivars has been reported for responsiveness to soil microbes, like mycorrhiza colonization, rhizobia nodulation, and in association with free-living bacteria (Wissuwa et al., 2009). Ion transporters might also be important targets for improving the acquisition of nitrate and for enhancing salt tolerance (Lynch, 2007). However, the incorporation of such traits that offer possibilities for improving nutrient acquisition capacity, plant–microbe interactions, and tolerance to abiotic and biotic soil stresses in breeding programs has been slow (Wissuwa et al., 2009). Reasons for this are (1) the lack of simple and efficient methods that allow for direct selection of large numbers of breeding materials for rhizosphere-related traits, and (2) the relatively slow progress in deciphering the genetics, physiology, and biochemistry behind the mechanisms of nutrient efficiency.

Therefore, future innovations that may increase the number of genotypes that can be screened for root growth and plant-microbe-soil interactions will be of great importance, especially for organic farming. Modern molecular tools, like marker-assisted selection, offer great potential to plant breeders in their endeavors (De Dorlodot et al., 2007; Johnson et al., 2000). Genomics-assisted selection (Varshney et al., 2005) combined with root modeling approaches (Wu et al., 2005) might further improve the understanding of these traits. The successful identification of the genes and processes involved will rely heavily on accurate phenotypic assessment of these rhizosphere-related traits under relevant farming conditions. Collaborative efforts among breeders, soil scientists, physiologists, agronomists, and farmers are needed to elucidate these complex interactions and to identify reliable screening methods for efficient selection.

Integration of Breeding and Agronomic Optimization

While considerable selection gain has been obtained for crops cultivated under high-input conditions in recent decades, these cultivars show little yield improvement under organic and low input conditions (Ceccarelli et al., 1998; Dawson et al., 2008a; Keneni and Imtiaz, 2010). Therefore, breeding efforts for these target environments need to be substantially enforced (Keneni and Imtiaz, 2010; Kotschi, 2009). As long as no large-scale screening methods are available for breeders to select for specific rhizosphere-related traits, they have to rely on selection conditions that are representative of organic farming conditions. This includes diverse biotic and abiotic stress factors such as limited N supply and/or release from organic fertilizer, weed interference, and seedling diseases in various combinations. Therefore, G×E interactions are expected to be more pronounced as yield differences among growing conditions increases (Dawson et al., 2008b; Reid et al., 2011). Breeders have several options of how to deal with G×E in their breeding process: (1) avoidance of G×E by selecting cultivars for broad adaptation to large geographic regions and management systems with little variation across environments, and (2) breeding for specific adaptation by trying to exploit G×E caused by predictable environmental variation such as site-specific pedo-climatic conditions (G×L) or management practice (G×M), while seeking adaptation to unpredictable seasonal variations, e.g., genotype × year interactions (G×Y; Annicchiarico, 2002). In general, target-specific adaptation allows for greater selection gain, however this is associated with higher relative costs compared to breeding for wide adaptation.

The frequently observed rank changes of cultivars under organic and conventional farming conditions highlights the need to select cultivars under conditions closely mimicking commercial organic farms. It is important for breeders to include traits important to organic farmers. Selection of traits with low G×M interaction might be possible under both organic and conventional farming, whereas in later generations the G×M interaction should be utilized by direct selection under organic farming to obtain maximum selection gain. For example, Baresel et al. (2008) suggested that wheat cultivars with improved nitrogen uptake efficiency during early growth stages would be more adapted to the timing of nitrogen mineralization on organically managed soils.

Breeding for yield stability across organic environments is also very challenging and requires multi-location trials representing different geographic regions but also the different types of organic farming. An alternative to yield stability across organic environments represents the decentralized participatory breeding approach allowing for local adaptation while maintaining cultivar diversity (see Chapter 6). The recent elucidation of environmentally induced epigenetic mechanisms might also boost interest in target-specific adaptation. Marker-assisted selection for utilizing G×E more efficiently might become an important tool in the near future (Backes and Østergard, 2008; Varshney et al., 2005; Zheng et al., 2010).

Finally, breeding efforts and improvement of organic management practices (e.g., reduced tillage and intercropping in organic farming) should be combined to utilize synergistic effects (Messmer et al., 2010; Keneni and Imtiaz, 2010). The successful integration of concomitant improvement of rice breeding and direct seeding mulch based cropping system was demonstrated by Vales et al. (2009). Considering nutrient efficiency issues on a holistic basis implies to develop nutrient-efficient crops and best-management practices that increase plant efficiency for utilization of applied organic fertilizers in the crop rotation. Breeding crops in and adapted to the unique conditions inherent in organic systems will be the key for eco-functional intensification and realization of the full potential of organic agriculture in different parts of the world.

References

Achat, D.L., M.R. Bakker, and P. Trichet. 2008. Rooting patterns and fine root biomass of *Pinus pinaster* assessed by trench wall and core methods. *Journal of Forest Research* 13:165–175.

Acuna, C.A., T.R. Sinclair, C.L. Mackowiak, A.R. Blount, K.H. Quesenberry, and W.W. Hanna. 2010. Potential root depth development and nitrogen uptake by tetraploid bahiagrass hybrids. *Plant and Soil* 334:491–499.

Almeida, M.M.T.B., A.T. Lixa, E.E. da Silva, P.H.S. de Azevedo, H. De-Polli, and R.D.D. Ribeiro. 2008. Legume fertilizers as alternative sources of nitrogen for organic lettuce production. *Pesquisa Agropecuaria Brasileira* 43:675–682.

Andrews, M., P.J. Lea, J.A. Raven, and K. Lindsey. 2004. Can genetic manipulation of plant nitrogen assimilation enzymes result in increased crop yield and greater N-use efficiency? An assessment. *Annals of Applied Biology* 145:25–40.

Annicchiarico, P. 2002. Genotype × environment interactions -Challenges and opportunities for plant breeding and cultivar recommendations. FAO plant production and protection paper No. 174. Rome, Italy: Food and Agriculture Organization of the United Nations.

Arp, W.J. 1991. Effects of source-sink relations on photosynthetic acclimation to elevated carbon dioxide. *Plant, Cell and Environment* 14:869–876.

Atlin, G.N., and K.J. Frey. 1990. Selecting oat lines for yield in low-productivity environments. *Crop Science* 30:556–561.

Backes, G., and H. Østergård. 2008. Molecular markers to exploit genotype-environment interactions of relevance in organic growing systems. *Euphytica* 163:523–531.

Badri, D.V., N. Quintana, E.G. El Kassis, H.K. Kim, Y.H. Choi, A. Sugiyama, R. Verpoorte, E. Martinoia, D.K. Manter, and J.M. Vivanco. 2009. An ABC transporter mutation alters root exudation of phytochemicals that provoke an overhaul of natural soil microbiota. *Plant Physiology* 151:2006–2017.

Baldani, J.I., V.M. Reis, V.L.D. Baldani, and J. Dobereiner. 2002. A brief story of nitrogen fixation in sugarcane: Reasons for success in Brazil. *Functional Plant Biology* 29:417–423.

Baresel, J.P., G. Zimmermann, and H.J. Reents. 2008. Effects of genotype and environment on N uptake and N partition in organically grown winter wheat (*Triticum aestivum* L.) in Germany. *Euphytica* 163:347–354.

Baudoin, E., A. Lerner, M.S. Mirza, H. El Zemrany, C. Prigent-Combaret, E. Jurkevich et al. 2010. Effects of *Azospirillum brasilense* with genetically modified auxin biosynthesis gene *ipdC* upon the diversity of the indigenous microbiota of the wheat rhizosphere. *Research in Microbiology* 161:219–226.

Berg, G. 2009. Plant-microbe interactions promoting plant growth and health: Perspectives for controlled use of microorganisms in agriculture. *Applied Microbiology and Biotechnology* 84:11–18.

Bhupinderpal-Singh, and Z. Rengel. 2007. The role of crop residues in improving soil fertility. *In*: P. Marschner, and Z. Rengel (eds.). Nutrient Cycling in Terrestrial Ecosystems. Berlin/Heidelberg, Germany: Springer.

Blum, A., G. Golan, J. Mayer, and B. Sinmena. 1997. The effect of dwarfing genes on sorghum grain filling from remobilized stem reserves, under stress. *Field Crops Research* 52:43–54.

Bonos, S.A., D. Rush, K. Hignight, and W.A. Meyer. 2004. Selection for deep root production in tall fescue and perennial Ryegrass. *Crop Science* 44:1770–1775.

Bourion, V., S.M.H. Rizvi, S. Fournier, H. de Larambergue, F. Galmiche, P. Marget, G. Duc, and J. Burstin. 2010. Genetic dissection of nitrogen nutrition in pea through a QTL approach of root, nodule, and shoot variability. *Theoretical & Applied Genetics* 121(1):71–86.

Brancourt-Hulmel, M., G. Doussinault, C. Lecomte, P. Berard, B. Le Buanec, and M. Trottet. 2003. Genetic improvement of agronomic traits of winter wheat cultivars released in France from 1946 to 1992. *Crop Science* 43:37–45.

Brancourt-Hulmel, M., E. Heumez, P. Pluchard, D. Beghin, C. Depatureaux, A. Giraud, and J. Le Gouis. 2005. Indirect versus direct selection of winter wheat for low-input or high-input levels. *Crop Science* 45:1427–1431.

Brennan, R., J.W. Bowden, M. Bolland, Z. Rengel, and D. Isbister. 2000. Is nutrition the answer to wheat after canola problems? Paper read at Crop Updates, February 2000, in Perth, Western Australia.

Brundrett, M.C. 2009. Mycorrhizal associations and other means of nutrition of vascular plants: Understanding the global diversity of host plants by resolving conflicting information and developing reliable means of diagnosis. *Plant and Soil* 320:37–77.

Burger, H., M. Schloen, W. Schmidt, and H.H. Geiger. 2008. Quantitative genetic studies on breeding maize for adaptation to organic farming. *Euphytica* 163:501–510.

Caba, J.M., C. Lluch, and F. Ligero. 1993. Genotypic differences in nitrogen assimilation in *Vicia faba*: Effect of nitrate. *Plant and Soil* 151:167–174.

Carr, P.M., H.J. Kandel, P.M. Porter, R.D. Horsley, and S.F. Zwinger. 2006. Wheat cultivar performance on certified organic fields in Minnesota and North Dakota. *Crop Science* 46:1963–1971.

Ceccarelli, S. 1996. Adaptation to low high input cultivation. *Euphytica* 92:203–214.

Ceccarelli, S., S. Grando, and A. Impiglia. 1998. Choice of selection strategy in breeding barley for stress environments. *Euphytica* 103:307–318.

Chloupek, O. 1972. The relationship between electric capacitance and some other parameters of plant roots. *Biologia Plantarum* 14:227–230.

———. 1977. Evaluation of size of a plant's root-system using its electrical capacitance. *Plant and Soil* 48:525–532.

Chloupek, O., B.P. Forster, and W.T.B. Thomas. 2006. The effect of semi-dwarf genes on root system size in field-grown barley. *Theoretical and Applied Genetics* 112(5):779–786.

Costigan, P.A. 1984. Critical concentrations of potassium required for maximum growth in lettuce and cabbage in the presence and absence of sodium. *Journal of the Science of Food and Agriculture* 35:296–296.

———. 1988. The placement of starter fertilizers to improve the early growth of drilled and transplanted vegetables. *In*: Proceedings of the International Fertilizer Society.

Crowley, D.E., and Z. Rengel. 1999. Biology and chemistry of nutrient availability in the rhizosphere. *In*: Rengel, Z. (ed.), Mineral nutrition of crops: Fundamental mechanisms and implications. pp. 1–40. New York: Food Products Press.

Crush, J.R., S.N. Nichols, and L. Ouyang. 2010. Adventitious root mass distribution in progeny of four perennial ryegrass (*Lolium perenne* L.) groups selected for root shape. *New Zealand Journal of Agricultural Research* 53:193–200.

Dawson, J.C., D.R. Huggins, and S.S. Jones. 2008. Characterizing nitrogen use efficiency in natural and agricultural ecosystems to improve the performance of cereal crops in low-input and organic agricultural systems. *Field Crops Research* 107:89–101.

Dawson, J.C., K.M. Murphy, and S.S. Jones. 2008. Decentralized selection and participatory approaches in plant breeding for low-input systems. *Euphytica* 160:143–154.

De Dorlodot, S., B. Forster, L. Pages, A. Price, R. Tuberosa, and X. Draye. 2007. Root system architecture: Opportunities and constraints for genetic improvement of crops. *Trends in Plant Science* 12:474–481.

de Faria, S.M., A.G. Diedhiou, H.C. de Lima, R.D. Ribeiro, A. Galiana, A.F. Castilho, and J.C. Henriques. 2010. Evaluating the nodulation status of leguminous species from the Amazonian forest of Brazil. *Journal of Experimental Botany* 61:3119–3127.

Delhaize, E., B.D. Gruber, and P.R. Ryan. 2007. The roles of organic anion permeases in aluminium resistance and mineral nutrition. *FEBS Letters* 581:2255–2262.

den Herder, G., G. Van Isterdael, T. Beeckman, and I. De Smet. 2010. The roots of a new green revolution. *Trends in Plant Science* 15:600–607.

Deng, W., K. Luo, Z. Li, Y. Yang, N. Hu, and Y. Wu. 2009. Overexpression of *Citrus junos* mitochondrial citrate synthase gene in *Nicotiana benthamiana* confers aluminum tolerance. *Planta* 230:355–365.

Diacono, M., and F. Montemurro. 2010. Long-term effects of organic amendments on soil fertility. A review. *Agronomy for Sustainable Development* 30:401–422.

———. 2010. Long-term effects of organic amendments on soil fertility. A review. *Agronomy for Sustainable Development* 30:401–422.

Drew, E.A., R.S. Murray, and S.E. Smith. 2006. Functional diversity of external hyphae of AM fungi: Ability to colonise new hosts is influenced by fungal species, distance and soil conditions. *Applied Soil Ecology* 32:350–365.

Dutta, S., and A.R. Podile. 2010. Plant growth promoting rhizobacteria (PGPR): The bugs to debug the root zone. *Critical Reviews in Microbiology* 36:232–244.

Eilers, K.G., C.L. Lauber, R. Knight, and N. Fierer. 2010. Shifts in bacterial community structure associated with inputs of low molecular weight carbon compounds to soil. *Soil Biology & Biochemistry* 42:896–903.

Fageria, N.K., V.C. Baligar, and Y.C. Li. 2008. The role of nutrient efficient plants in improving crop yields in the twenty first century. *Journal of Plant Nutrition* 31:1121–1157.

Falconer, D.S., and T.F.C. Mackay. 1996. Introduction to quantitative genetics. 4th ed. Essex, UK: Longman, Harlow.

Feddermann, N., R. Finlay, T. Boller, and M. Elfstrand. 2010. Functional diversity in arbuscular mycorrhiza – The role of gene expression, phosphorous nutrition and symbiotic efficiency. *Fungal Ecology* 3:1–8.

Frantz, J.M., G.E. Welbaum, Z.X. Shen, and R. Morse. 1998. Comparison of cabbage seedling growth in four transplant production systems. *HortScience* 33:976–979.

Friedel, J.K., S. Jakupaj, M. Gollner, R. Hrbek, C. Flamm, M. Oberforster, E. Zechner, A. Kinastberger, and F. Löschenberger. 2008. Mycorrhization of winter wheat cultivars in organic farming. Paper read at "Cultivating the Future Based on Science." Second Scientific Conference of the ISOFAR. June 16–20, 2008, Modena.

Gahoonia, T.S., and N.E. Nielsen. 2004. Root traits as tools for creating phosphorus efficient crop varieties. *Plant and Soil* 260:47–57.

———. 2004. Barley genotypes with long root hairs sustain high grain yields in low-P field. *Plant and Soil* 262:55–62.

Gahoonia, T.S., O. Ali, A. Sarker, M.M. Rahman, and W. Erskine. 2005. Root traits, nutrient uptake, multi-location grain yield and benefit-cost ratio of two lentil (*Lens culinaris*, Medikus.) varieties. *Plant and Soil* 272:153-161.

Gahoonia, T.S., R. Ali, R.S. Mallhotra, A. Jahoor, and M.M. Rahman. 2007. Variation in root morphological and physiological traits and nutrient uptake of chickpea genotypes. *Journal of Plant Nutrition* 30:829–841.

Garnett, T., V. Conn, and B.N. Kaiser. 2009. Root based approaches to improving nitrogen use efficiency in plants. *Plant, Cell and Environment* 32:1272–1283.

George, T.S., A.E. Richardson, P.A. Hadobas, and R.J. Simpson. 2004. Characterization of transgenic *Trifolium subterraneum* L. which expresses phyA and releases extracellular phytase: Growth and P nutrition in laboratory media and soil. *Plant, Cell and Environment* 27:1351–1361.

George, T.S., R.J. Simpson, P.A. Hadobas, and A.E. Richardson. 2005. Expression of a fungal phytase gene in *Nicotiana tabacum* improves phosphorus nutrition of plants grown in amended soils. *Plant Biotechnology Journal* 3:129–140.

George, T.S., P.J. Gregory, P. Hocking, and A.E. Richardson. 2008. Variation in root-associated phosphatase activities in wheat contributes to the utilization of organic P substrates in vitro, but does not explain differences in the P-nutrition of plants when grown in soils. *Environmental & Experimental Botany* 64:239–249.

George, T.S., A.E. Richardson, S.S. Li, P.J. Gregory, and T.J. Daniell. 2009. Extracellular release of a heterologous phytase from roots of transgenic plants: Does manipulation of rhizosphere biochemistry impact microbial community structure? *FEMS Microbiology Ecology* 70:433–445.

Ghaouti, L., and W. Link. 2009. Local vs. formal breeding and inbred line vs. synthetic cultivar for organic farming: Case of *Vicia faba* L. *Field Crops Research* 110:167–172.

Guarda, G., S. Padovan, and G. Delogu. 2004. Grain yield, nitrogen-use efficiency and baking quality of old and modern Italian bread-wheat cultivars grown at different nitrogen levels. *European Journal of Agronomy* 21:181–192.

Guppy, C.N., and M.J. McLaughlin. 2009. Options for increasing the biological cycling of phosphorus in low-input and organic agricultural systems. *Crop & Pasture Science* 60:116–123.

Gustafson, G.M., E. Salomon, S. Jonsson, and S. Steineck. 2003. Fluxes of K, P, and Zn in a conventional and an organic dairy farming system through feed, animals, manure, and urine – A case study at Ojebyn, Sweden. *European Journal of Agronomy* 20:89–99.

Hartmann, A., M. Schmid, D. Van Tuinen, and G. Berg. 2009. Plant-driven selection of microbes. *Plant and Soil* 321:235–257.

Hauggaard-Nielsen, H., and E.S. Jensen. 2005. Facilitative root interactions in intercrops. *Plant and Soil* 274:237–250.

Hetrick, B.A.D., G.W.T. Wilson, and T.S. Cox. 1992. Mycorrhizal dependence of modern wheat varieties, landraces, and ancestors. *Canadian Journal of Botany* 70:2032–2040.

———. 1993. Mycorrhizal dependence of modern wheat cultivars and ancestors – A synthesis. *Canadian Journal of Botany-Revue Canadienne De Botanique* 71:512–518.

Hildermann, I., M. Messmer, P. Kunz, A. Pregitzer, T. Boller, A. Wiemken, and P. Mäder. 2009. Sorte × Umwelt-Interaktionen von Winterweizen im biologischen Landbau – Cultivar × site interaction of winter wheat under diverse organic farming conditions. Tagungsband der 60. Tagung der Vereinigung der Pflanzenzüchter und Saatgutkaufleute Österreichs, November 24–26, 2009, pp. 163–166. Austria: Raumberg-Gumpenstein.

Hildermann, I., A. Thommen, D. Dubois, T. Boller, A. Wiemken, and P. Mäder. 2009. Yield and baking quality of winter wheat cultivars in different farming systems of the DOK long-term trials. *Journal of the Science of Food and Agriculture* 89:2477–2491.

Hildermann, I., M. Messmer, D. Dubois, T. Boller, A. Wiemken, and P. Mader. 2010. Nutrient use efficiency and arbuscular mycorrhizal root colonisation of winter wheat cultivars in different farming systems of the DOK long-term trial. *Journal of the Science of Food and Agriculture* 90:2027–2038.

Horneburg, B., and H.C. Becker. 2008. Crop adaptation in on-farm management by natural and conscious selection: A case study with lentil. *Crop Science* 48:203–212.

Hund, A., Y. Fracheboud, A. Soldati, and P. Stamp. 2008. Cold tolerance of maize seedlings as determined by root morphology and photosynthetic traits. *European Journal of Agronomy* 28:178–185.

Hund, A., S. Trachsel, and P. Stamp. 2009. Growth of axile and lateral roots of maize: I development of a phenotyping platform. *Plant and Soil* 325:335–349.

Hungria, M., R.J. Campo, E.M. Souza, and F.O. Pedrosa. 2010. Inoculation with selected strains of *Azospirillum brasilense* and A. lipoferum improves yields of maize and wheat in Brazil. *Plant & Soil* 331:413–425.

Jiang, Z.C., W.M. Sullivan, and R.J. Hull. 2000. Nitrate uptake and nitrogen use efficiency by Kentucky bluegrass cultivars. *HortScience* 35:1350–1354.

Johnson, W.C., L.E. Jackson, O. Ochoa, R. van Wijk, J. Peleman, D.A. St Clair, and R.W. Michelmore. 2000. Lettuce, a shallow-rooted crop, and *Lactuca serriola*, its wild progenitor, differ at QTL determining root architecture and deep soil water exploitation. *Theoretical and Applied Genetics* 101:1066–1073.

Jones, D.L., A. Hodge, and Y. Kuzyakov. 2004. Plant and mycorrhizal regulation of rhizodeposition. *New Phytologist* 163:459–480.

Jones, D.L., C. Nguyen, and R.D. Finlay. 2009. Carbon flow in the rhizosphere: Carbon trading at the soil-root interface. *Plant and Soil* 321:5–33.

Jones, H., S. Clarke, Z. Haigh, H. Pearce, and M. Wolfe. 2010. The effect of the year of wheat variety release on productivity and stability of performance on two organic and two non-organic farms. *Journal of Agricultural Science* 148:303–317.

Kamara, A.Y., A. Menkir, I. Kureh, and L.O. Omoigui. 2006. Response to low soil nitrogen stress of S-1, maize breeding lines, selected for high vertical root-pulling resistance. *Maydica* 51:425-433.

Kaspar, T.C., H.M. Taylor, and R.M. Shibles. 1984. Taproot-elongation rates of soybean cultivars in the glasshouse and their relation to field rooting depth. *Crop Science* 24:916–920.

Keneni, G., and M. Imtiaz. 2010. Demand-driven breeding of food legumes for plant-nutrient relations in the tropics and the sub-tropics: Serving the farmers; not the crops! *Euphytica* 175:267–282.

Kohler, J., F. Caravaca, M. del Mar Alguacil, and A. Roldan. 2009. Elevated CO_2 increases the effect of an arbuscular mycorrhizal fungus and a plant-growth-promoting rhizobacterium on structural stability of a semiarid agricultural soil under drought conditions. *Soil Biology & Biochemistry* 41:1710–1716.

Kotschi, J. 2009. The role of organic farming for global food security. *Gaia-Ecological Perspectives for Science and Society* 18:200–204.

Kristensen, H.L., and K. Thorup-Kristensen. 2004. Uptake of N-15 labeled nitrate by root systems of sweet corn, carrot and white cabbage from 0.2–2.5 meters depth. *Plant and Soil* 265:93–100.

———. 2004. Root growth and nitrate uptake of three different catch crops in deep soil layers. *Soil Science Society of America Journal* 68:529–537.

Lainé, P., A. Ourry, J. Macduff, J. Boucaud, and J. Salette. 1993. Kinetic parameters of nitrate uptake by different catch crop species: Effects of low temperatures or previous nitrate starvation. *Physiologia Plantarum* 88:85–92.

Lambers, H., C. Mougel, B. Jaillard, and P. Hinsinger. 2009. Plant-microbe-soil interactions in the rhizosphere: An evolutionary perspective. *Plant and Soil* 321:83–115.

Lammerts van Bueren, E.T., P.C. Struik, and E. Jacobsen. 2002. Ecological concepts in organic farming and their consequences for an organic crop ideotype. *Netherlands Journal of Agricultural Science* 50:1–26.

Lammerts van Bueren, E.T., S.S. Jones, L. Tamm, K.M. Murphy, J.R. Myers, C. Leifert, and M.M. Messmer. 2011. The need to breed crop varieties suitable for organic farming, using wheat, tomato and broccoli as examples: A review. *NJAS Wageningen Journal of Life Sciences* 58:193–205.

Landi, P., M.C. Sanguineti, S. Salvi, S. Giuliani, M. Bellotti, M. Maccaferri, S. Conti, and R. Tuberosa. 2005. Validation and characterization of a major QTL affecting leaf ABA concentration in maize. *Molecular Breeding* 15:291–303.

Leifeld, J., and J. Fuhrer. 2010. Organic farming and soil carbon sequestration: What do we really know about the benefits? *Ambio* 39:585–599.

Liao, H., X.L. Yan, G. Rubio, S.E. Beebe, M.W. Blair, and J.P. Lynch. 2006. Genetic mapping of basal root gravitropism and phosphorus acquisition efficiency in common bean. *Functional Plant Biology* 33:207.

Lin, S.F., D. Grant, S. Cianzio, and R. Shoemaker. 2000. Molecular characterization of iron deficiency chlorosis in soybean. *Journal of Plant Nutrition* 23:1929–1939.

Liu, Y., G.H. Mi, F.J. Chen, J.H. Zhang, and F.S. Zhang. 2004. Rhizosphere effect and root growth of two maize (*Zea mays* L.) genotypes with contrasting P efficiency at low P availability. *Plant Science* 167:217–223.

Lorenzana, R.E., and R. Bernardo. 2008. Genetic correlation between corn performance in organic and conventional production systems. *Crop Science* 48:903–910.

Löschenberger, F., A. Fleck, H. Grausgruber, H. Hetzendorfer, G. Hof, J. Lafferty, M. Marn, A. Neumayer, G. Pfaffinger, and J. Birschitzky. 2008. Breeding for organic agriculture: The example of winter wheat in Austria. *Euphytica* 163:469–480.

Lugtenberg, B., and F. Kamilova. 2009. Plant-growth-promoting rhizobacteria. *Annual Review of Microbiology* 63:541–556.

Lynch, J. 1998. The role of nutrient-efficient crops in modern agriculture. *In*: Rengel, Z. (ed.), Nutrient use in crop production. New York: The Haworth Press.

Lynch, J.P., and K.M. Brown. 2001. Topsoil foraging – An architectural adaptation of plants to low phosphorus availability. *Plant and Soil* 237:225–237.

Lynch, J.P. 2007. Roots of the second green revolution. *Australian Journal of Botany* 55:493–512.

Ma, Q.F., Z. Rengel, and J. Kuo. 2002. Aluminium toxicity in rye (*Secale cereale*): Root growth and dynamics of cytoplasmic Ca^{2+} in intact root tips. *Annals of Botany* 89:241–244.

Mäder, P., D. Hahn, D. Dubois, L. Gunst, T. Alfoldi, H. Bergmann, M. Oehme et al. 2007. Wheat quality in organic and conventional farming: Results of a 21-year field experiment. *Journal of the Science of Food and Agriculture* 87:1826–1835.

Mäder, P., F. Kaiser, A. Adholeya, R. Singh, H.S. Uppal, A.K. Sharma, R. Srivastava et al. 2011. Inoculation of root microorganisms for sustainable wheat-rice and wheat-black gram rotations in India. *Soil Biology & Biochemistry* 43:609–619.

Maj, D., J. Wielbo, M. Marek-Kozaczuk, and A. Skorupska. 2010. Response to flavonoids as a factor influencing competitiveness and symbiotic activity of *Rhizobium leguminosarum*. *Microbiological Research* 165:50–60.

Mandal, N.P., P.K. Sinha, M. Variar, V.D. Shukla, P. Perraju, A. Mehta, A.R. Pathak et al. 2010. Implications of genotype × input interactions in breeding superior genotypes for favorable and unfavorable rain-fed upland environments. *Field Crops Research* 118:135–144.

Manjarrez, M., M. Wallwork, S.E. Smith, F. Andrew Smith, and S. Dickson. 2009. Different arbuscular mycorrhizal fungi induce differences in cellular responses and fungal activity in a mycorrhiza-defective mutant of tomato (rmc). *Functional Plant Biology* 36:86–96.

Manske, G.G.B., J.I. Ortiz-Monasterio, R.M. van Ginkel, S. Rajaram, and P.L.G. Vlek. 2002. Phosphorus use efficiency in tall, semi-dwarf and dwarf near-isogenic lines of spring wheat. *Euphytica* 125(1):113–119.

Marinari, S., A. Lagomarsino, M.C. Moscatelli, A. Di Tizio, and E. Campiglia. 2010. Soil carbon and nitrogen mineralization kinetics in organic and conventional three-year cropping systems. *Soil & Tillage Research* 109:161–168.

Marschener, H. 1998. Role of root growth, arbuscular mycorrhiza, and root exudates for the efficiency in nutrient acquisition. *Field Crops Research* 56:203–207.

Marschner, H. 1995. Mineral nutrition of higher plants. 2nd ed. London: Academic Press.

Marschner, P., and Z. Rengel. 2003. Contributions of rhizosphere interactions to soil biological fertility. *In*: Soil biological fertility: A key to sustainable land use in agriculture. Dordrecht, The Netherlands: Kluwer Academic Publishers.

Marschner, P., D.E. Crowley, and C.H. Yang. 2004. Development of specific rhizosphere bacterial communities in relation to plant species, nutrition and soil type. *Plant and Soil* 261:199–208.

Marschner, P., and S. Timonen. 2005. Interactions between plant species and mycorrhizal colonization on the bacterial community composition in the rhizosphere. *Applied Soil Ecology* 28:23–36.

Marschner, P., Z. Solaiman, and Z. Rengel. 2005. Growth, phosphorus uptake, and rhizosphere microbial-community composition of a phosphorus-efficient wheat cultivar in soils differing in pH. *Journal of Plant Nutrition and Soil Science* 168:343–351.

———. 2006. Rhizosphere properties of Poaceae genotypes under P-limiting conditions. *Plant and Soil* 283:11–24.

———. 2007. Brassica genotypes differ in growth, phosphorus uptake and rhizosphere properties under P-limiting conditions. *Soil Biology & Biochemistry* 39:87–98.

Marschner, P., and Z. Rengel. 2010. The effects of plant breeding on soil microbes. *In*: Dixon, G.R., and E.L. Tilston (eds.), Soil microbiology and sustainable crop production. Dordrecht, The Netherlands: Springer.

Mattoo, A.K., and J.R. Teasdale. 2010. Ecological and genetic systems underlying sustainable horticulture. *In*: Janick, J. (ed.), Horticultural reviews, volume 37. Hoboken, NJ: John Wiley & Sons, Inc. doi: 10.1002/9780470543672.

McCaig, T.N., and J.A. Morgan. 1993. Root and shoot dry-matter partitioning in near-isogenic wheat lines differing in height. *Canadian Journal of Plant Science* 73:679–689.

Messmer, M., H. Burger, W. Schmidt, and H. Geiger. 2009. Importance of appropriate selection environments for breeding maize adapted to organic farming systems. Tagungsband der 60. Jahrestagung der Vereinigung der Pflanzenzüchter und Saatgutkaufleute Österreichs, November 24–26, 2009, pp. 49–52. Austria: Raumberg-Gumpenstein.

Messmer, M.M., A. Berner, M. Krauss, J. Jansa, T. Presterl, W. Schmidt, and P. Mäder. 2010. Genetic variation for nutrient use efficiency in maize under different tillage and fertilization regimes with special emphasis to plant microbe interaction. *In*: Goldringer, I., J.C. Dawson, A. Vettoretti, F. Rey (eds.) Breeding for resilience: A strategy for organic and low-input farming systems? Proceedings of EUCARPIA conference: Section organic and Low-input Agriculture, December 1–3, 2010, pp. 72–75. Paris, France: INRA and ITAB.

Meyers, L.L., M.P. Russelle, and J.A.F.S. Lamb. 1996. Fluridone reveals root elongation differences among alfalfa germplasms. *Agronomy Journal* 88:67–72.

Miralles, D.J., G.A. Slafer, and V. Lynch. 1997. Rooting patterns in near-isogenic lines of spring wheat for dwarfism. *Plant and Soil* 197:79–86.

Morandi, D., C. le Signor, V. Gianinazzi-Pearson, and G. Duc. 2009. A *Medicago truncatula* mutant hyper-responsive to mycorrhiza and defective for nodulation. *Mycorrhiza* 19:435–441.

Murphy, K., and S.S. Jones. 2007. Genetic assessment of the role of breeding wheat for organic systems. *In*: Buck, H.T., J.E. Nisi, and N. Salomon (eds.). Wheat production in stressed environments series: Developments in plant breeding, volume 12. pp. 217–222. New York: Springer.

Muurinen, S., G.A. Slafer, and P. Peltonen-Sainio. 2006. Breeding effects on nitrogen use efficiency of spring cereals under northern conditions. *Crop Science* 46:561–568.

Neumann, G., T.S. George, and C. Plassard. 2009. Strategies and methods for studying the rhizosphere – The plant science toolbox. *Plant and Soil* 321:431–456.

Nuruzzaman, M., H. Lambers, M.D.A. Bolland, and E.J. Veneklaas. 2006. Distribution of carboxylates and acid phosphatase and depletion of different phosphorus fractions in the rhizosphere of a cereal and three grain legumes. *Plant and Soil* 281:109–120.

Olesen, J.E., M. Askegaard, and I.A. Rasmussen. 2009. Winter cereal yields as affected by animal manure and green manure in organic arable farming. *European Journal of Agronomy* 30:119–128.

Osborne, L.D., and Z. Rengel. 2002. Screening cereals for genotypic variation in efficiency of phosphorus uptake and utilisation. *Australian Journal of Agricultural Research* 53:295–303.

Parker, D.R., and W.A. Norvell. 1999. Advances in solution culture methods for plant mineral nutrition research. *Advances in Agronomy* 65:151–213.

Presterl, T., G. Seitz, M. Landbeck, E.M. Thiemt, W. Schmidt, and H.H. Geiger. 2003. Improving nitrogen-use efficiency in European maize: Estimation of quantitative genetic parameters. *Crop Science* 43:1259–1265.

Przystalski, M., A. Osman, E. Thiemt, B. Rolland, L. Ericson, H. Østergård, L. Levy et al. 2008. Comparing the performance of cereal varieties in organic and non-organic cropping systems in different European countries. *Euphytica* 163:417–433.

Raun, W.R., and G.V. Johnson. 1999. Improving nitrogen use efficiency for cereal production. *Agronomy Journal* 91:357–363.

Reid, T.A., R.C. Yang, D.F. Salmon, and D. Spaner. 2009. Should spring wheat breeding for organically managed systems be conducted on organically managed land? *Euphytica* 169:239–252.

Reid, T.A., R.C. Yang, D.F. Salmon, A. Navabi, and D. Spaner. 2011. Realized gains from selection for spring wheat grain yield are different in conventional and organically managed systems. *Euphytica* 177:253–266.

Rengel, Z., and R.D. Graham. 1995. Importance of seed Zn content for wheat growth on Zn-deficient soil. II. Grain yield. *Plant and Soil* 173:267–274.

Rengel, Z. 2001. Genotypic differences in micronutrient use efficiency in crops. *Communications in Soil Science and Plant Analysis* 32:1163–1186.

———. 2002. Breeding for better symbiosis. *Plant and Soil* 245:147–162.

———. 2005. Breeding crops for adaptation to environments with low nutrient availability. *In*: Ashraf, M., and P.J.C. Harris (eds.) Abiotic stresses: Plant resistance through breeding and molecular approaches. New York: The Haworth Press.

Rengel, Z., and P. Marschner. 2005. Nutrient availability and management in the rhizosphere: Exploiting genotypic differences. *New Phytologist* 168:305–312.

Richardson, A.E., P.A. Hadobas, and J.E. Hayes. 2000. Acid phosphomonoesterase and phytase activities of wheat (*Triticum aestivum* L.) roots and utilization of organic phosphorus substrates by seedlings grown in sterile culture. *Plant, Cell and Environment* 23:397–405.

———. 2001. Extracellular secretion of Aspergillus phytase from Arabidopsis roots enables plants to obtain phosphorus from phytate. *Plant Journal* 25:641–649.

Richardson, A.E. 2009. Regulating the phosphorus nutrition of plants: Molecular biology meeting agronomic needs. *Plant and Soil* 322:17–24.

Richardson, A.E., J.M. Barea, A.M. McNeill, and C. Prigent-Combaret. 2009. Acquisition of phosphorus and nitrogen in the rhizosphere and plant growth promotion by microorganisms. *Plant and Soil* 321:305–339.

Roelofs, R.F.R., Z. Rengel, G.R. Cawthray, K.W. Dixon, and H. Lambers. 2001. Exudation of carboxylates in Australian Proteaceae: Chemical composition. *Plant, Cell and Environment* 24:891–903.

Ryan, J., S. Masri, S. Ceccarelli, S. Grando, and H. Ibrikci. 2008. Differential responses of barley landraces and improved barley cultivars to nitrogen-phosphorus fertilizer. *Journal of Plant Nutrition* 31:381–393.

Sadeghzadeh, B., Z. Rengel, and C. Li. 2009. Differential zinc efficiency of barley genotypes grown in soil and chelator-buffered nutrient solution. *Journal of Plant Nutrition* 32:1744–1767.

Sadeghzadeh, B., Z. Rengel, C. Li, and H. Yang. 2010. Molecular marker linked to a chromosome region regulating seed Zn accumulation in barley. *Molecular Breeding* 25:167–177.

Sadeghzadeh, B., and Z. Rengel. 2011. Zinc in soils and crop nutrition. *In*: Hawkesford, M.J., and P. Barraclough (eds.). The molecular basis of nutrient use efficiency in crops. London: John Wiley & Sons, pp. 335–376.

Salvagiotti, F., K.G. Cassman, J.E. Specht, D.T. Walters, A. Weiss, and A. Dobermann. 2008. Nitrogen uptake, fixation and response to fertilizer N in soybeans: A review. *Field Crops Research* 108:1–13.

Schmid, O., S. Padel, N. Halberg, M. Huber, I. Darnhofer, C. Micheloni, C. Koopmans, C. Stopes, H. Bügel, H. Willer, M. Schlüter, E. Cuoco. 2009. Strategic research agenda for organic food and farming. TP Organics, Brüssel. Accessed April 13, 2011. http://orgprints.org/16694/.

Sinebo, W., R. Gretzmacher, and A. Edelbauer. 2002. Environment of selection for grain yield in low fertilizer input barley. *Field Crops Research* 74:151–162.

Smit, A.L., and J. Groenwold. 2005. Root characteristics of selected field crops: Data from the Wageningen Rhizolab (1990–2002). *Plant and Soil* 272:365–384.

Solaiman, Z., P. Marschner, D. Wang, and Z. Rengel. 2007. Growth, P uptake and rhizosphere properties of wheat and canola genotypes in an alkaline soil with low P availability. *Biology and Fertility of Soils* 44:143–153.

Steele, K.A., D.S. Virk, R. Kumar, S.C. Prasad, and J.R. Witcombe. 2007. Field evaluation of upland rice lines selected for QTLs controlling root traits. *Field Crops Research* 101:180–186.

Stockdale, E.A., and P.C. Brookes. 2006. Detection and quantification of the soil microbial biomass: Impacts on the management of agricultural soils. *Journal of Agricultural Science* 144:285–302.

Stone, L.R., D.E. Goodrum, M.N. Jaafar, and A.H. Khan. 2001. Rooting front and water depletion depths in grain sorghum and sunflower. *Agronomy Journal* 93:1105–1110.

Svecnjak, Z., and Z. Rengel. 2006. Nitrogen utilization efficiency in canola cultivars at grain harvest. *Plant and Soil* 283:299–307.

Tarafdar, J.C, and A. Jungk. 1987. Phosphatase activity in the rhizosphere and its relation to the depletion of soil organic phosphorus. *Biology and Fertility of Soils* 3:199–204.

Thorup-Kristensen, K., and R. van den Boogaard. 1998. Temporal and spatial root development of cauliflower (*Brassica oleracea* L. var. *botrytis* L.). *Plant and Soil* 201:37–47.

Thorup-Kristensen, K. 1998. Root growth of green pea (*Pisum sativum* L.) genotypes. *Crop Science* 38:1445–1451.

Thorup-Kristensen, K., and R. van den Boogaard. 1999. Vertical and horizontal development of the root system of carrots following green manure. *Plant and Soil* 212:145–153.

Thorup-Kristensen, K. 2001. Are differences in root growth of nitrogen catch crops important for their ability to reduce soil nitrate-N content, and how can this be measured? *Plant and Soil* 230:185–195.

———. 2006. Root growth and nitrogen uptake of carrot, early cabbage, onion and lettuce following a range of green manures. *Soil Use and Management* 22:29–38.

———. 2006. Effect of deep and shallow root systems on the dynamics of soil inorganic N during 3-year crop rotations. *Plant and Soil* 288:233–248.

Thorup-Kristensen, K., M.S. Cortasa, and R. Loges. 2009. Winter wheat roots grow twice as deep as spring wheat roots, is this important for N uptake and N leaching losses? *Plant and Soil* 322:101–114.

Unkovich, M., and J. Baldock. 2008. Measurement of asymbiotic N-2 fixation in Australian agriculture. *Soil Biology & Biochemistry* 40:2915–2921.

Vales, M., L. Seguy, S. Bouzinac, and J. Taillebois. 2009. Improvement of cropping systems by integration of rice breeding: A novel genetic improvement strategy. *Euphytica* 167:161–164.

van Beem, J., M.E. Smith, and R.W. Zobel. 1998. Estimating root mass in maize using a portable capacitance meter. *Agronomy Journal* 90:566–570.

Varshney, R.K., A. Graner, and M.E. Sorrells. 2005. Genomics-assisted breeding for crop improvement. *Trends in Plant Science* 10:621–630.

Vlachostergios, D.N., and D.G. Roupakias. 2008. Response to conventional and organic environment of thirty-six lentil (*Lens culinaris* Medik.) varieties. *Euphytica* 163:449–457.

Werth, M., and Y. Kuzyakov. 2010. C-13 fractionation at the root-microorganisms-soil interface: A review and outlook for partitioning studies. *Soil Biology & Biochemistry* 42:1372–1384.

Whitelaw, M.A., T.J. Harden, and K.R. Helyar. 1999. Phosphate solubilisation in solution culture by the soil fungus *Penicillium radicum*. *Soil Biology and Biochemistry* 31:655–665.

Whitelaw, M.A. 2000. Growth promotion of plants inoculated with phosphate-solubilizing fungi. *Advances in Agronomy* 69:99–151.

Wissuwa, M., and N. Ae. 2001. Genotypic variation for tolerance to phosphorus deficiency in rice and the potential for its exploitation in rice improvement. *Plant Breeding* 120:43–48.

Wissuwa, M., M. Mazzola, and C. Picard. 2009. Novel approaches in plant breeding for rhizosphere-related traits. *Plant and Soil* 321:409–430.

Wojciechowski, T., M.J. Gooding, L. Ramsay, and P.J. Gregory. 2009. The effects of dwarfing genes on seedling root growth of wheat. *Journal of Experimental Botany* 60:2565–2573.

Wu, L., M.B. McGechan, C.A. Watson, and J.A. Baddeley. 2005. Developing existing plant root system architecture models to meet future, agricultural challenges. *Advances in Agronomy* 85:181–219.

Yadav, R.S., and J.C. Tarafdar. 2003. Phytase and phosphatase producing fungi in arid and semi-arid soils and their efficiency in hydrolyzing different organic P compounds. *Soil Biology and Biochemistry* 35:1–7.

Yazdanbakhsh, N., and J. Fisahn. 2009. High throughput phenotyping of root growth dynamics, lateral root formation, root architecture and root hair development enabled by PlaRoM. *Functional Plant Biology* 36:938–946.

Zheng, B.S., J. Le Gouis, M. Leflon, W.Y. Rong, A. Laperche, and M. Brancourt-Hulmel. Using probe genotypes to dissect QTL × environment interactions for grain yield components in winter wheat. *Theoretical and Applied Genetics* 121:1501–1517.

Zhu, Y.G., S.E. Smith, A.R. Barritt, and F.A. Smith. 2001. Phosphorus (P) efficiencies and mycorrhizal responsiveness of old and modern wheat cultivars. *Plant and Soil* 237:249–255.

3 Pest and Disease Management in Organic Farming: Implications and Inspirations for Plant Breeding

Thomas F. Döring, Marco Pautasso, Martin S. Wolfe, and Maria R. Finckh

Introduction

The co-evolution of plants with their pests and diseases is a major driving force in evolution in nature. As a consequence, many pests and pathogens have multiple functions involved in survival on host populations. As a result of this continuous co-evolution, plant pests and pathogens have been selected for high reproduction rates, because of the low probability of an individual being able to find or infect a compatible host plant. For their part, host plants in natural ecosystems are often interspersed among other plant species and show wide and changing variation among individuals for genes affecting specific or non-specific resistance to each of the many pathogens that may attack them. Throughout the plant world there are dynamic and unstable equilibria between hosts and pathogens that are easily perturbed.

A major form of perturbation is agriculture, particularly in the form of monoculture, which creates massive exposure of a single host genotype to the pathogen population, raising, by orders of magnitude, the probability of successful infection by individual spores or pests and creating an opportunity for epidemics. Such a fundamental ecological shift is countered by either a technological response (breeding resistant varieties or developing interventions such as pesticides) or by mimicking or building on natural systems (diversifying host stands). The view of organic agriculture is that the greater holism of the second approach should make more tools available on the host side of the co-evolution while avoiding the risks associated with a more linear, technological approach.

We begin this chapter with two case studies that exemplify the central importance of plant diversity for plant protection. We then briefly review the principles and practices of plant protection in organic agriculture. With this context laid out, we next turn to some key questions that are directly related to plant breeding for organic plant protection: What are the target areas of organic plant protection for breeding, i.e., what should be the priorities of organic plant breeding? What are suitable breeding goals for organic plant protection? Which breeding approaches can be used to directly target pests and diseases? What are possible approaches of plant breeding that indirectly contribute to plant protection? And finally, what are suitable selection strategies from the specific perspective of organic plant protection?

Late Blight of Potato

Late blight is one of the most notorious plant diseases, with a dramatic history and global economic importance. Caused by the Oomycete *Phytophthora infestans* (Montagne) de Bary, the disease affects leaves, stems, and tubers of the potato plant. When late blight is left uncontrolled it can spread very rapidly (within a few weeks) through the potato crop, destroying all foliage and leaving behind a brown stinking mass. This reduces photosynthesis and tuber growth, and leaves the soil exposed to wind and water erosion and to late season weed development. Sporangia of the pathogen can be washed off the plants by rain into the soil where they may infect tubers.

Originating from Mexico, *P. infestans* was first recorded in the United States in 1843. However, it had its greatest historical impact from the mid-1840s in Ireland, where it contributed, along with other socio-economic factors, to the Great Irish Famine, with millions of people leaving the country or starving to death (Turner, 2005). This was due largely to the narrow genetic base of potatoes and the widespread use of a single variety (cv. Lumper; Finckh and Wolfe, 2006; Fowler and Mooney, 1990; MacLeod et al., 2010).

While organic farmers usually perceive late blight as a major problem in potato production (Tamm et al., 2004), yields are often limited by nutrient availability rather than by late blight itself (e.g., Möller et al., 2007; but see Nemecek et al., 2011). Similarly, variations in water supply can override effects of late blight on potato yields. Moreover, in the special case of seed potato production, late defoliation by blight can help prevent late season virus transmission to potato tubers and keep the tubers below the size threshold for seed.

Organic farmers use rotational breaks between potato crops to provide adequate plant nutrition and to reduce soil-borne inoculum of late blight and other pathogens. They are also advised to maximize the distance between potato fields to reduce inoculum spread. Although Aylor (2003) showed that the critical gap for *P. infestans* dispersal is about 35 to 50 km, there is evidence that reducing landscape connectivity by deploying resistant cultivars is a promising control strategy (Skelsey et al., 2010). In addition, growing potatoes in diversified systems such as strip intercropping has been shown to reduce overall disease (Bouws and Finckh, 2008). Pre-sprouting of tubers is recommended for speeding up plant development so that yield losses due to late blight are minimized (Karalus and Rauber, 1997). Where permitted, copper used as a contact fungicide, in combination with forecasting services, can reduce late blight (Finckh et al., 2008b). However, copper is a toxic heavy metal and has detrimental effects in the environment.

One of the most important tools for late blight control is cultivar choice. Organic growers tend to use more resistant and earlier maturing varieties than other growers (Tamm et al., 2004). Although they may be susceptible, early varieties often escape blight epidemics (Finckh et al., 2008). The susceptibility of potatoes to *P. infestans* depends not only on the potato variety and the amount of pathogen inoculum available, but also on the strains of the pathogen present. This issue became more important in Europe after Mexican potatoes imported in the late 1970s led to the establishment of the second mating type, A2, which allowed *P. infestans* to reproduce sexually (Fry et al., 1993). This resulted in greater genetic diversity of the pathogen (Shattock, 2002), increasing its potential to overcome host resistance and fungicides. The likelihood that a virulent strain of a pathogen establishes a large population and becomes a problem increases with the frequency of susceptible host plants. Therefore, diversification of the crop has been repeatedly suggested as a strategy to reduce late blight levels (Phillips et al., 2005; Finckh et al., 2008a). Success will depend on resistance management, i.e., not deploying resistant varieties on large areas (both within regions and worldwide) so as to minimize selection for pathogen strains that can overcome the resistance. This could be helped by participatory plant breeding as in the Netherlands (see Chapter 14), which encourages development and use of a range of potato genotypes at the farm level to help in buffering against environmental variation.

In summary, late blight, if and when it does pose a problem in organic potato production, is not always easy to manage with cultivation measures and control interventions. Cultivar choice and improvement also play a crucial and complementary role; their success will depend on a high degree of genetic diversity in the resistances used. While targeted breeding for late blight resistance is clearly desirable from an organic viewpoint, the availability of resistant cultivars should not lead to complacency; host resistance to late blight needs to be integrated into a complementary package of control measures.

Yellow Rust in Wheat

Yellow rust in wheat (also called stripe rust), is caused by the fungus *Puccinia striiformis* Westend. f. sp. *tritici* Eriks., and is characterized by stripes of small yellowish pustules on the leaves. Until recently, it was regarded as important mainly in cool climates, because the pathogen can survive cold winters in infected host tissue. Epidemics usually start in the autumn, but symptoms develop only slowly over winter. In late spring, infected leaves can desiccate rapidly when the weather is warm and dry. As an obligate pathogen, *P. striiformis* requires green plant material to survive and over-summers on volunteer plants or early autumn-sown crops. During the growing season, the fungus produces multiple asexual generations during which new races may emerge through asexual recombination.

P. striiformis has been shown to travel long distances, even between continents, with subsequent sweeping changes to the genetic make-up and ecological potential of regional rust populations. For instance, new isolates from yellow rust in the United States showed adaptation to warmer climates and were also more aggressive (Milus et al., 2009), underlining the potential of genetically diverse pathogen populations to adapt to new conditions. Trying to contain yellow rust by resistance breeding is a race between breeders and the evolving pathogen. It results in the dilemma indicated above: If a variety bred for race-specific resistance is grown over a large area, the probability rises for emergence of a rust strain able to overcome the resistance.

Diversification of disease resistance is therefore seen as the key to this boom and bust problem (McDonald and Linde, 2002). For example, Hovmøller et al. (2002) concluded that "the risk of periodic outbreaks of pan-European epidemics of yellow rust would be lessened if wheat breeders had access to more diverse sources of resistance . . . and if different sources of resistance were used by different companies, preferably in different countries." However, this requires coordination and continuous production of new differently resistant varieties. An additional form of resistance management which has been successful in recent decades has been the use of multilines and cultivar mixtures (Mundt, 2002; Wolfe, 1985); such mixtures usually protect crops from more than one target disease (Finckh and Wolfe, 2006).

While yellow rust resistance in wheat is considered as important for organic breeding (Osman et al., 2008) there are few data available on the importance of this disease. In a comparison between conventional and organic conditions in Germany, Münzing et al. (2004) found that yellow rust levels and frequency were lower under organic management. In organic wheat trials on two sites in England, yellow rust levels were mostly zero (Döring and Wolfe, unpublished). Rust diseases appear not to play a major role in organic cereals, probably due to lower nitrogen availability and plant densities. A further factor might be the application of organic soil amendments; for example, Rodgers-Gray and Shaw (2000) showed that wheat fertilized with cow manure was less susceptible to brown rust caused by *Puccinia triticina*.

In summary, organic systems have several strategies already in place for reducing the impact of wheat yellow rust. As in the case of late blight, breeding for diversity needs to complement these approaches.

Plant Protection in Organic Farming

What Is Organic Plant Protection?

Here, we define organic plant protection as the management of plant pests and diseases under the conditions of organic farming as regulated by national and international organic farming standards. We exclude weed control because it is dealt with in Chapter 4 of this book, and we do not cover abiotic plant health problems, such as nutrient deficiencies (see Chapter 2).

Organic plant protection resembles integrated pest management (IPM). Dent (1995) defines IPM as "a pest management system that in the socio-economic context of farming systems, the associated environment and the population dynamics of the pest species, utilizes all suitable techniques in as compatible a manner as possible and maintains the pest population levels below those causing economic injury." (In this definition of IPM, pests include both animal pests and plant pathogens such as viruses, fungi, and bacteria.)

According to the U.S. Environmental Protection Agency (U.S. EPA, 2011), organic plant production, "applies many of the same concepts as IPM but limits the use of pesticides to those that are produced from natural sources, as opposed to synthetic chemicals." In our view, this definition misses an important point; by merely mentioning the replacement of substances it tells us only little about the range of specific organic approaches. In contrast, we propose that organic plant protection can be seen as one IPM system, in which the "socio-economic context" of the farming system is determined by the organic standards, regulations, and current practices. However, these regulations go beyond plant protection (regulation of substances available for control of pests and diseases), in that they refer to nutrient management, treatment of the soil, animal husbandry, and other issues.

Importantly, these areas, though not directly related to plant protection, impact upon the crop production system, particularly on the design of rotations, and so determine indirectly which plant protection methods are suitable. An example is the use of leguminous plants in organic rotations, which are mainly, but not exclusively, grown to make nitrogen available to plants and livestock. The inclusion of legumes in organic rotations means that the frequency of cereals is reduced in comparison with conventional systems. Organic rotations thus reduce the potential for soil-borne cereal diseases, by breaking the life-cycle of pathogens such as *Gaeumannomyces graminis* (take-all) or *Oculimacula spp.* (eyespot). At the same time, the high proportion of legumes in the rotation necessitates management of health in the legumes themselves, although rotation also helps here.

Beyond the local organic standards, organic farming is also guided by internationally agreed principles of health, ecology, fairness, and care (IFOAM, 2009). These principles form a second framework within which ideal ecological plant protection systems can be devised, though this may be restricted by current economic realities. In the following section we indicate plant breeding strategies for organic plant protection within existing systems, together with ways in which plant breeding can help encompass the more general principles.

Targets, Aims, and Principles of Organic Plant Protection

In 1919, the citizens of Enterprise (Alabama, USA) put up a monument to the cotton boll weevil (*Anthonomus grandis*), to commemorate "that the weevil attacks had led to the diversification of agriculture and the resulting prosperity of the area, which previously had been a dangerous one-crop (cotton) economy. Weevil attacks also made cotton difficult to grow and maintain its price" (Ordish, 1976, caption to plate 5). This reminds us of the value of diversified agricultural production systems, but it also raises a fundamental question about which criteria should be used to assess the effects of pests and diseases.

In many cases, researchers use pest or pathogen-related measures such as insect abundance or pathogen-induced lesions on plant tissues to gauge the success of plant protection activities. However, crop yields might not always be affected negatively by the presence of a pest or pathogen, due to the ability of plants to (over-)compensate for biotic stresses. In addition, as for the boll weevil, pest control programs that reduce local yield losses might not necessarily be successful from a regional economic point-of-view. It is therefore advisable to define the overarching goals of plant protection in a given system before embarking on a plant protection strategy. The general aims of high, stable, and well-distributed yields, as well as good quality of products at low costs to the environment and to human and animal health, can often be conflicting. It is therefore essential to design processes that can achieve agreement on optimal solutions, for example (1) to apply precautionary principles when the situation is uncertain; (2) to keep as many ways of development open as possible; (3) to view the problem from different sides (including consideration of multiple trade-offs) in a multi- or interdisciplinary fashion; and (4) to view actions and effects with the whole system in mind, considering potential interactions, non-linear effects, and complexities (Chellemi, 2010).

In organic systems, plant protection cannot be considered as an isolated discipline just to reduce pest and disease levels. It has to be integrated into the system of the whole farm and entire regional economy and ecology. Solutions, whether brought about by breeding or other methods, must be designed and negotiated in a holistic way.

Direct Strategies for Organic Plant Protection

In both organic and non-organic plant production systems, strategies of pest and disease management can be grouped in the three categories that follow. The strategies of (non-chemical) plant protection have been reviewed extensively elsewhere (e.g., Altieri, 1995; Dent, 1995; Zehnder et al., 2007), so we give only a brief overview on measures to strengthen the plant, to weaken the pest or pathogen, and to avoid coincidence of plant and pest in space or time (see table 3.1).

It is essential to combine and integrate these strategies, preferably in functionally complementary ways. Specific strategies can often be optimized by pest or disease prognosis systems for appropriate timing of planting or applications.

Organic Practices Affecting Pests and Diseases

Organic practices have many functions and simultaneously serve several strategies of plant protection. Importantly, their primary function might not be related to plant protection at all, but may be rooted in other areas of plant production such as plant nutrition. We review two overlapping areas of organic practices: (1) plant diversity; and (2) organic rotations, plant nutrition, and manure application.

Table 3.1 Overview of plant protection measures for organic agriculture

Strengthening the plant
- Minimizing abiotic stress by choosing crop species and varieties that are well-adapted to the site conditions, but also by finding or altering abiotic conditions that favor the crop
- Optimizing plant nutrition and soil condition
- Using resistant varieties and enhancing resistance induction (see text)
- Increasing competitive ability of the crop against weeds
- Using certified healthy seed or planting material
- Using plant derived plant strengtheners, e.g., applied as spray preparations.

Weakening the pest or pathogen
- Removing the pest or pathogen, e.g., by using glue traps, using seed cleaning machinery, and/or removal of inoculum in the field by roguing infected plants
- Making abiotic conditions unfavorable to the pest or pathogen by cultural methods, e.g., by drying stored products
- Using organic pesticides such as copper, sulfur, or hot water treatments for fungal control, or, e.g., Bt (*Bacillus thuringiensis*) preparations, Quassia, or Neem against insects
- Using biological control agents against pests and pathogens
- Encouraging the pests' natural enemies such as carabid beetles by enhancing landscape structure or providing habitats and improving availability of food resources
- Encouraging microbial antagonists of pathogens in the soil by using suppressive composts and other methods to increase soil organic matter and microbial activity
- Breaking the life cycle of the pest or pathogen, e.g., by removal of secondary hosts
- Disrupting the mating behavior of pests by using pheromones
- Waiting until the pathogen is dead or has lost its viability in store

Avoiding coincidence of plant and pest or pathogen in space or time
- Moving to a site with conditions unfavorable to the pest (e.g., where disease vectors are infrequent due to long, cold winters)
- Increasing the distance between fields with the same crop beyond the dispersal ability of the pest or pathogen
- Maximizing the distance to known sources of inoculum
- Spreading risk by staggered planting schedules or using a seasonal time window that is known for low pest or disease pressure
- Accelerating crop growth to escape the pest or pathogen through short time to maturity or harvest
- Disturbing host finding and colonization, e.g., by using repellents affecting visual or olfactory senses of insect pests
- Luring the pest away from the target crop to a preferred host as a trap crop
- Using physical barriers to dispersal and crop colonization such as covers, tall or sticky crops, hedges or tree belts, water ditches, or even forest belts

(1) *Enhancing plant diversity* is a key characteristic of organic agriculture, with wide-ranging implications for the management of plant pests and diseases. In agricultural contexts plant diversity can refer to cultivated or wild plants and is determined by the number of elements (e.g., plant species or varieties that grow together), the evenness of their frequencies, and their similarity. The higher the number of elements, the more even their distribution; and the more dissimilar they are, the greater the diversity. Similarity can be defined genetically (within or between species) or functionally. Over recent years, the importance of functional diversity for the provision of ecosystem services such as the production of biomass (Balvanera et al., 2006) and decomposition of organic matter (Hättenschwiler et al., 2005) has been recognized; particular interest was generated by major analyses of grassland ecosystems (Kennedy et al., 2002; Tilman et al., 1997).

In comparison to conventional agriculture, organic systems showed greater species richness (i.e., higher number of different elements; Bengtsson et al., 2005) and higher evenness (Crowder et al., 2010), but information on similarity regarding plant diversity is scarce in comparisons between conventional and organic systems. Plant diversity can be used at the field level as variety mixtures

(Mundt, 2002; Wolfe, 1985) or as genetically highly diverse plant populations (Döring et al., 2010; Chapter 5 in this book). Diversity can also be enhanced by inter-cropping different crop species. For example, organic farmers in temperate climates often under-sow grass clover green manure (leys) in a growing cereal crop. Also, grain legumes can be mixed with small grain spring cereals in numerous combinations, e.g., faba beans and spring oats, or peas and spring barley. In the tropics, inter-cropping is widespread in organic and non-organic systems (Altieri, 1995). At the farm level, plant diversity is increased by the use of more crop varieties or species, and by maintenance and conservation of wildlife and landscape structures such as hedges, ponds, or permanent grass strips. Diversity is also increased at the regional or landscape level where different farms grow different varieties or crops (Jarvis et al., 2008; Keesing et al., 2010). Organic farmers use longer rotations than conventional, with different fields at different stages, and the diversity of cultivated plants is usually greater. In addition, the diversity of wild plants, e.g., arable weeds, is higher on organic than on conventional farms because of differences in crop rotations and weed management (Bengtsson et al., 2005; Boutin et al., 2008).

Increased plant diversity affects interactions with pests and diseases in multiple ways (Finckh and Wolfe, 2006; Keesing et al., 2006; Simon et al., 2010). Generally, high plant diversity reduces pest (Altieri and Nicholls, 2004, Bianchi et al., 2008; Gurr et al., 2003) and disease problems (Finckh and Wolfe, 2006). For example, on rice in China, mixed cropping systems reduced rice blast incidence significantly compared with monocultures (Zhu et al., 2000). Similarly, variety mixtures of barley in the former German Democratic Republic reduced mildew levels until monocultures were re-introduced after re-unification (Wolfe et al., 1992). Also, inter-cropping in vegetables reduced insect pest populations compared to non-intercropped controls (Finch et al., 2003; Tahvanainen and Root, 1972; Theunissen and Schelling, 2000).

Many mechanisms underlie the effects of plant biodiversity on pests and diseases, knowledge of which can be important for optimization and breeding. In mixed cropping, one crop may serve as a structural support for another, e.g., when peas attach their tendrils to cereal stems, so the peas grow more upright under lower humidity, reducing the probability of fungal infection. Within diverse plant stands, non-host crops can also act as physical barriers to the dispersal of spores or insects. In a diverse stand, specific races of pathogens encounter a lower density of hosts they can each infect, and they therefore spread more slowly. Pathogens may even be filtered out almost completely, where large and dense physical barriers such as hedgerows are used. One crop may also camouflage another, chemically or optically, thereby disrupting host searching by pests, or by acting as a trap crop. An intriguing example is the relationship between barley and the cereal aphid *Rhopalosiphum padi*. Ninkovic and Åman (2009) found that some barley genotypes produce volatiles that induce other genotypes of barley to become more resistant to aphid attack. Greater plant diversity may support natural enemies as antagonists of a pest or pathogen, through provision of alternative or complementary food. For example, flower strips adjacent to crops have been shown to increase the abundance of aphid-attacking parasitoids and hoverflies, which benefit from nectar provided by the flowers (Tylianakis et al., 2004).

Increasing plant diversity within and between farms does have limits. For instance, registration and trade in genetically diverse plant populations is currently not possible in many countries (see Chapter 5). Practical problems associated with increasing plant diversity include competition and quality aspects. While increased diversity can effectively reduce the abundance of pests and disease incidence or severity, competition between the inter-cropped plants may reduce overall yield. This is particularly problematic when the biomass of the inter-crop partner is not used. Variety mixtures of cereals for bread-making have not penetrated the market, because large-scale millers and bakers have rejected lots composed of varieties with potentially different processing qualities. Composing inter-crops therefore requires considerations beyond effects on the pathosystem. The most serious limit to making use of increased biodiversity, however, is the loss of agro-biodiversity itself: We can only mix what is extant. The conservation of cultivated plants (e.g., through maintenance breeding) and of wild (or not yet cultivated) species is therefore of highest priority.

(2) *Rotation design* forms the backbone of organic production systems. On the one hand it is a tool for nutrient cycling and balancing and, on the other, it can be crucial in reducing pest, disease, and weed problems (Curl, 1963; Evans et al., 2010). As a measure to increase diversity over time (Lampkin, 1994), rotations also play their part in enhancing biodiversity on the farm.

Rotations break the life cycles of pests and pathogens by creating a host-free period. For example, plant pathogenic nematodes can be managed by alternating host plants with non-hosts in the rotation (Flint and Roberts, 1988). However, the life cycles of some pests and pathogens may be longer than a typical rotation: For example, root-feeding larvae of the click beetle *Agriotes* persist up to six years in the soil. In addition, problems can arise in rotations with high proportions of legumes, as in many organic production systems. Here, the potential build-up of legume specific diseases or pests is critical and needs attention from plant breeding (see next section).

The use of break or cover crops such as mustard or other Brassicaceae species can have direct effects on plant diseases, notably when so-called biofumigation crops are used (Kirkegaard, 2009). Plant breeding efforts can be directed at improving the content of secondary plant metabolites responsible for disease suppression.

Further, organic rotations (and the application of animal manures) typically lead to only moderate amounts of plant available nitrogen, in contrast to conventional systems in which readily available mineral nitrogen is regularly applied. Lower levels of mineralized nitrogen and the slower decomposition of organically bound nitrogen facilitate plant protection, because many insect pests and fungal pathogens thrive on high levels of mineral nitrogen (e.g., aphids, Van Emden and Bashford, 1969; mildew, Shaner and Finney, 1977).

Finally, to create a healthy and biologically active soil, organic agriculture is concerned with promotion of microbial communities (e.g., *mycorrhizae*) for plant nutrition, which may also be important in relation to disease expression. There is also strong reliance on organic soil amendments such as farmyard manure or composts, which can have suppressive effects on plant pathogens (Hoitink and Fahy, 1986). Similarly, Garratt et al. (2010) showed that aphids were more abundant on barley plants that had received conventional (mineral) fertilizer than on organically fertilized plants.

An important consequence of this rotational and plant nutritional context is that plant breeding should ideally take place within an organic rotation and under relevant nutrient conditions (e.g., comparatively low nitrogen availability). In summary, the practices directly or indirectly related to organic plant protection interact with biodiversity at all levels. Diversity therefore, needs to be a breeding goal itself.

Key Target Areas of Plant Breeding for Organic Plant Protection

Breeding for organic plant protection needs clear targets for prioritization. We believe these should include seed-borne pathogens, legumes (because of their central role in nitrogen supply), perennial plants, and crops used as unprocessed food such as vegetables or table fruit. For these areas, alternative approaches in plant protection are currently underdeveloped. However, we argue that in the longer term, prioritization of only a few target crop species will be insufficient, as the maintenance and development of crop diversity at the species level can be achieved only with continued diverse breeding work.

Seed-Borne Diseases

Seed production is of special concern for organic agriculture. In the European Union (EU), organic regulations prescribe that organic growers have to use certified organic seed. However, because of the small market, and issues of market predictability and transparency, organic seed of many crop species or varieties is not always available. This situation is aggravated by the absence of effective resistance to many seed pathogens. Together with low tolerance levels for diseases in seed, this means that the organic sector faces difficulties in producing enough seed to meet the legal quality standards for seed trading.

In Europe, where organic seed is unavailable, farmers can apply for derogations to use non-organic seed. Despite efforts to reduce the dependency of the organic sector on conventional seed, the number of derogations in the EU for use of non-organic seed is only slowly decreasing. Although the system in the United States is different (growers have to demonstrate to their certifier that they checked availability of the variety in organic form with at least three seed companies), dependence on conventional seed is still high for choice of varieties, and seed quality. The organic sector therefore faces a credibility problem. Moreover, while derogations for cheaper non-organic seed are available, there is no incentive to commercial breeders and seed producers to target the organic market and to invest in resistance breeding for seed-borne diseases.

Indeed, seed-borne diseases have received only limited attention in conventional plant breeding in recent decades because of the dominance of fungicide use for seed disease control (e.g., Mathre et al., 2001). This dependency could be reduced by improving the health of organically produced seed. In the past, research on seed health under organic conditions focused on non-chemical seed treatments, using either hot water treatment or plant extracts for disease control. Although progress has been made in developing these approaches (e.g., Schmitt et al., 2009; Tinivella et al., 2009), they still carry the disadvantage of relatively high costs, underlining the importance of plant breeding for healthy organic seed.

A key area for development is breeding for resistance against the smuts and bunts (*Ustilago spp.* and *Tilletia spp.*), the major seed-borne diseases of cereals. *Tilletia* spp. causing bunt was almost eliminated by using seed treatment developed in the 1950s, but it can still be a major problem for farmers using organic seed (Wolfe et al., 2008). It can be controlled effectively through seed treatments with hot water, hot air, or yellow mustard-powder, but loose smut (*Ustilago tritici*) is more recalcitrant, so that the need for resistance development is even greater.

Perennial Plants

From an agro-ecological point of view, plants growing for more than one year provide several benefits over annuals. Perennials (1) reduce soil erosion from wind and water; (2) build up more soil organic matter; (3) sequester carbon, reducing agriculture's net contribution to greenhouse gas emissions; (4) provide structural features that support farm wildlife; (5) make use of untapped nutrient resources with their deeper rooting systems; and (6) provide alternative food resources, thereby increasing food diversity and security. Therefore, it is widely agreed that, to develop more resource-efficient forms of agriculture, the proportion of perennial plants needs to be increased.

However, perennials can be vulnerable to plant diseases and pests because of their potential build-up over time. One of the most powerful tools of organic plant protection, the use of rotations, is not an immediate or direct option for perennials. However, pest and disease control can be improved through breeding. For example, DeHaan et al. (2007) observed that perennial grains show higher levels of resistance against pests and diseases than their annual relatives, possibly because the continuous selection exerted by these biotic stressors in perennial plants has led to evolution of more stable resistance.

Key examples of perennials that do suffer from unsolved pest and disease problems are grapevines and apples. The development of cultivars resistant to apple diseases (e.g., scab, *Venturia inaequalis*; brown rot, *Monilinia fructigena*) is important for organic production, to help reduce fungicide use (Holb et al., 2011). Such a reduction is worth pursuing partly for environmental reasons (Page, 2011), but also because fungicides permitted in organic apple production (e.g., sulfur) can be detrimental, for example, to predatory mites, which are essential in biological control of insects.

Deployment of apple diversity (e.g., mixtures of cultivars at the farm, landscape, and regional level) and restricting the area planted with the same resistant cultivar, are sensible approaches to reduce potential plant health problems. This implies that conservation of rare and ancient apple varieties, as well as of the germplasm of populations in wild habitats (e.g., *Malus sylvestris*), is important in maintaining a potential pool of genes for resistance to diseases and pests. Conversely, there is the possibility that apple trees in wild habitats may harbor pathogen diversity which may then threaten cultivated resistant varieties, although this was not the case for apple scab in France (Lê Van et al., 2011). Research in Switzerland has highlighted the importance of consumer attitudes in the efforts to introduce disease resistance in apples (Weibel and Leder, 2004). Because consumers showed little inclination to purchase varieties with unknown names, novel varieties with scab resistance or tolerance were difficult to market. The introduction of a new apple classification based on broad taste categories made it possible to reconcile consumer preferences with the need for disease resistance and for the conservation of varietal diversity.

In organic wine production, a major problem is infection of vine plants by downy mildew, caused by *Plasmopara viticola*. As with apple, it is possible to avoid the use of permitted fungicides in vineyards by employing alternative management strategies, from novel abiotic compounds (e.g., bicarbonates, milk, plant oils; Sivčev et al., 2010) to biological antagonists (Crisp et al., 2006), and by improving the timing of applications, e.g., by using Web-based decision support systems (Kuflik et al., 2009). Unfortunately, recent research showed that cultivar mixtures of grapevines did not consistently reduce severity of *Plasmopara* (Matasci, 2008). Breeding for resistance therefore has high priority in organic wine production (Basler and Pfenniger, 2003), but acceptance of resistant cultivars is slow because of the long-term nature of viticulture, and because organic and non-organic vine growers use the same susceptible varieties (Willer and Zanoli, 2000).

Fruit and Vegetables as Non-processed Food Plants

As the organic sector promotes fresh organic food for health and well-being, produce sold to the consumer must have an attractive appearance. Currently, because the organic sector relies on a more direct interaction between producers and consumers, deviations from the ideal can be better explained to the consumer than in a supermarket. However, expansion of organic product range and volume in supermarkets means that high standards for presentation need to be met, with direct implications for plant protection. For instance, consumers do not tolerate insect pests on purchased products: Although a few aphids on a lettuce do not affect yield, they can lead to product rejection by the consumer. Similarly, visible fungal infections, such as sooty blotch on apples, can reduce marketability even if the lesions are superficial and have no impact on yield or taste. Here, the difficulty for breeders is to develop high levels of resistance because the threshold of accepted damage is so low. This is only possible to achieve when resistance is integrated with other measures.

Leguminous Plants

Legumes are integral to organic production because of their ability to fix nitrogen. In livestock farming, the common legume-grass mixtures grown for fodder or pasture also serve general soil fertility by providing a period without tillage in which organic matter is accumulated in the soil. In stockless farming systems, grain legumes are used more frequently because they produce a direct income.

However, grain legumes especially, but also many forage species, can suffer from severe disease and pest problems. The most serious diseases are soil-borne and often caused by generalist pathogens such as *Phoma medicaginis, Aphanomyces euteiches, Rhizoctonia solani, Sclerotinia sclerotiorum*, or *S. minor*. Most of these are able to attack many legumes and, in the case of *A. euteiches* and the *Sclerotinia spp.*, even a number of non-legumes. There are also more host-specific species such as *Mycosphaerella spp.* and the various forms of *Fusarium solani* and *F. oxysporum*, which can often survive for long periods in the soil, making them difficult to control by rotation.

The less specialized pathogens can often be reduced by enhancing soil microbial activity and soil organic matter: For example, by applying suppressive composts or by growing and incorporating green manures from other plant families. Resistances to many of the more specialized pathogens such as *M. pinodes* and *F. solani* and *F. oxysporum* are known. Unfortunately, legume breeding has declined severely in favor of programs concentrating on the major cash crops. This leaves organic growers often with only few legume species and varieties from which to choose.

Considering that there are thousands of annual and perennial legume species, often with potential use as animal or human food, it should be possible to identify additional species with different complex resistances to be substituted as break crops for the currently used legumes. These novel species might also help to close functional gaps left by currently available legume species, e.g., in terms of decomposition rate, palatability, or micronutrient uptake.

A Wider Range of Crop Species

The major pests and diseases are foremost because agriculture and the food industry concentrate on only a handful of host crops. The diversity that is available could be better deployed (variety mixtures, populations), but there is no doubt that an overall increase, particularly among species, is essential. In addition to pest and disease management, this would help in buffering against climate change and global market variation, and also in extending crop seasonality and local diet range. New options are available from (1) recent history – there are many crops that could be reconsidered and reselected; (2) introduction of novel crops from similar climate zones; (3) introduction of novel crops from areas with conditions similar to those predicted to develop in an existing zone; and (4) new crop ranges relevant to new cropping systems such as perennials, and annuals, used, for example, in agroforestry.

The breeding, selection, and seed production costs involved in making such introductions will be high, but offset to some extent in that each additional crop is likely to be less at risk from pests and diseases, particularly if used in a system of management of diversity.

Breeding Goals for Ecological Plant Protection

A fundamental question in breeding is whether to pursue wide or narrow (specific) adaptation (Annicchiarico, 2009). Widely adapted genotypes (cultivars, mixtures, or populations) perform similarly in different environments whereas narrowly adapted genotypes perform well in specific environments (better than the widely adapted genotypes), but poorly in other environments. While the term "environment" often refers to climatic and edaphic factors, it can also relate to biotic factors, such as pests and pathogens. Genotypes can be widely resistant to many different pathogen races (or species), or narrowly resistant to some races (or species) but susceptible to others.

The answer will depend largely on the variability and predictability of the environment: If variability is large and conditions are likely to change rapidly or in unknown ways, more widely adapted genotypes that buffer against environmental fluctuation will be preferred. With global climate change, local and regional weather is predicted to become more variable, i.e., less predictable and more extreme. At the same time, global trade is extending, unpredictably, the dispersal of pathogens and pests, which can contribute to greater genetic diversity and flexibility of those organisms, as noted above for late blight and yellow rust.

Consequently, wide adaptation will need to play a more important role in the future, both for organic and non-organic agriculture. In addition, because organic systems show greater environmental variability than conventional systems, plant breeding for organic systems needs to focus specifically on wide adaptation, to provide stability of crop performance through genetics rather than pesticide inputs.

However, breeding also needs to produce genotypes specifically adapted to organic conditions. With respect to plant protection this means, for example, development of cereal varieties that are resistant to smut and bunt, but not by focusing on resistance to one or few races of the pathogen. We therefore propose that breeding for organic systems will have to develop breeding goals and strategies on a case-by-case basis, considering the flexibility of the pathogen (McDonald and Linde 2002) and using approaches of "wide-within-narrow" (WWN) or "narrow-within-wide" (NWW) adaptation. This could mean, for example, that a variety mixture designed specifically for rust resistance is based on a wide range of resistance genes targeting different rust species and races (WWN); or that in a genetically diverse plant population that buffers against unpredictable stress factors such as drought, all genotypes are uniformly adapted to cope with a particular stress factor, such as a regular occurring pest species (NWW). Such approaches will require future research to separate unpredictable from predictable variability in organic systems (e.g., large but stable differences among localities).

Beyond stability, there is also growing recognition that the genotype and the system in which it is grown need to be resilient. They need to respond dynamically to changing environments by adapting to change; thus, they remain functional under stress by *adjusting* to the stressful conditions. Experimental evidence shows that resilience is promoted by crop diversity (Döring et al., 2010).

Finally, a general breeding goal to be developed is the genotype's system compatibility, allowing it to be integrated into a specific organic system, such as the ability to mix well with other crops (see next section).

Plant Breeding Approaches Directly Targeting Pests or Diseases

Breeding for Resistance

Breeding for resistance against pests and diseases is the most direct contribution of breeding to plant protection. Resistance can be based on a large range of mechanisms, such as hardness of tissues, roughness of the leaf surface, leaf hairiness, thickness of the husk, chemical defense mechanisms through secondary metabolites in the plant cells, repellent volatiles, or even the ability of plants to attract natural enemies in response to attack by herbivores. In many cases, high levels of resistance are correlated with a comparatively low yield potential under unstressed conditions. One reason could be that it is physiologically costly for the plant to produce secondary metabolites involved in resistance. Alternatively, Foyer et al. (2007) argue that such costs are not significant for the plant but that breeding has reduced the content of unpalatable or toxic plant compounds to make plants more attractive to human tastes, which has incidentally decreased their resistance to pests. This link between taste and resistance stresses the importance of the market and the need to integrate breeding activities with appropriate marketing strategies (e.g., see the apple example given previously).

Resistance can be fleeting, because the ability of pests and pathogens to evolve allows them to overcome the plant's resistance (e.g., Kolmer, 1996). Durability of resistance in crops planted over large areas is therefore an important aim of resistance breeding. However, in conventional agriculture, large breeding companies are geared to a rapid turnover of new varieties which allows them to accept rapid loss of resistances and to forsake durability of resistance (Michelmore, 2003).

An exceptionally durable resistance is conveyed by the Lr34 gene for leaf rust resistance in wheat. Despite wheat cultivars with Lr34 occupying "more than 26 million ha in various developing countries alone ... no evolution of increased virulence toward Lr34 has been observed for more than 50 years" (Krattinger et al., 2009). Also, some forms of qualitative resistance based on the loss of function in a gene are generally difficult to overcome (Pavan et al., 2010). These tend to be recessive and do not confer a hypersensitive response as opposed to dominant hypersensitivity genes.

However, as Michelmore (2003) warns, successful breeding for resistance may also have detrimental effects because concentration on few resistance genotypes may accelerate the disappearance of genetic material thus increasing genetic vulnerability. By outcompeting a large number of varieties or landraces in just one trait (the resistance against a major pest or disease), a "successful" variety therefore threatens to eliminate the diversity present in several other ecologically or nutritionally important traits.

As a rule of thumb, quantitative resistance (QR), providing imperfect resistance against many pathogen races is often more stable over time than major gene resistance (MGR), which provides a higher level of resistance but may also be overcome more easily (but see Johnson, 1984 for examples of durable MGR). However, as McDonald and Linde (2002) argue, the reproductive biology of the pathogen and its genetic population structure are as important for the durability of resistance as the nature of the plant's resistance gene. Based on the criteria of pathogen genetic diversity and the amount of gene flow among pathogen populations, these authors suggest four breeding strategies to cope with the risk of resistance break down: The use of single MGR for low pathogen diversity with low gene flow; MGR pyramids for high gene flow in less diverse pathogens; QR and regional use of MGR for highly diverse pathogens with low gene flow; and a combination of QR and the intensive management of MGR with multilines and cultivar mixtures for the most risky group, pathogens with high genetic diversity and high gene flow.

Induced Resistance

Resistance reactions in plants can be divided roughly into constitutive resistance, which is constantly present and induced resistance (IR), which is triggered. The two best known types of IR are systemic acquired resistance (SAR) and induced systemic resistance (ISR). SAR is based on the production of pathogenicity related (PR) proteins and salicylic acid while in ISR, jasmonic acid and ethylene are involved (Vallad and Goodman, 2004). IR may be based not only on the triggering of defense mechanisms but also on the alteration of plant quality, e.g., making plants less attractive to pests (e.g., Zehnder et al., 2001). There are many different mechanisms involved in plant defense which, potentially, may all be involved in IR (Lebeda et al., 2008). IR responses may compete with each other and either enhance or decrease susceptibility to different pests and pathogens (Vallad and Goodman, 2004).

IR in nature is usually triggered by pests and pathogens or by beneficial microorganisms. Thus, if an avirulent race of a pathogen lands on a plant that is resistant to that race, the plant's resistance reaction will be induced. Once induced, the resistance will not only defend the plant against the avirulent race that caused the initial resistance reaction but often more generally, e.g., against attack from other pathogens or races that arrive later. In addition, IR can also be triggered by various chemicals or parts of pathogens. For example, a fraction of a protein from the pathogen *Xanthomonas oryza pv. oryzicola*, the causal agent of bacterial leaf streak in rice, is involved in the triggering of a hypersensitive response in rice. The so-called harpin is capable of inducing strong resistance in rice against several important diseases. In addition, treating rice with the protein fragment also enhances plant growth resulting in yield increases in field experiments.

While the true dimensions of naturally induced resistance in practical agriculture are not known, a number of examples suggest that IR plays a much more important role in plant defense than previously thought (Walters, 2009). For example, Calonnec et al. (1996) demonstrated that infection efficiency (in the laboratory) and disease severity in the field due to *P. striiformis* were reduced by more than 40% on a susceptible wheat cultivar when first challenged with an avirulent race: This could be a useful trait to select for to increase natural protection in the field.

To be of interest for breeding, however, several conditions have to be met: (1) there has to be clearly identifiable genetic variation for this trait in the host; (2) the variation has to be stable across environments; (3) selection for the trait should be practicable. There is, by now, ample evidence that the first condition holds true and that inducibility of resistance is genotype-specific as suggested by Walter et al. (2005) and recently demonstrated by Chen et al. (2008) for rice, or by Sharma et al. (2009) for the *P. infestans*-tomato pathosystem. Thus, it is selectable. However, conditions (2) and (3) are more complicated. Besides being genotype-specific, IR also depends on the inducer used and it may be strongly affected by environmental conditions (e.g., Walters et al., 2005). In addition, at least in the tomato–*P. infestans* system, IR is also isolate specific, i.e., resistance is not necessarily induced against all isolates of a pathogen (Sharma et al., 2009), which complicates its identification and applicability in general agriculture. Unfortunately, this could also lead to development of plants inducible by a specific chemical and when grown in a particular substrate, thus reducing flexibility and potentially creating unwanted dependencies for growers.

An alternative to breeding for inducibility of resistance of single genotypes may be to work on general, or better, natural resistance induction by adapting the growing system. This can be achieved by enhancing (1) the resistance diversity in the system and (2) the microbial diversity at various levels. This was demonstrated, for example, by Calonnec et al. (1996) in experiments with wheat yellow rust. Interestingly, not only avirulent pathogens but also the combination of different virulent isolates can also enhance IR as was shown for the tomato–*P. infestans* system (Sharma et al., 2009). They also found that the isolate specific IR due to different inducers and tomato accessions largely disappeared when isolate mixtures were used.

In addition to pathogen diversity, microbial diversity in the rhizosphere, phyllosphere, and endosphere has been shown to enhance plant resistance (e.g., Lindow, 2007; Rodriguez et al., 2009) and this is being followed especially as a potential solution to the huge number of disease problems faced in banana production (e.g., Kavino et al., 2008).

The greater the host and microbial diversity in a system the greater are the chances for effective triggering of induced resistance. As each plant genotype is associated with its own specific set of associated microorganisms (e.g., Mazzola et al., 2002), diversification is probably the most efficient approach to enhance IR within agricultural systems rather than by expensive and difficult selection for single genotypes with enhanced inducibility.

Tolerance

Although "resistance" and "tolerance" to plant pathogens are often used as synonyms, they refer to different phenomena. Being resistant to a pathogen indicates that a plant can withstand attempted infection to a greater or lesser extent. This results in cessation of infection or reductions in infection rate, efficiency, and lesion size, leading to reduced pathogen propagation. The resistance mechanisms invoked are likely to require energy and may thus be associated with reduced productivity (but see the previous reference to Foyer et al., 2007). Some resistances have been associated with yield losses in the absence of disease, making them less attractive for use. This was the case initially in exploitation of the *mlo* gene conveying resistance to powdery mildew (caused by *Blumeria graminis*) in barley. It took several years of breeding to reduce the yield losses associated with the hypersensitive responses due to that resistance. The same happened with the commercial resistance inducer against barley powdery mildew "BION®": If used too late in the season it caused yield losses in addition to IR.

In contrast to resistance, tolerance refers to the capability of a plant to be infected and yet remain productive, as with tolerance to salt, cold, heat, or drought. Two varieties, one with disease resistance, the other with disease tolerance, may produce yields similar to each other in both the absence and presence of disease. However, the variety with resistance is more desirable in practice because it generates less pathogen inoculum. The ideal breeding objective would thus be to combine the two characteristics, but this is often difficult to achieve.

Plant Breeding Approaches with Indirect Effects on Plant Health

Plant breeding can also contribute to plant protection indirectly, through the integration into systems which achieve pest or disease control non-genetically. In this breeding approach, the aim is to develop crops that are better suited to these systems, rather than to select for direct or inducible resistance to pests or diseases. Here we propose five different indirect breeding approaches with relevance for (organic) plant protection.

(1) Plant architecture. The relative sizes of the plant organs and their spatial organization are important features for organic plant breeding, e.g., for conveying high competitive ability of the crop against weeds. For protection against plant diseases, the importance of plant architecture is highlighted in those fungal diseases of cereals, where a large distance between the leaves and the ear can contribute to disease reduction in the ear. Another example is the effect of the openness of canopy on pests in apples (Simon et al., 2006).

(2) Rotational breeding. Plant breeding for organic agriculture should always consider implications throughout the rotation. In rotational breeding, crops are selected for their properties as sanitation crops where there is a beneficial effect on the subsequent crop via restriction of plant pests or diseases. A good example is *Brassicaceae* cover crops, which can reduce soil-borne plant

pathogenic nematodes. Breeding can then exploit within-species variation and select genotypes best suited as biofumigation crops.

(3) Co-breeding for intercropping. It is well established that intercropping systems in which two (or more) different plant species are grown together in the same field (often in alternating strips) can reduce pest and disease severity in a target crop (see section 2 of this volume). Often however, there is still room for improvement regarding the way in which the intercropped plants interact and in selecting for appropriate crop genotypes. One possible approach is through observation of morphologically or physiologically diverse plant populations that evolve over time; i.e., seed from a population grown as an intercrop is re-sown annually with the best fitted genotypes contributing more seed to the following generation. Similarly, plant breeding can target push-pull systems, which reduce pests by manipulating host selection behavior. In these systems, the target crop has a property that "pushes" pests away (e.g., through repellent odors), while a trap crop grown around the target is more attractive than the target and "pulls" the pest away from it. Here, plant breeding can attempt to select for both greater repulsion and attraction.

(4) Breeding for natural biological control. Flower-rich strips of non-cultivated or cultivated plants can be sown around crops to provide nectar resources for natural enemies of pests. However, while this resource can increase the fitness of beneficial insects such as parasitoid wasps, they can also provide additional food resources for pests such as butterflies with herbivorous larvae. Because nectar composition varies among plants and different insects prefer nectars of different composition, some plant species provide a better balance for pest control than others (Winkler, 2005). Although the first step is the selection of plant species best suited for providing selective resources for beneficial organisms, breeding could make this approach even more efficient.

(5) Matched breeding for diversity. As we have argued throughout this chapter, plant diversity is the indispensable basis for successful plant protection in organic systems. However, in many areas of the food industry, diversity of the raw product poses a problem as the processing properties of different genotypes diverge. Although efforts should be made to overcome objections to diversity at the processing end, plant breeding can also attempt to address the issue by creating matched genotypes that are uniform in their processing qualities (e.g., regarding baking quality), but that are diverse in traits relevant for plant protection (such as resistance against diseases).

In all these approaches, it is necessary to consider the practicability and costs of breeding; in effect, the process sets the limits to the complexity that can be handled.

Discussion and Conclusions

From our review of plant breeding in the context of organic pest and disease management, four main messages emerge. The first is the tremendous importance of plant diversity for developing resilient agricultural production systems. So much crop diversity has been lost over the last five to six decades (Fowler and Mooney, 1990), and current agriculture is so dominated by monocultures, that promoting the increased use and protection of plant diversity is imperative to slow down and reverse these dangerous developments.

The second message is the importance of the economic dimensions for plant breeding in plant protection contexts. Globalization of agricultural trade contributes to pest and disease problems. Moreover, ecologically sound solutions in plant protection are often impeded by current market forces which favor large structures over small, thereby contributing to a decrease and homogenization of agricultural diversity. In addition, the short–term thinking in current economic systems hampers the development of breeding and breeding research, as long-term investment leaves resources only for a few major crops, thereby aggravating pest and disease problems. Breeding for plant protection in organic systems therefore needs to engage with the consumers and other players in the market in order to find and establish viable alternatives.

Third, plant protection in organic systems poses different challenges to plant breeding than does conventional agriculture. In many cases, organic systems suffer less from pests and diseases than their conventional counterparts because preventive measures are already in place. In other cases (such as pests and diseases of perennial plants, or seed-borne diseases), plant breeding urgently needs to step in and develop durable solutions. We expect that both conventional and specifically organic plant breeding may support the improvement of plant protection in organic agriculture. Conversely, efforts to solve existing and emerging problems in conventional systems will also benefit from breeding for organic plant protection.

Finally, breeding for plant protection needs to consider the dynamic and unpredictable nature of our constantly evolving agricultural systems. Current science is strongly dominated by deterministic philosophies whose proponents proclaim that, with a mechanistic understanding of underlying principles, everything can be manipulated and brought under human control. This hubris is neither supported by scientific evidence (Prigogine, 1997) nor by practitioners' experiences. Agricultural science and practice need to accept the inherent indeterminism of living systems. As plant breeding for plant protection has so often attempted, and failed, to bring pathosystems under tight control, it needs to come to terms with its own inherent limitations. This is particularly relevant in organic contexts, where environments are highly variable. Still, breeding can help to solve organic plant protection problems. To achieve this, it needs to enthusiastically embrace approaches based on high diversity.

References

Altieri, M.A. 1995. Agroecology – The science of sustainable agriculture. 2nd edition. Boulder, CO: Westview Press.
Altieri, M.A., and C.I. Nicholls. 2004. Biodiversity and pest management in agroecosystems. Binghamton, NY: Haworth Press.
Annicchiarico, P. 2009. Coping with and exploiting genotype-by-environment interactions. *In*: Ceccarelli, S., E.P. Guimarães, and E. Weltzien (eds.) Plant breeding and farmer participation. pp. 519–564. Rome, Italy: FAO.
Aylor, D.E. 2003. Spread of plant disease on a continental scale: Role of aerial dispersal of pathogens. *Ecology* 84:1989–1997.
Balvanera, P., A.B. Pfisterer, N. Buchmann, J.S. He, T. Nakashizuka, D. Raffeaelli, and B. Schmid. 2006. Quantifying the evidence for biodiversity effects on ecosystem functioning and services. *Ecology Letters* 9:1146–1156.
Basler, P., and H. Pfenninger. 2003. Disease-resistant cultivars as a solution for organic viticulture. *Acta Horticulturae* 603: 681–685.
Bengtsson, J., J. Ahnström, and A.C. Weibull. 2005. The effects of organic agriculture on biodiversity and abundance: A meta-analysis. *Journal of Applied Ecology* 42:261–269.
Bianchi, F.J.J.A., C.J.H. Booij, and T. Tscharntke. 2008. Sustainable pest regulation in agricultural landscapes: A review on landscape composition, biodiversity and natural pest control. *Proceedings of the Royal Society* B 273:1715–1727.
Boutin, C., A. Baril, and P.A. Martin. 2008. Plant diversity in crop fields and woody hedgerows of organic and conventional farms in contrasting landscapes. *Agriculture, Ecosystems and Environment* 123:185–193.
Bouws, H., and M.R. Finckh. 2008. Effects of strip-intercropping of potatoes with non-hosts on late blight severity and tuber yield in organic production. *Plant Pathology* 57:916–927.
Calonnec, A., H. Goyeau, and C. de Vallavieille-Pope. 1996. Effects of induced resistance on infection efficiency and sporulation of *Puccinia striiformis* on seedlings in varietal mixtures and on field epidemics in pure stands. *European Journal of Plant Pathology* 102:733–741.
Chellemi, D.O. 2010. Back to the future: Total system management (organic, sustainable). *In*: Gisi, U. et al. (eds.) Recent developments in management of plant diseases. pp. 285–292. Berlin, Germany: Springer.
Chen, L., S.J. Zhang, S.S. Zhang, S. Qu, X. Ren, J. Long, Q. Yin et al. 2008. A fragment of the *Xanthomonas oryzae* pv. *oryzicola* harpin HpaGXooc reduces disease and increases yield of rice in extensive grower plantings. *Phytopathology* 98:792–802.
Crisp, P., T.J. Wicks, M. Lorimer, and E.S. Scott. 2006. An evaluation of biological and abiotic controls for grapevine powdery mildew. 1. Greenhouse studies. *Australian Journal of Grape and Wine Research* 12:192–202.

Crowder, D.W., T.D. Northfield, M.R., Strand, and W.E. Snyder. 2010. Organic agriculture promotes evenness and natural pest control. *Nature* 466:109–112.

Curl, E. 1963. Control of plant diseases by crop rotation. *The Botanical Review* 29:413–479.

DeHaan, L.R., T.S. Cox, D.L. Van Tassel, and J.D. Glover. 2007. Perennial grains. *In*: Scherr, S.J., and J.A. McNeely (eds.) Farming with nature: The science and practice of eco-agriculture. Washington, D.C.: Island Press.

Dent, D. (ed.) 1995. Integrated pest management. London: Chapman and Hall.

Döring, T.F., M. Wolfe, H. Jones, H. Pearce, and J. Zhan. 2010. Breeding for resilience in wheat – Nature's choice. *In*: Goldringer, I., and E.T. Lammerts van Bueren. (eds.) Breeding for resilience: A strategy for organic and low-input farming systems? EUCARPIA 2nd Conference of the Organic and Low-Input Agriculture Section. pp. 45–48. December 1–3, 2010. Paris, France.

Evans, M.L., G.J. Hollaway, J.I. Dennis, R. Correll, and H. Wallwork. 2010. Crop sequence as a tool for managing populations of *Fusarium pseudograminearum* and *F. culmorum* in south-eastern Australia. *Australasian Plant Pathology* 39:376–382.

Finch, S., H. Billiald, and R.H. Collier. 2003. Companion planting – do aromatic plants disrupt host-plant finding by the cabbage root fly and the onion fly more effectively than non-aromatic plants? *Entomologia Experimentalis et Applicata* 109:183–195.

Finckh, M.R. 2008. Integration of breeding and technology into diversification strategies for disease control in modern agriculture. *European Journal of Plant Pathology* 121:399–409.

Finckh, M.R., and M.S. Wolfe. 2006. Diversification strategies. *In*: Cooke, B.M., D.G. Jones, and B. Kaye (eds.), The epidemiology of plant diseases, 2nd ed. pp. 269–307. Berlin: Springer.

Finckh, M.R., M.S. Wolfe, and E.T. Lammerts van Bueren. 2008a. The canon of potato science: Variety mixtures and diversification strategies. *Potato Research* 50:335–339.

Finckh, M.R., F. Hayer, E. Schulte-Geldermann, and C. Bruns. 2008b. Diversität, Pflanzenernährung und Prognose: Ein integriertes Konzept zum Management der Kraut-und Knollenfäule in der ökologischen Landwirtschaft [Diversity, plant nutrition, and prognosis: An integrated concept for the management of late blight in organic potato production] German with English abstract. *Gesunde Pflanzen* 60:159–170.

Flint, M.L., and P.A. Roberts. 1988. Using crop diversity to manage pest problems: Some California examples. *American Journal of Alternative Agriculture* 3:163–167.

Fowler, C., and P.R. Mooney. 1990. *In*: Shattering: Food, politics, and the loss of genetic diversity. Tucson, AZ: University of Arizona Press.

Foyer, C.H., G. Noctor, and H.F. van Emden. 2007. An evaluation of the costs of making specific secondary metabolites: Does the yield penalty incurred by host plant resistance to insets result from competition for resources? *International Journal of Pest Management* 53:175–182.

Fry, W.E., S.B. Goodwin, A.T. Dyer, J.M. Matuszak, A. Drenth, P.W. Tooley, L.S. Sujkowski et al. 1993. Historical and recent migrations of *Phytophthora infestans*: Chronology, pathways, and implications. *Plant Disease* 77:653–661.

Garratt, M.P.D., D.J. Wright, and S.R. Leather. 2010. The effects of organic and conventional fertilizers on cereal aphids and their natural enemies. *Agricultural and Forest Entomology* 12:307–318.

Gurr, G.M., S.D. Wratten, and J.M. Luna. 2003. Multi-function agricultural biodiversity: Pest management and other benefits. *Basic and Applied Ecology* 4:107–116.

Hättenschwiler, S., A.V. Tiunov, and S. Scheu. 2005. Biodiversity and litter decomposition in terrestrial ecosystems. *Annual Review of Ecology and Systematics* 36:191–218.

Hoitink, H.A.J., and P.C. Fahy. 1986. Basis for the control of soilborne plant pathogens with composts. *Annual Review of Phytopathology* 24:93–114.

Holb, I.J., B. Balla, F. Abonyi, M. Fazekas, P. Lakatos, and J.M. Gáll. 2011. Development and evaluation of a model for management of brown rot in organic apple orchards. *European Journal of Plant Pathology* 129:469–483.

Hovmøller, M.S., A.F. Justesen, and J.K.M. Brown. 2002. Clonality and long-distance migration of *Puccinia striiformis* f.sp. *tritici* in north-west Europe. *Plant Pathology* 51:24–32.

IFOAM. 2009. The principles of organic agriculture. Accessed February 25, 2011. http://www.ifoam.org/about_ifoam/principles/index.html.

Jarvis, D.I., A.H. Brown, P.H. Cuong, L. Collado-Panduro, L. Latournerie-Moreno, S. Gyawali, T. Tanto et al. 2008. A global perspective of the richness and evenness of traditional crop-variety diversity maintained by farming communities. *Proceedings of the National Academy of Sciences of the USA* 105:5326–5331.

Johnson, R. 1984. A critical analysis of durable resistance. *Annual Review of Phytopathology* 22:309–330.

Karalus, W., and R. Rauber. 1997. Effect of presprouting on yield of main crop potatoes (*Solanum tuberosum*) in organic farming. *Journal of Agronomy and Crop Science* 179:241–249.

Kavino, M., S. Harish, N. Kumar, D. Saravankumar, T. Damodaran, K. Soorianathasundaram, and R. Samiyappan. 2008. Biohardening with plant growth promoting rhizosphere and endophytic bacteria induces systemic resistance against banana bunchy top virus. *Applied Soil Ecology* 39:187–200.

Keesing, F., R.D. Holt, and R.S. Ostfeld. 2006. Effects of species diversity on disease risk. *Ecology Letters* 9:485–498.

Keesing, F., L.K. Belden, P. Daszak, A. Dobson, C.D. Harvell, R.D. Holt, P. Hudson et al. 2010. Impacts of biodiversity on the emergence and transmission of infectious diseases. *Nature* 468:647–652.

Kennedy, T.A., S. Naeem, K.M. Howe, J.M.H. Knops, D. Tilman, and P. Reich. 2002. Biodiversity as a barrier to ecological invasion. *Nature* 417:636–638.

Kirkegaard, J. 2009. Biofumigation for plant disease control – from the fundamentals to the farming system. *In*: Walters, D. (ed.) Disease control in crops: Biological and environmentally friendly approaches. pp. 172–195. Oxford, UK: Wiley-Blackwell.

Kolmer, J.A. 1996. Genetics of resistance to wheat leaf rust. *Annual Review of Phytopathology* 34:435–455.

Krattinger, S.G., E.S. Lagudah, W. Spielmeyer, R.P. Singh, J. Huerta-Espino, H. McFadden, E. Bossolini, L.L. Selter, and B. Keller. 2009. A putative ABC transporter confers durable resistance to multiple fungal pathogens in wheat. *Science* 323:1360–1363.

Kuflik, T., D. Prodorutti, A. Frizzi, Y. Gafni, S. Simon, and I. Pertot. 2009. Optimization of copper treatments in organic viticulture by using a web-based decision support system. *Computers and Electronics in Agriculture* 68:36–43.

Lampkin, N. 1994. Organic farming. Ipswich, UK: Farming Press.

Lê Van, A., C.E. Durel, B. Le Cam, and V. Caffier. 2011. The threat of wild habitat to scab resistant apple cultivars. *Plant Pathology*, 60:621–630.

Lebeda, A., M. Selarovy, M. Petrivalsky, and J. Prokopova. 2008. Diversity of defence mechanisms in plant-oomycete interactions: a case study of *Lactuca* spp. and *Bremia lactucae*. *European Journal of Plant Pathology* 122:71–89.

Lindow, S.E. 2007. Phyllosphere microbiology: A perspective. *In*: Bailey, D.J. et al. (eds.) Microbial ecology of aerial plant surfaces. pp. 1–20. Oxford, UK: Oxford University Press.

MacLeod, A., M. Pautasso, M.J. Jeger, and R. Haines-Young. 2010. Evolution of the international regulation of plant pests and challenges for future plant health. *Food Security* 2:49–70.

Matasci, C.L. 2008. An examination of the effects of grapevine cultivar mixtures and organic fungicide treatments on the epidemiology and population structure of the grapevine downy mildew *Plasmopara viticola*. PhD Thesis ETH Zurich. 153 pp.

Mathre, D.E., R.H. Johnston, and W.E. Grey. 2001. *Small grain cereal seed treatment. The plant health instructor.* doi: 10.1094/PHI-I-2001-1008-01.

Mazzola, M., P.K. Andrews, J.P. Reganold, C.A. and Lévesque. 2002. Frequency, virulence, and metalaxyl sensitivity of *Pythium* spp. isolated from apple roots under conventional and organic production systems. *Plant Disease* 86:669–675.

McDonald, B.A., and C. Linde. 2002. The population genetics of plant pathogens and breeding strategies for durable resistance. *Euphytica* 124:163–180.

Michelmore, R.W. 2003. The impact zone: Genomics and breeding for durable disease resistance. *Current Opinion in Plant Biology* 6:1–8.

Milus, E.A., K. Kristensen, and M.S. Hovmøller. 2009. Evidence for increased aggressiveness in a recent widespread strain of *Puccinia striiformis* f. sp. *tritici* causing stripe rust of wheat. *Phytopathology* 99:89–94.

Möller, K., J. Habermeyer, V Zinkernagel, and H.J. Reents. 2007. Impact and interaction of nitrogen and *Phytophthora infestans* as yield-limiting and yield-reducing factors in organic potato (*Solanum tuberosum* L.) crops. *Potato Research* 49:281–301.

Mundt, C.C. 2002. Use of multiline cultivars and cultivar mixtures for disease management. *Annual Review of Phytopathology* 40:381–410.

Münzing, K., D. Meyer, D. Rentel, and J. Steinberger. 2004. Vergleichende Untersuchung über Weizen aus ökologischem und konventionellem Anbau. *Getreidetechnologie* 58:6–12.

Nemecek, T., D. Dubois, O. Huguenin-Elie, and G. Gaillard. 2011. Life cycle assessment of Swiss farming systems: I. Integrated and organic farming. *Agricultural Systems* 104:217–232.

Ninkovic, V., and I. Åman. 2009. Aphid acceptance of *Hordeum* genotypes is affected by plant volatile exposure and is correlated with aphid growth. *Euphytica* 169:177–185.

Ordish, G. 1976. The constant pest – A short history of pests and their control. London, UK: Peter Davies.

Osman, A., L. van den Brink, and E.T. Lammerts van Bueren. 2008. Comparing organic and conventional VCU testing for spring wheat in the Netherlands. *In*: Rey F., L. Fontaine, A. Osman, and J. van Waes (eds.) Proceedings of the COST ACTION 860 – SUSVAR and ECO-PB Workshop on Value for Cultivation and Use testing of organic cereal varieties. What are the key issues? pp. 37-41. Brussels, Belgium, February 28–29, 2008, SUSVAR, COST, ECO-PB. Paris, France: ITAB. Full proceedings available at www.eco-pb.org.

Page, G., T. Kelly and M. Minor 2011. Modelling sustainability: What are the factors that influence sustainability of organic fruit production systems in New Zealand? *Organic Agriculture* 1:55–64.

Pavan, S., E. Jacobsen, R.G.F. Visser, and Y. Bai. 2010. Loss of susceptibility as a novel breeding strategy for durable and broad-spectrum resistance. *Molecular Breeding* 25:1–12.

Phillips, S.L., M.W. Shaw, and M.S. Wolfe. 2005. The effect of potato variety mixtures on epidemics of late blight in relation to plot size and level of resistance. *Annals of Applied Biology* 147:245–252.

Prigogine, I.R. 1997. The end of certainty: Time, chaos and the new laws of nature. New York, NY: The Free Press.

Rodgers-Gray, B.S., and M.W. Shaw. 2000. Substantial reductions in winter wheat diseases caused by addition of straw but not manure to soil. *Plant Pathology* 49:590–599.

Rodriguez, R.J., J.F. White, Jr., A.E. Arnold, and R.S. Redman. 2009. Fungal endophytes: Diversity and functional roles. *New Phytologist* 182:314–330.

Schmitt, A., E. Koch, D. Stephan, C. Kromphardt, M. Jahn, H.J. Krauthausen, G. Forsberg et al. 2009. Evaluation of non-chemical seed treatment methods for the control of *Phoma valerianellae* on lamb's lettuce seeds. *Journal of Plant Diseases and Protection* 116:200–207.

Shaner, G., and R.E. Finney. 1977. The effect of nitrogen fertilization on the expression of slow-mildewing resistance in Knox wheat. *Phytopathology* 67:1051–1056.

Sharma, K., C. Bruns, and M.R. Finckh. 2009. Isolate mixtures increase the effectiveness of plant strengtheners in inducing resistance in tomatoes against *Phytophthora infestans*. *Canadian Journal of Plant Pathology* 31, 497 (abstract).

Shattock, R.C. 2002. *Phytophthora infestans*: Populations, pathogenicity and phenylamides. *Pest Management Science* 58:944–950.

Simon, S., P.E. Lauri, L. Brun, H. Defrance, and B. Sauphanor. 2006. Does fruit-tree architecture manipulation affect the development of pests and pathogens? A case study in an organic apple orchard. *Journal of Horticultural Science and Biotechnology* 81:765–773.

Simon, S., J-C. Bouvier, J-F. Debras, and B. Sauphanor. 2010. Biodiversity and pest management in orchard systems. A review. *Agronomy for Sustainable Development* 30:139–152.

Sivčev, B.V., I.L. Sivčev, and Z.Z. Ranković-Vasić. 2010. Plant protection products in organic grapevine growing. *Journal of Agricultural Sciences* 55:103–122.

Skelsey, P., W.A.H. Rossing, G.J.T. Kessel, and W. van der Werf. 2010. Invasion of *Phytophthora infestans* at the landscape level: How do spatial scale and weather modulate the consequences of spatial heterogeneity in host resistance? *Phytopathology* 100:1146–1161.

Tahvanainen, J.O., and R.B. Root. 1972. The influence of vegetational diversity on the population ecology of a specialized herbivore, *Phyllotreta cruciferae* (Coleoptera: Chrysomelidae). *Oecologia* 10:321–346.

Tamm, L., B. Smit, M. Hospers, B. Janssens, J. Buurma, J.P. Molgaard, P.E. Larke et al. 2004. Assessment of the socio-economic impact of late blight and state of the art management in EU organic production systems. Forschungsinstitut für biologischen Landbau, Frick. (unpublished).

Theunissen, J., and G. Schelling. 2000. Undersowing carrots with clover: Suppression of carrot rust fly (*Psila rosae*) and cavity spot (*Pythium spp.*) infestation. *Biological Agriculture and Horticulture* 18:67–76.

Tilman, D., J. Knops, D. Wedin, P. Reich, M. Ritchie, and E. Siemann. 1997. The influence of functional diversity and composition on ecosystem processes. *Science* 277:1300–1302.

Tinivella, F., L.M. Hirata, M.A. Celan, S.A.I. Wright, T. Amein, A. Schmitt, E. Koch et al. 2009. Control of seed-borne pathogens on legumes by microbial and other alternative seed treatments. *European Journal of Plant Pathology* 123:139–151.

Turner, R.S. 2005. After the famine: Plant pathology, *Phytophthora infestans* and the late blight of potatoes, 1845–1960. *Historical Studies in Physical and Biological Sciences* 34:341–370.

Tylianakis, J.M., R.K. Didham, and S.D. Wratten. 2004. Improved fitness of aphid parasitoids receiving resource subsidies. *Ecology* 85:658–666.

U.S. Environmental Protection Agency (USEPA). 2011. Accessed February 2011. http://www.epa.gov/opp00001/factsheets/ipm.htm.

Vallad, G.E., and R.M. Goodman. 2004. Systemic acquired resistance and induced systemic resistance in conventional agriculture. *Crop Science* 44:1920–1934.

Van Emden, H.F. and M.A Bashford. 1969. A comparison of the reproduction of *Brevicoryne brassicae* and *Myzus persicae* in relation to soluble nitrogen concentration and leaf age (leaf position) in the Brussels sprout plant. *Entomologia Experimentalis et Applicata* 12:351–364.

Walters, D.R. 2009. Are plants in the field already induced? Implications for practical disease control. *Crop Protection* 28:459-465.

Walters, D., D. Walsh, A. Newton, and G. Lyon. 2005. Induced resistance for plant disease control: Maximizing the efficacy of resistance elicitors. *Phytopathology* 95:1368–1373.

Weibel, F., and A. Leder. 2004. Consumer reaction to the "Flavour Group Concept" to introduce scab resistant apple varieties into the market. "Variety-Teams" as a further development of the concept. *In*: Proceedings of the 11th International Conference on Cultivation Technique and Phytopathological Problems in Organic Fruit-Growing. pp. 196–201, February 3–5, 2004, Weinsberg, Germany.

Willer, H., and R. Zanoli. 2000. Organic viticulture in Europe. *In*: Willer, H., and U. Meier (eds.), Proceedings of the 6th International Congress on Organic Viticulture. August 25–26, 2000. pp. 163–165. Convention Center Basel/Helga Willer; Urs Meier. – Bad Dürkheim. Stiftung Ökologie und Landbau.

Winkler, K. 2005. Assessing the risks and benefits of flowering field edges – Strategic use of nectar sources to boost biological control. PhD thesis. Wageningen, The Netherlands: Wageningen University.

Wolfe, M.S. 1985. The current status and prospects of multiline cultivars and variety mixtures for disease resistance. *Annual Review of Phytopathology* 23:251–273.

Wolfe, M.S., U.E. Braendle, B. Koller, E. Limpert, J. McDermott, K. Mueller, and D. Schaffner. 1992. Barley mildew in Europe: Population biology and host resistance. *Euphytica* 63:125–139.

Wolfe, M.S., J.P. Baresel, D. Desclaux, I. Goldringer, S.P. Hoad, G. Kovacs, F. Löschenberger, T. Miedaner, H. Østergård, and E.T. Lammerts van Bueren. 2008. Developments in breeding cereals for organic agriculture. *Euphytica* 163:323–346.

Zehnder, G., G.M. Gurr, S. Kühne, M.R. Wade, S.D. Wratten, and E. Wyss. 2007. Arthropod pest management in organic crops. *Annual Review of Entomology* 52:57–80.

Zehnder, G.W., J.F. Murphy, E.J. Sikora, and J.W. Kloepper. 2001. Application of rhizobacteria for induced resistance. *European Journal of Plant Pathology* 107:39–50.

Zhu, Y., H. Chen, J. Fan, Y. Wang, Y. Li, J. Chen, J.X. Fan et al. 2000. Genetic diversity and disease control in rice. *Nature* 406:718–722.

4 Approaches to Breed for Improved Weed Suppression in Organically Grown Cereals

Steve P. Hoad, Nils-Øve Bertholdsson, Daniel Neuhoff, and Ulrich Köpke

Background

Poor weed control is one of the main limitations in organic agriculture (Bond and Grundy, 2001) and results in yield loss and the build-up of weeds and the weed seed bank in the soil (Beveridge and Naylor, 1999). Efficient weed management is essential for successful organic crop production, and crop species and cultivars that confer a high degree of competitive ability against weeds are highly desirable (Mason and Spaner, 2006). Crops with high weed suppression, especially against aggressive weeds, will have value as part of cultural control to reduce the build-up of weeds in organic agriculture, as well as in conventional high input systems (Mason and Spaner, 2006).

Complete weed control is not always encouraged in organic farming, as it is recognized that a low weed population can be advantageous because it provides food and habitats for a range of beneficial organisms (Aebischer, 1997; Clements et al., 1994; van Elsen, 2000; Fuller, 1997). However, above critical thresholds, weed control is required in organic (Bulson et al., 1996), as well as conventional (Cussans, 1968; Hewson, et al., 1973; Wilson and Wright, 1990) farming systems.

Our knowledge of crop competitive ability against weeds and its application to organic farming and plant breeding is most advanced in cereals. This chapter focuses on cereal species, but the principles of crop competitiveness and plant traits for weed suppression that have been developed in cereals can be applied to a wide range of other crops. Indeed, the findings presented herein complement agronomic and physiological studies of crop-weed interactions in other species such as soybean, *Glycine max* L. (Vollmann et al., 2010), potato, *Solanum tuberosum* L. (Hutchinson et al., 2011; Love et al., 1995; Nelson and Thoreson, 1981), tomato, *Lycopersicon esulentum* Mill., pepper, *Capsicum annuum* L. (González-Ponce et al., 1996), and other tropical cereals such as sorghum, *Sorghum bicolor* L. Moench (Wu et al., 2010).

Selection of crops or cultivars suited to organic agriculture requires a different approach to that used in conventional high input systems. This is because there are fewer opportunities to compensate for limitations to yield imposed by diseases, low nutrients, and weeds in organic agriculture compared to conventional agriculture, as well as a requirement to adapt to highly variable environmental conditions across a diversity of organic growing systems (Wolfe et al., 2008).

Organic Crop Breeding, First Edition. Edited by Edith T. Lammerts van Bueren and James R. Myers.
© 2012 John Wiley & Sons, Inc. Published 2012 by John Wiley & Sons, Inc.

Research such as the European Union funded project on Strategies of Weed Control in Organic Farming (Neuhoff et al., 2005) has been essential in providing an understanding of crop-weed interactions and highlighting the potential usefulness of cereal traits, which can be targeted in plant breeding. Selection of new cereal cultivars for competitiveness against weeds under organic conditions relies on the identification of plant traits or crop characteristics and the development of routine methodologies to indicate their potential usefulness. Competitive ability is usually not attributed to a single characteristic, either within or between varieties (Korres and Froud-Williams, 2002; Lemerle et al., 2001a), but the interaction between a series of desirable characteristics is important (Eisele and Köpke, 1997b; Mason and Spaner, 2006; Seavers and Wright, 1999).

Crop competitive ability in reducing weed growth has been recognized in a wide range of cereal species (Bastiaans et al., 1997; Begna et al., 2001; Caton et al., 2001; Cosser et al., 1996; Didon, 2002; Doll, 1997). This knowledge is important for breeders in the selection and evaluation of cultivars for weed suppression, especially for organic farmers and farmers practicing integrated farming methods. Among temperate cereal species, the most suppressive or competitive are oats (*Avena sativa* L.) and winter rye (*Secale cereale* L.), followed by triticale, barley (*Hordeum vulgare* L.), and wheat (*Triticum aestivum* L.; Davies and Welsh, 2002; Lemerle et al., 1995).

Although plant breeders record a wide range of morphological and developmental features, weed suppression is unlikely to have been a high priority in conventional plant breeding (Lammerts van Bueren et al., 2002). However, a large body of research activity has made significant progress in understanding the morphological and physiological basis of competitive ability and tolerance to weeds (Lemerle et al., 1996, 2006; Mason and Spaner, 2006; Watson et al., 2006) and this can be applied to plant breeding. Both plant traits (e.g., height) and crop characteristics (e.g., ground cover) are recognized as important selection criteria. Large genotype by environment interactions may cause difficulties in selecting for competitiveness (Coleman et al., 2001), and selection for competitiveness could be at the expense of other important criteria (Brennan et al., 2001). Nevertheless, within a given climatic zone there appears to be sufficient genetic variation in crop competitive ability (Acciaresi et al., 2001; Coleman et al., 2001) for such selection to be introduced into breeding programs.

Crop Competitiveness Against Weeds

There has been recognition for some time that cereal cultivars differ in their competitiveness with weeds. Lemerle et al. (2001a) reported the extent of, and some reasons for, wheat cultivar variation in competitive ability. More specifically, Moss (1985) showed that some cultivars of winter wheat and barley withstood *Alopecurus myosuroides* competition better than others. Richards (1989) and Richards and Davies (1991) found differences in weed cover across spring barley and wheat cultivars, which were related to early crop ground cover and subsequent ease of their control in non-organic experimental plots treated with herbicides. Blackshaw (1994) found that semi-dwarf cultivars of wheat were less competitive with *Bromus tectorum* than tall cultivars, a result that was attributed in part to competition for light. Lemerle et al. (2001b) examined 250 genotypes of wheat in competition with *Lolium rigidum* and found a difference in yield response dependent on cultivar. Older genotypes are often more competitive than recent introductions (Bertholdsson, 2005; Lemerle et al., 2001b), especially if this is associated with high early biomass accumulation, large numbers of tillers and tallness with an extensive leaf canopy.

Cosser et al. (1997) suggested that tall cultivars of wheat such as 'Maris Widgeon' (a UK cultivar) could be beneficial in organic systems when high weed growth was anticipated. Gooding et al. (1997) evaluated *Rht* genes for dwarfing in organically grown wheat. A shortening of plants increased weed infestation with *Alopecurus myosuroides,* which was related to a reduction in shading ability as assessed by penetration of photosynthetically active radiation and/or height differences between cultivars. Verschwele and Niemann (1996) confirmed the potential of wheat cultivar variation for weed suppression, and attributed shading ability to high crop cover and increased crop canopy height. However, Eisele and Köpke (1997a) stated that while tallness was important, other traits that provided good overall shading ability were essential. For example, they confirmed that in organically grown wheat, planophile rather than erectophile leaf inclination gave increased ground shading during growth, which could significantly decrease weed biomass. Huel and Hucl (1996) found that in spring wheat tallness, large seedling ground cover, and flag leaf length were associated with differences in competitiveness against *Avena fatua* and *Brassica juncea*, while Acciaresi et al. (2001) examined competition between wheat cultivars and *Lolium multiflorum*, and Coleman et al. (2001) and Mokhtari et al. (2002) with *Lolium rigidum,* found differences in competitiveness related to plant height and leaf length or size.

A crop's competitiveness against weeds is also strongly influenced by agronomic factors (Lemerle et al., 2001a) including seed rate and crop spacing (e.g., plant-row width), as well as soil and climatic conditions, which influence the growth and development of both crop and weeds. In organic systems where nutrient availability is very different compared to conventional systems, the identification of characteristics that allow the crop to give adequate shading, particularly early in the season before soil nitrogen release fully develops, is important. Especially in organic agriculture early crop growth in winter crops will depend on crop nutrient uptake capacity and nitrogen use efficiency because growth is often limited by nitrogen availability in the early spring (Eisele and Köpke, 1997a). Besides cultivar choice, narrowing plant row width was shown to increase early crop ground cover considerably. This could be a consequence of improved plant spacing, enhanced root growth, and higher uptake capacity for soil resources.

Other environmental influences on competitiveness, such as water availability, have been alluded to in the literature as discussed in Lemerle et al. (2001a). In the future, it would be beneficial to consider cultivar differences in crop characteristics such as leaf canopy development across different environments, especially in relation to obtaining nutrients early in the season, or in dry environments where plant traits for moisture scavenging ability are important. This would identify cultivars with plant structures that are usefully shading weeds even with a low soil nutrient or moisture status.

Crop Traits Involved in Weed Suppression

Table 4.1 summarizes a broad range of crop characteristics that are essential or desirable in weed suppression. Each has been identified in the literature as affecting weed growth. Some of these traits will determine whole crop characteristics such as crop ground cover, leaf canopy size, and light interception. Although there are a large number of potentially useful traits, high crop ground cover is universally important. Crop ground cover comprises a broad range of plant traits (Didon and Hansson, 2002; Huel and Hucl, 1996; Ogg and Seefeldt, 1999). For example, the importance of leaf inclination for cultivar specific competitiveness has been widely reported (Drews et al., 2009; Huel and Hucl, 1996; Korres and Froud-Williams, 2002; Lemerle et al., 1996; Niemann, 1992). Subsequently, this has led to idea of desirable plant habits for weed suppression (see next section).

Although some cultivars have higher weed suppression than others, this is usually not attributed to a single characteristic (Eisele and Köpke, 1997b). Generally, traits conferring high shading ability determine the competitive ability of a cultivar (Cosser et al., 1997; Eisele and Köpke, 1997a) and

Table 4.1 Relative value of cereal plant traits and crop characteristics for weed suppression

Trait	Value of Trait	Sources
Good plant establishment	Essential	1,2,3
High early season ground cover	Essential	4,5,6
High tillering ability	Essential	1,2,3,7,8
Increased plant height	Desirable	1,2,3,9,10,11
Rapid growth rate	Desirable	3,6,7,11
Planophile leaf habit	Highly desirable	10,12,13,14,15
High leaf area index	Highly desirable	12,13,14,15
Wide leaf laminae	Desirable	13,15
High yield potential	Desirable	16
Allelopathic activity	Desirable	6,17

Sources: 1. Wicks et al., 1986; 2. Korr et al., 1996; 3. Didon and Hansson, 2002; 4. Richards and Davies, 1991; 5. Froud-Williams, 1997; 6. Bertholdsson, 2005; 7. Lemerle et al., 2001a; 8. Korres and Froud-Williams, 2002; 9. Gooding et al., 1993; 10. Drews et al., 2009; 11. Ogg and Seefeldt, 1999; 12. Niemann, 1992; 13. Heul and Hucl, 1996; 14. Seavers and Wright, 1999; 15. Lemerle et al., 1996; 16. Hoad et al., 2008; 17. Bertholdsson, 2007.

it is the identification of these traits and their interactions that is important for incorporating weed competition in breeders' selection strategies.

To a large extent, the evaluation of above-ground traits for weed suppression has been considered in relation to maximizing light interception by the crop (Cosser et al., 1997; Eisele and Köpke, 1997a; Gooding et al., 1993; Neuhoff et al., 2005), though the relative importance of the different morphological characteristics involved in canopy development has not always been clarified in the literature. In particular, the potential for substitution of structures allowing for differing genotypic variation to give a similar shading end result has not been fully explored.

Sometimes there may be only small differences in ground shading of contrasting cultivar types, though even small differences in shading ability, or the percentage of light intercepted, can have a significant effect on weed growth (Eisele and Köpke, 1997b). High light interception at the top of crop stands, especially in tall, planophile, genotypes would appear to offer a distinct competitive advantage (Amesbauer and Hartl, 1999; Christensen, 1995).

Assessing the value of below-ground traits in weed suppression is more challenging for plant breeding, though potentially very beneficial. Allelopathic differences among cultivars has been alluded to in the literature, derived from in vitro testing (Belz and Hurle, 2001; Wu et al., 2000), but less is known about in vivo behavior under field conditions. Competition for nutrients and allelopathic effects of the crop on weeds are likely to be key areas for further investigation.

The following section describes the most important plant traits and crop characteristics in more detail. These findings come mainly from research, but they also come from field observations of plant breeders. Some traits may become more or less important, depending on the climatic region. Furthermore, the strengths in some characteristics may compensate for weaknesses in others.

Selection of Traits and Their Evaluation in Plant Breeding Programs

The Importance of Crop Ground Cover as a Composite Trait in Plant Selection

Crop ground cover is the most important crop characteristic for competing against weeds (Davies et al., 2004; Hoad et al., 2006a, 2006b; Lemerle et al., 1996; Richards and Whytock, 1993). When

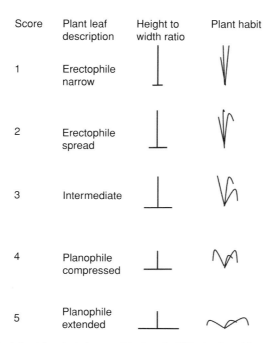

Figure 4.1 Plant growth habits defined by plant shape and leaf angle. This simple guide provides a means to score plants in breeders' selection programs and variety trials (modified from an earlier version described in Hoad et al., 2006a).

measured from directly above the leaf canopy, crop ground cover is a good indicator of shading characteristics and can be used as a guide in cultivar selection. Total leaf area index or green area index are often correlated with crop ground cover, shading ability, and weed suppression (Neuhoff et al., 2005).

When a cultivar competes well against weeds it is usually associated with a relatively high fractional light interception. Some research has also shown that high weed suppression is associated with relatively high light interception in the upper leaf canopy of tall, planophile cultivars (Amesbauer and Hartl, 1999; Christensen, 1995). As stated earlier, crop ground cover integrates several plant and crop characteristics. It is also influenced by agronomic factors such as plant spacing, plant and tiller density and soil nitrogen availability, as well as by plant traits such as leaf inclination. Although seasonal variations in plant establishment and shoot production will influence a crop's ground cover, it is possible to describe general plant growth habits that are likely to confer varying degrees of competitive ability (see fig. 4.1 and next section).

Plant Growth Habit – General Features

The main determinant of plant growth habit is leaf inclination or angle. Figure 4.1 illustrates how plants with different leaf angles may be described, or scored, in variety trials. The scheme is based on a 1 to 5 scale for assessing erectophile, planophile, or intermediate growth habits (Hoad et al., 2006a).

The importance of leaf inclination for cultivar specific competitive ability was confirmed in the findings of several authors (Drews et al., 2009; Huel and Hucl, 1996; Korres and Froud-Williams, 2002; Lemerle et al., 1996; Niemann, 1992). High weed suppression ability in planophile cultivars has been associated with large flag leaves (Huel and Hucl, 1996; Niemann, 1992; Seavers and Wright, 1999) and increased fractional light interception (Eisele and Köpke, 1997a, 1997b) compared to erectophile cultivars. However, the erectophile habit has been the long established ideotype for high yields in cereals. This habit can be an advantage when weed levels are low, but it is a high-risk phenotype when weed establishment is high, especially early in the growing season. If an erectophile habit is desired, then increased plant height may be of greater value. The erectophile growth habit can be more competitive than the planophile type as long as traits such as plant height and high tillering are the main features of competitive ability (Korres and Froud-Williams, 2002).

A planophile habit is most useful where weed growth continues throughout the cropping season (i.e., through to leaf canopy closure and beyond, such as in cool-temperate regions). By contrast, in a Mediterranean climate, late season weed growth is often reduced by hot, dry conditions. As most cultivars in this region are erectophile at present, large leaves and plant height become relatively more important characters in maintaining shading ability.

Early Crop Vigor and Rapid Early Season Growth

Early crop vigor is associated with increased competitive ability (Acciaresi et al., 2001; Bertholdsson, 2005; Lemerle et al., 2001a, 2001b; Rebetzke and Richards, 1999). Traits conferring early vigor include seed size, seedling growth rate, and coleoptile tiller development (Liang and Richards, 1994, 1999; Qingwu and Staugaard, 2002) and seed embryo size (López-Cartaneda et al., 1996). Various measures of rapid, or early, growth rate are useful for organic cereal breeding programs (Bertholdsson, 2005; Jönsson et al., 1994). For example, initial shoot (and root) growth rates are important indicators of competitive ability (Christensen, 1995; Didon, 2002; Froud-Williams, 1997), as is the rate of stem extension (Didon and Hansson, 2002; Ogg and Seefeldt, 1999). Leaf width is also an indicator of potential crop ground cover (Huel and Hucl, 1996; Lemerle et al., 1996). Selections in variety trials should also consider early season ground cover in relation to plant growth habit, e.g., prostrate growth (Richards and Davies, 1991; Froud-Williams, 1997) and erectophile habit (Korres and Froud-Williams, 2002; Oberforster et al., 2003), or as defined in figure 4.1.

Although the literature presents conflicting views about the relative importance of early season ground cover on weed suppression (Cosser et al., 1997; Davies et al., 2004; Huel and Hucl, 1996; Lemerle et al., 1996; Richards and Whytock, 1993), it is clear that rapid early growth allows the crop to maintain a light interception lead over rapidly growing weeds and to shade newly emerging weeds. Rapidly developing ground cover (e.g., during tillering) is evident across all growth habits, though erectophiles may need to compensate by producing more tillers or longer and larger leaves (Korres and Froud-Williams, 2002; Oberforster et al., 2003). Crop ground cover at the end of tillering has a strong, negative correlation with weed suppression throughout the season (Davies et al., 2004; Richards and Davies, 1991).

High Tillering Capacity

A crop's ability to produce or retain a high number of shoots is an important determinant of ground coverage and shading ability (Didon and Hansson, 2002; Korr et al., 1996; Wicks et al., 1986). Tiller or shoot population density is a function of plant number per unit of ground area and the ability of a plant to produce and maintain tillers in a given environment. Cultivars may vary in either or both of these characteristics. In practice, high tillering ability is likely to be a key trait in compensating for low plant establishment or plant population density. For example, in cool-temperate regions high tillering ability would maintain shoot numbers under low plant populations, e.g., less than 150 to 160 plants per square meter.

Increasing Plant Height

Plant height is widely reported as a desirable trait for increasing crop competitiveness (Didon and Hansson, 2002; Drews et al., 2009; Gooding et al., 1993; Korr et al., 1996; Oberforster et al., 2003; Wicks et al., 1986), with seasonal development of crop canopy height and final plant height both being important (Didon and Hansson, 2002; Korr et al., 1996; Oberforster et al., 2003; Wicks et al., 1986). Taller varieties are likely to be more competitive than shorter ones as competition for light increases (Hashem et al., 1998; Cudney et al., 1991).

Plant height, however, must also be considered as part of overall shading ability in suppressing weeds (Cosser et al., 1997; Eisele and Köpke, 1997a). This is because the importance of height decreases if a range of other traits ensures competitiveness against weeds. For instance, Seavers and Wright (1999) assessed higher weed suppression ability of a shorter wheat cultivar *Avalon*, which additionally had planophile leaves, higher leaf area index, and faster development compared to a taller cultivar *Spark*.

Although the ability to grow above the weed population may be independent of leaf inclination and amount of light intercepted (Blackshaw, 1994), tallness associated with high overall shading ability is an important target (Eisele and Köpke, 1979a), as conferred by a planophile leaf habit in the upper canopy (Amesbauer and Hartl, 1999; Christensen, 1995). Tall cultivars may be most important for the suppression of certain weed species; for example, tall-growing or scrambling weeds including grasses such as *Apera-spica venti* and *Avena fatua* (Balyan et al., 1991; Blackshaw, 1994; Cousens et al., 2003; Ogg and Seefeldt, 1999).

Tall or very tall cultivars would appear to be most competitive when established at moderate to high plant population densities, or if they have a planophile leaf habit in the upper canopy (Amesbauer and Hartl, 1999; Christensen, 1995). The examples above are consistent with the idea that combinations of factors contribute to weed suppression (Eisele and Köpke, 1997a; Seavers and Wright, 1999). This is further illustrated by wheat cultivars such as 'Maris Widgeon', which in variety trials have developed high ground cover by being very tall and developing a more planophile habit during stem extension.

Below-Ground Competition for Resources and Allelopathy

There will be circumstances where plant competition for soil resources is more important than competition for light, as reviewed by Wilson (1988). Particularly in early growth stages root traits are often more important for competitiveness than shoot traits, as shown for ryegrass grown with white clover (Martin and Field, 1984). Early crop growth in winter crops, especially in organic agriculture, is dependent on crop nutrient uptake capacity and nitrogen-use efficiency, because growth is often limited by nitrogen (Eisele and Köpke, 1997a).

Ideally, selection criteria for above-ground traits should be developed alongside those for below-ground characteristics such as root competition and allelopathy (Bertholdsson, 2004, 2005, 2007, 2010). However, characteristics such as allelopathy are less well understood and may be less practical for use by plant breeders in their selections. For example, a vigorous root system confers increased competitive ability for water and nutrients, though there is no correlation between root size and allelopathic activity (Bertholdsson, 2004; Jensen et al., 2001).

Variation in allelopathic effects on weeds has been identified from in vitro testing in wheat (Belz and Hurle, 2001; Bertholdsson, 2010; Wu et al., 2000), barley (Bertholdsson, 2004, 2005), and rice (Olofsdotter et al., 2001), but much less is known about in vivo behavior.

It is possible to find cultivars and breeding lines with high allelopathic activity, but breeding for this trait is complex (Bertholdsson, 2007), as it is difficult to separate allelopathy from other characteristics of crop competitive ability (Bertholdsson, 2005). Consequently there may be as yet unexplored potential for the selection of varieties showing a high allelopathic activity against weeds (Olofsdotter et al., 2002; Wu et al., 2000). So far, most efforts have been made in rice breeding (Kong et al., 2006) and in spring wheat with several lines under evaluation (Bertholdsson, 2010).

In oat (*Avena sativa* L.) and barley (*Hordeum vulgare* L.) several allelochemicals have been identified (Kato-Noguchi et al., 2010; Kremer and Ben-Hammouda, 2009). Interestingly, in Scandinavian landraces and old cultivars of barley the allelopathic potentials show a decreasing trend with introduction of new cultivars onto the market (Bertholdsson, 2004).

One of the first crops to be exploited for allelopathy was rice (*Oryza sativa* L.; Olofsdotter, 2001; Olofsdotter et al., 2002). For example, the highly allelopathic rice cultivar Huagan-1 suppressed growth of *Echinochloa crus-gali* L. (banyard grass; Kong et al., 2006). Rye (*Secale cereale* L.) is another crop known for its allelopathic proprieties and has also been used in cover-crop residue management systems for weed control (Kruidhof et al., 2009). Hybrids of wheat and rye (e.g., triticale, *Triticosecale*, Wittm.) can also have high allelopathic activity, though other traits such as baking quality may be compromised (Merker, 1992). Recent studies have shown that the weed suppression ability of wheat could be improved by 60% if early vigor and allelopathy are improved to the level of triticale (Bertholdsson, 2010).

Selection Strategies

Methods for quantifying weed suppression need to be practical and sufficiently robust for use in breeders' selections and in the evaluation of new lines and cultivars by testing authorities. Selection for weed suppression can be carried out directly in the presence of weeds (Cosser et al., 1997; Froud-Williams, 2002; Hoad et al., 2006a; Korres and Lemerle et al., 2006b), or indirectly by selecting for traits associated with competitive ability (Bertholdsson, 2005; Gibson et al., 2003; Zhao et al., 2006).

Direct selection can be undertaken when crop growth and yield is assessed against natural weed populations, or sown-in weed species and weed mixtures. By contrast, indirect selection is carried out on the basis of selecting traits, or trait combinations, of known value. Knowledge of the latter is most likely to come from research activity, rather than breeding itself.

Direct selection for high weed suppression should be undertaken across as many trials as possible so that rank order in competitive ability can be considered in relation to yield and overall cultivar robustness, or stability in yield (Hoad et al., 2008). This is because cultivars for use in organic farming, especially those bred conventionally, will need to cope with large seasonal differences in weed populations compared to conventional farming, as well as other potential sources of high variability such as soil fertility and crop establishment.

Some traits such as plant height and rate of development are relatively easy to measure and relatively stable, while other characteristics such as leaf area or shading ability integrate several traits and are less easy to select. Selection for competitiveness needs to take into account a broad spectrum of traits and assessments to optimize all potentially useful factors (Caton et al., 2001), for example in the identification of early vigor or early season growth.

Traits such as early vigor and rapid early growth rate can be screened for under controlled conditions, including hydroponics (Bertholdsson and Brantestam, 2009; Jönsson et al., 1994), by measuring the width of the first leaf (Rebetzke et al., 2008) or the embryo size (López-Castañeda et al., 1996). This type of screening has the advantage in that it can be done in early generations thus reducing the risk of losing useful cultivars if selection had started in later generations.

Environmental conditions affect the competitive level of both crop and weeds. For example, Coleman et al. (2001) indicated seasonal effects on the relative importance of different crop shading factors. Lemerle et al. (2001b) found that some cultivars showed environment-specific competitive advantage while others were relatively poorly competitive in some situations. This has particular relevance for organic farming situations where the availability of nutrients is more variable. Cultivars that show greater stability in response to variations in the environment, in terms of their ability to suppress weeds, will be more beneficial to organic farmers. However, relatively few studies have examined cultivar responses across a wide environmental range in organically grown crops (Cosser et al., 1997; Eisele and Köpke, 1997a; Gooding et al., 1997; Hoad et al., 2008).

Agronomic factors such as seed rate and row (drilling) width strongly influence crop characteristics, especially crop ground cover. For example, narrowing the row width or increasing plant density will increase ground cover. Furthermore, the relative benefit of some traits such as a planophile growth habit is likely to increase when row spacing is widened. Modifying agronomic factors could also change the rank order of cultivars or their traits: E.g., tillering ability. Therefore, plant and crop assessments should use a broad range of cultivars (Caton et al., 2001) to identify optimal cultivar performance.

The timing of crop and weeds assessments, e.g., by crop growth stage, is important as this will influence the relative value of different plant traits, or the scoring of overall weed suppression ability (Hoad et al., 2008). The fact that weeds tend to cluster in patches (Dessaint et al., 1991) must also be considered to avoid under- or over-estimating the effects of weeds on crop performance. Variability in weed distribution or growth patterns can be taken into account by increased replication or modifying sampling procedures, as in the presence and absence scoring described by Gold et al. (1996).

Much of the research cited in the previous sections has been carried out either with grass weeds that have a similar growth pattern with the crop or with individual weed species, with relatively little observation on the background flora (e.g., annual dicotyledonous species), which may be important under farming conditions. Therefore, a key development in any selection strategy would be to understand how cultivar performance varies with changes in the weed species population. Significant levels of cultivar by competition interaction have been observed (de Vida et al., 2006) and anecdotal evidence from field trials also suggests that cultivar performance varies depending on the amount of weed growth and the species of weeds present (Hoad et al., 2008).

Various analytical methods to quantify genotype by environment interactions (Truberg and Huhn, 2002) can be used to help understand genotypic responses to changes in weed growth. Furthermore, both weed suppression ability and its sensitivity to weed growth could be introduced in cultivar selection and variety testing systems for organic agriculture (Hoad et al., 2008). Ideally, new cultivars would be selected on the basis of high weed suppression and/or low sensitivity to changes in weed growth. Accounting for genotype variation will be most important when genotype by environment interactions are high or when differences between environments are large or extreme, as is often found in organic agriculture (Wolfe et al., 2008).

Understanding Crop-Weed Interactions to Assist Plant Breeding

For plant breeding it is most useful if plant traits are of generic value. However, crop-weed associations can be highly specific, meaning that particular traits or crop characteristics have benefits under certain conditions. It is important to consider the competitive nature of individual weed species or the weed population as a whole. This is partly because of the direct effects of different weed species on resource capture (light and soil resources), but mostly because of their differential tolerance to shading and responses to the availability of soil moisture and nitrogen.

According to Mohler (1996), in early growth stages many crops including cereals have a size advantage over weeds due to a relatively large seed and seedling. Thus, earlier or faster crop growth and development provides a basis for a competitive advantage of crops over weeds at the onset of competition for light (Liebman and Davis, 2000). However, if the early growth of a winter cereal is comparatively slow or limited by nitrogen availability, then weed flora can become more competitive.

Later emerging weeds have a shorter period of time to compensate the growth advantage of crop plants. Consequently late emerging weed plants are clearly less competitive against crops compared to earlier emerged plants, e.g., potentially competitive tall-growing or climbing species like *Chenopodium album* and *Polygonum convolvulus* can be strongly suppressed by winter wheat competition. The dominant effect of weed emergence time for weed competitiveness was confirmed for *Avena fatua* in wheat and barley (O'Donovan et al., 1985) and for *Chenopodium album* in barley Håkansson (1979). According to the latter, delay of a few days in weed emergence resulted in significantly shorter weed plants and lower weed dry matter at flowering time of barley, as a result of the relative competitive advantage of the crop.

For some crop-weed relationships, it may be that reliable weed control through crop shading ability is predominantly limited to short-statued weed types (e.g., *Aphanes arvensis*) and to late spring emerging weeds (e.g., *Polygonum convolvulus*). Suppression of early emerging or tall-growing or climbing annual weeds (e.g., *Vicia hirsuta*) may be influenced by other factors, such as dry soil conditions in spring, to favor crop growth compared to weed growth. Under favorable growth conditions for weeds, e.g., cold and rainy springs, weed species such as *Vicia hirsuta* or *Galium aparine* are able to overtop crops, thus escaping from shading, with consequences for crop grain yield (Iqbal and Wright, 1999; Willey and Holliday, 1971).

Some studies indicate that cereal shading ability may not offer adequate control of highly competitive perennials such as *Cirsium arvense* or *Agropyron repens* (Rauber and Böttger, 1984), though other work has shown that even these species (e.g., *Cirsium arvense*) are sensitive to light competition (Dau et al., 2004; Moore, 1975). Shoot and rhizome growth of species such as *Agropyron repens* may also be susceptible to crop shading, but at later cereal growth stages (e.g., grain filling), when the leaf canopy is in decline, its growth is strongly accelerated as a result of increasing light interception (Rauber and Böttger, 1984).

The crop-weed association will also depend on the ability of both the crop and weeds to tolerate competition. Cultivars vary in their tolerance to weeds, i.e., maintain yield under the presence of competition, as defined by Lemerle et al. (2006). Weeds also vary in their tolerance to shading. Ground cover of shaded short-statued weeds is not always reduced, which suggests tolerance to shading. For example, *Stellaria media* has a high degree of tolerance to shading (Jørnsgård et al., 1996). Physiological tolerance to shading depends on each species and its specific ability to adapt to shading (Stoller and Myers, 1989). For example, in contrast to *Stellaria media,* ground cover of shaded *Ranunculus sardous* and *Vicia hirsuta* can be significantly reduced by crop ground cover and light interception.

Weeds also vary in their competitive ability as influenced by soil nitrogen availability. Competitiveness of nitrophilous weed species such as *Stellaria media* and *Gallium aparine* is enhanced by high soil nitrogen availability (Grundy and Froud-Williams, 1993; Rooney et al., 1990). By contrast, other species such as *Anagallis arvensis, Apera spica-venti,* and *Matricaria recutita* can be negatively correlated with crop growth under the same high-nitrogen conditions. The effect of soil nitrogen on crop-weed relations is also influenced by weed tolerance to shading ability. Jørnsgård et al. (1996) investigated the interaction between nitrogen availability and biomass production of nitrophilous weed species in cereals species. Higher nitrogen availability enhanced crop growth and crop shading ability, but the response of weed species differed according to their tolerance to shading. For example, the biomass of less tolerant species was suppressed by enhanced crop shading ability, and they did not benefit from increasing nitrogen availability.

Concluding Remarks and Wider Perspectives

Cereal cultivars conferring a high degree of crop competitive ability are highly beneficial in organic farming as well as other farming systems that aim to limit the use of herbicides. Breeding and evaluation of new cultivars in relation to weed suppression should consider characteristics such as growth habit, speed of early development, plant height, and tillering ability. A crop's competitiveness against weeds is also determined by agronomic factors, including seed rate and crop spacing (e.g., row width), as well as soil and climatic condition that influence the growth and development of both crop and weeds.

Many of the important weed suppression traits of value in cereals can be applied to the breeding of other crops, including *Solanum tuberosum* (potato), *Glycine max* (soybean), *Phaseolus* (beans), *Pisum sativum* (pea), *Cicer arietinum* (chickpeas), *Medicago sativa* (alfalfa), *Beta vulgaris* (sugar beet) and *Gossypium* species (cotton), in which other plant traits such as leaf size and branching patterns will be important for conferring high shading ability.

Generally, a high season-long crop ground cover is important. Weed suppression cannot be attributed to a single characteristic. Instead, the interaction between a series of desirable characteristics is important, with cultivars compensating for weakness in certain traits with strengths in others. Further research is needed to understand some of these interactions and especially the potential benefits of below-ground traits (e.g., allelopathic effects and root competition).

Breeders' selection for plant growth habits is based on their value under different soil and cropping conditions, or locations in organic farming. The balance between different characteristics for weed suppression will determine the value of the cultivar for early, late, and season-long weed control. A key factor for all these characteristics is a robust consistency in weed suppression and yield required for organic cultivars.

Competitive ability is only one of a number of desirable characteristics that need to be considered when selecting new cultivars, or when devising new methods to assist in the selection process. It is an essential characteristic for cultivars use in organic farming, as long as it can be integrated with other desirable traits for organic agriculture. Otherwise, selection of weed suppression alone could be considered too narrow, especially if other selection criteria were neglected or if there were trade-offs between different selection criteria (Wolfe et al., 2008).

In selecting cultivars suited to organic farming it is also important to measure the consistency, or stability, in general performance across environments. In the future, it would be useful to screen for weed suppression ability across different levels of weed growth and composition, or even under different selection strategies, as reviewed in Wolfe et al. (2008).

Breeding for improved competitive ability against weeds would bring added benefit to organic cultivars and organic farming. Although many of these desired traits have not received enough priority in conventional breeding systems (Lammerts van Bueren et al., 2002), it is clear than certain traits could be selected for and incorporated into variety recommended lists. This would allow farmers to identify cultivars to maximize weed suppression under their particular growing conditions and farming system.

References

Acciaresi, H.A., H. Chidichimo, and S.J. Sarondon. 2001. Traits related to competitive ability of wheat (*Triticum aestivum*) varieties against Italian ryegrass (*Lolium multiflorum*). *Biological Agriculture and Horticulture* 19:275–286.

Aebischer, N.J. 1997. Effects of cropping practices on declining farmland birds during the breeding season. *In*: British Crop Protection Council (ed.), Proceedings of the Brighton Crop Protection Conference – Weeds. pp. 915–922. Brighton, UK: British Crop Protection Council, Hampshire.

Amesbauer, W., and W. Hartl. 1999. Lichtkonkurrenz in Winterweizenbeständen – eine Überlegung zur Sortenwahl im biologischen Landbau. Beiträge zur 5. Wissenschaftstagung zum Ökologischen Landbau, Berlin. pp. 505–508.

Balyan, R.S., R.K. Malik, R.S. Panwar, and S. Singh. 1991. Competitive ability of winter wheat cultivars with wild oat (*Avena ludoviciana*). *Weed Science* 39:154–158.

Bastiaans, L., M.J. Kropff, N. Kempuchetty, A. Rajan, and T.R. Migo. 1997. Can simulation models help design rice cultivars that are more competitive against weeds? *Field Crops Research* 51:101–111.

Begna, S.H., R.I. Hamilton, L.M. Dwyer, D.W. Stewart, D. Cloutier, L. Assemat, K. Foroutan-Pour, and D.L. Smith. 2001. Morphology and yield response to weed pressure by corn hybrids differing in canopy architecture. *European Journal of Agronomy* 14:293–302.

Belz, R., and K. Hurle. 2001. Tracing the source – Do allelochemicals in root exudates of wheat correlate with cultivar-specific weed-suppressing ability? *In*: Proceedings of the Brighton Crop Protection Conference – Weeds. pp. 317–320.

Bertholdsson, N-O. 2004. Variation in allelopathic activity over one hundred years of barley selection and breeding. *Weed Research* 44:78–86.

Bertholdsson, N-O. 2005. Early vigour and allelopathy – Two useful traits for enhanced barley and wheat competitiveness against weeds. *Weed Research* 45:94–102.

Bertholdsson, N-O. 2007. Varietal variation in allelopathic activity in wheat and barley and possibilities for use in plant breeding. *Allelopathy Journal* 19:193–201.

Bertholdsson, N-O. 2010. Breeding spring wheat for improved allelopathic potential. *Weed Research* 50:49–57.

Bertholdsson, N-O., and A.K. Brantestam. 2009. A century of Nordic barley breeding – Effects on early vigour root and shoot growth, straw length, harvest index and grain weight. *European Journal of Agronomy* 30:266–274.

Beveridge, L.E., and R.E.L. Naylor. 1999. Options for organic weed control – What farmers do. *In*: Proceedings of the Brighton Crop Protection Conference – Weeds. pp. 939–944.

Blackshaw, R.E. 1994. Differential competitive ability of winter wheat cultivars against downy brome. *Agronomy Journal* 86:649–654.

Bond, W., and A.C. Grundy. 2001. Non-chemical weed management in organic farming systems. *Weed Research* 41:383–405.

Brennan, J.P., D. Lemerle, and P. Martin. 2001. Economics of increasing wheat competitiveness as a weed control weapon. Contributed paper presented to the 45th Annual Conference of the Australian Agricultural and Resource Economics Society. Adelaide, Australia: Australian Agricultural and Resource Economics Society.

Bulson, H., J. Welsh, C. Stopes, and L. Woodward. 1996. Agronomic viability and potential economic performance of three organic four year rotations without livestock, 1988–1995. Aspects of Applied Biology 47: Rotations and Cropping Systems. pp. 277–286.

Caton, B.P., A.M. Mortimer, T.C. Foin, J.E. Hill, K.D. Gibson, and A.J. Fischer. 2001. Weed shoot morphology effects on competiveness for light in direct-seeded rice. *Weed Research* 41:155–163.

Christensen, S. 1995. Weed suppression ability of spring barley varieties. *Weed Research* 35:241–249.

Clements, D.R., F.W. Stephan, and C.J Swanton. 1994. Integrated weed management and weed species diversity. *Phytoprotection* 75:1–18.

Coleman, R.D., G.S. Gill, and G.J. Rebetzke. 2001. Identification of quantitative trait loci for traits conferring weed competitiveness in wheat (*Triticum aestivum*). *Australian Journal of Agricultural Research* 52:1235–1246.

Cosser, N.D., J.M. Gooding, and R.J. Froud-Williams. 1996. The impact of wheat cultivar, sowing date and grazing on the weed seed bank of an organic farming system. Aspects of Applied Biology 47: Rotations and Cropping Systems. pp. 429–432.

Cosser, N.D., J.M. Gooding, A.J. Thompson, and R.J. Froud-Williams. 1997. Competitive ability and tolerance of organically grown wheat cultivars to natural weed infestations. *Annals of Applied Biology* 130:523–535.

Cousens, R.D., A.G. Barnett, and G.C. Barry. 2003. Dynamics of competition between wheat and oats: 1. Effects of changing the time of phenological events. *Agronomy Journal* 95:1293–1304.

Cudney, D.W., L.S. Jordan, and A.E. Hall. 1991. Effect of wild oat (*Avena fatua*) infestations on light interception and growth rate of wheat (*Triticum aestivum*). *Weed Science* 39:175–179.

Cussans, G.E. 1968. The growth and development of *Agropyron repens* (L.) Beauv. In competition with cereals, field beans and oilseed rape. *In*: Proceedings 9th British Weed Control Conference. pp. 131–136.

Davies, D.H.K., and J.P. Welsh. 2002. Weed control in organic cereals and pulses. *In:* Younie, D., B.R. Taylor, J.P. Welsh, and J.M. Wilkinson (eds.). Organic Cereals & Pulses. pp. 77–114. Lincoln, UK: Chalcombe Publications.

Davies, D.K.H., S.P. Hoad, P.R. Maskell, and K. Topp. 2004. Looking at cereal varieties to help reduce weed control inputs. *In*: Proceedings Crop Protection Northern Britain. pp. 159–163. Dundee, UK.

Dau, A., B. Wassmuth, H-H. Steinmann, and B. Gerowitt. 2004. Keimung und Entwicklung von *Cirsium arvense* unter Lichtkonkurrenz – ein Modellversuch. Zeitschrift für Pflanzenkrankheiten und Pflanzenschutz, Sonderheft XIX. pp. 169–176.

Dessaint, F., R. Chadoeuf, and G. Barralis. 1991. Spatial pattern analysis of weed seeds in the cultivated soil seed bank. *Journal of Applied Ecology* 28:721–730.

Didon, U.M.E. 2002. Variation between barley cultivars in early response to weed competition. *Journal of Agronomy and Crop Science* 188:176–184.

Didon, U.M.E., and M.K. Hansson. 2002. Competition between six spring barley (*Hordeum vulgare* ssp. *vulgare* L.) cultivars and two weed flora in relation to interception of photosynthetic active radiation. *Biological Agriculture and Horticulture* 20:257–273.

Doll, H. 1997. The ability of barley to compete with weeds. *Biological Agriculture and Horticulture* 14:43–51.

Drews, S., D. Neuhoff, and U. Köpke. 2009. Weed suppression ability of three winter wheat varieties at different row spacing under organic farming conditions. *Weed Research* 49:526–533.

Eisele, J.A., and U Köpke. 1997a. Choice of variety in organic farming: New criteria for winter wheat ideotypes. 1: Light conditions in stands of winter wheat affected by morphological features of different varieties. *Pflanzenbauwissenschaften* 1:19–24.

Eisele, J.A., and U. Köpke. 1997b. Choice of cultivars in Organic Farming: New criteria for winter wheat ideotypes. 2: Weed competitiveness of morphologically different cultivars. *Pflanzenbauwissenschaften* 1:84–89.

Froud-Williams, R.J. 1997. Varietal selection for weed suppression. Aspects of Applied Biology 50: Optimising Cereal Inputs: Its Scientific Basis. pp. 355–360.

Fuller, R.J. 1997. Response of birds to organic arable farming: Mechanisms and evidence. *In*: Proceedings Brighton Crop Protection Conference – Weeds. pp. 897–906,

Gibson, K.D., A.J. Fischer, T.C. Foin, and J.E. Hill. 2003. Crop traits related to weed suppression in water-seeded rice (*Oryza sativa* L.). *Weed Science* 51:87–93.

Gold, H.J., J. Bay, and G.G. Wilkerson. 1996. Scouting for weeds based on the negative binomial distribution. *Weed Science* 44:504–510.

González-Ponce, R.G., C. Zancada, M. Verdugo, and L. Salas. 1996. Plant height as a factor in competition between black nightshade and two horticultural crops (tomato and pepper). *Journal of Horticultural Science* 71:453–460.

Gooding, M.J., A.J. Thompson, and W.P. Davies. 1993. Interception of photosynthetically active radiation, competitive ability and yield of organically grown wheat varieties. Aspects of Applied Biology 34: Physiology of Varieties. pp. 355–362.

Gooding, M.J., N.D. Cosser, A.J. Thompson, W.P. Davies, and R.J. Froud-Williams. 1997. The effect of cultivar and Rht genes on the competitive ability, yield and bread-making qualities of organically grown wheat. *In*: Proceedings of the 3rd ENOF workshop: Resource use in organic farming. pp. 113–121. Ancona, 1997.

Grundy, A.C., and R.J. Froud-Williams. 1993. The use of cultivar, crop seed rate and nitrogen level for the suppression of weeds in winter wheat. *In*: Brighton Crop Protection Conference – Weeds. pp. 997–1002.

Håkansson, S. 1979. Basic research in crop production. II. Factors of importance for plant establishment, competition and production in crop-weed stands. Cited in Håkansson, S. 2003. Weeds and weed management on arable land. Cambridge, MA: CABI Publishing.

Hashem, A., S.R. Radosevich, and M.L. Roush. 1998. Effect of proximity factors on competition between winter wheat (*Triticum aestivum*) and Italien ryegrass (*Lolium multiflorum*). *Weed science* 64:181–190.

Hewson, R.T., H.A. Roberts, and W. Bond. 1973. Weed competition in spring-sown broad beans. *Horticultural Research* 13:25–32.

Hoad, S.P., D.H.K. Davies, and C.F.E. Topp. 2006a. How to select varieties for organic farming: Science and practice. Aspects of Applied Biology, 79. What will organic farming deliver? COR 2006, pp. 117–120. Edinburgh, Scotland: Heriot-Watt University. September 18–20, 2006. Association of Applied Biologists, c/o Warwick HRI, Wellesbourne, Warwick

Hoad, S.P., D.H.K. Davies, and C.F.E. Topp. 2006b. Designing crops for low input and organic systems: Enhancing wheat competitive ability against weeds. *In*: Proceedings Crop Protection in Northern Britain. pp. 157–162.

Hoad, S.P., C.F.E. Topp, and D.H.K. Davies. 2008. Selection of cereals for weed suppression in organic agriculture: A method based on cultivar sensitivity to weed growth. *Euphytica* 163:355–366.

Huel, D.G., and P. Hucl. 1996. Genotype variation for competitive ability in spring wheat. *Plant Breeding* 115:325–329.

Hutchinson, P.J.S., B.R. Beutler, and J. Farr. 2011. Hairy nightshade (*Solanum sarrachoides*) competition with two potato varieties. *Weed Science* 59:37–42.

Iqbal, J., and D. Wright. 1999. Effects of weed competition of flag leaf photosynthesis and grain yield of spring wheat. *Journal of Agricultural Science* 132:23–30.

Jensen, L.B., B. Courtois, L.S. Shen, Z.K. Li, M. Olofsdotter, and R.P. Mauleon. 2001. Locating genes controlling allelopathic effects against barnyard grass in upland rice. *Agronomy Journal* 93:21–26.

Jönsson, R., N-O. Bertholdsson, G. Enqvist, and I. Åhman. 1994. Plant characters of importance in ecological farming. *Journal of the Swedish Seed Association* 104:137–148.

Jørnsgård, B., K. Rasmussen, J. Hill, and J.L. Christiansen. 1996. Influence of nitrogen on competition between cereals and their natural weed populations. *Weed Research* 36:461–470.

Kato-Noguchi, H., M. Hasegawa, T. Ino, K. Ota, and H. Kujime. 2010. Contribution of momilactone A and B to rice allelopathy. *Journal of Plant Physiology* 167:787–791.

Kong, C.H., H.B. Li, F. Hu, X.H. Xu, and P. Wang. 2006. Allelochemicals released by rice roots and residues in soil. *Plant and Soil* 288:47–56.

Korr, V., F-X. Maidl, and G. Fischbeck. 1996. Auswirkungen direkter und indirekter Regulierungsmaßnahmen auf die Unkrautflora in Kartoffeln und Weizen. Zeitschrift für Pflanzenkrankheiten und Pflanzenschutz, Sonderheft XV. pp. 349–358.

Korres, N.E., and R.J. Froud-Williams. 2002. Effects of winter wheat cultivars and seed rate on the biological characteristics of naturally occurring weed flora. *Weed Research* 42:417–428.

Kremer, R.J., and M. Ben-Hammouda. 2009. Allelopathic Plants. 19. Barley (*Hordeum vulgare* L). *Allelopathy Journal* 24:225–241.

Kruidhof, H.M., L. Bastiaans, and M.J. Kropff. 2009. Cover crop residue management for optimizing weed control. *Plant and Soil* 318:169–184.

Lammerts van Bueren, E.T., Struik, P.C., and E. Jacobsen. 2002. Ecological concepts in organic farming and their consequences for an organic crop ideotype. *Netherlands Journal of Agricultural Science* 50:1–26.

Lemerle, D., B. Verbeek, and N.E. Coombes. 1995. Losses and grain yield of winter crops from *Lolium rigidum* competition depend on crop species, cultivar and season. *Weed Research* 35:503–509.

Lemerle, D., B. Verbeek, R.D. Cousens, and N.E. Coombes. 1996. The potential for selecting wheat varieties strongly competitive against weeds. *Weed Research* 36:505–513.

Lemerle, D., G.S. Gill, C.E. Murphy, S.R. Walker, R.D. Cousens, S. Mokhtari, S.J. Peltzer, R. Coleman, and D.J. Luckett. 2001a. Genetic improvement and agronomy for enhance wheat competitiveness with weeds. *Australian Journal of Agricultural Research* 52:527–548.

Lemerle, D., B. Verbeek, and B. Orchard. 2001b. Ranking the ability of wheat varieties to compete with *Lolium rigidum*. *Weed Research* 41:197–209.

Lemerle, D., A. Smith, B. Verbeek, E. Koetz, P. Lockley, and P. Martin. 2006. Incremental crop tolerance to weeds: A measure for selecting competitive ability in Australian wheats. *Euphytica* 149:85–95.

Liang, Y.L., and R.A. Richards. 1999. Seedling vigor characteristics among Chinese and Australian wheats. *Communications in Soil Science and Plant Analysis* 30:159–165.

Liang, Y.L., and R.A. Richards. 1994. Coleoptile tiller development is associated with fast early vigor in wheat. *Euphytica* 80:119–124.

Liebman, M., and A.S.Davis. 2000. Integration of soil, crop and weed management in low-external-input farming systems. *Weed Research* 40:27–47.

López-Cartandena, C., R.A. Richards, G.D. Fraquhar, and R.E. Williamson. 1996. Seed and drilling characteristics contributing to variation in early vigor among temperate cereals. *Crop Science* 36:1257–1266.

Love, S. L., C.V. Eberlein, J.C. Stark, and W.H. Bohl. 1995. Cultivar and seed piece spacing effects on potato variety competitiveness with weeds. *American Journal of Potato Research* 72:197–213.

Martin, M.P.L.D., and R.J. Field. 1984. The nature of competition between perennial ryegrass and white clover. *Grass and Forage Science* 39:247–253.

Mason, H.E., and D. Spaner. 2006. Competitive ability of wheat in conventional and organic management systems; a review of the literature. *Canadian Journal of Plant Science* 86:333–343.

Merker, A. 1992. The *Triticeae* in cereal breeding. *Hereditas* 116:277–280.

Mohler, C.L. 1996. Ecological bases for the cultural control of annual weeds. *Journal of Production Agriculture* 9:468–474.

Mokhtari, S., N.W. Galwey, R.D. Cousens, and N. Thurling. 2002. The genetic basis of variation among wheat lines in tolerance to competition by ryegrass (*Lolium rigidum*). *Euphytica* 124:355–364.

Moore, R.J. 1975. The Biology of Canadian Weeds. 13. *Cirsium arvense* (L.) Scop. *Canadian Journal Plant Science* 55: 1033–1048.

Moss, S. 1985. The influence of crop variety and seed rate on *Alopecurus myosuroides* competition in winter cereals. *In*: Proceedings British Crop Protection Conference – Weeds. pp. 701–708.

Nelson, D.C., and M.C. Thoreson. 1981. Competition between potatoes (*Solanum tuberosum*) and weeds. *Weed Science* 29: 672–677.

Neuhoff, D., S. Hoad, U. Köpke, K. Davies, S. Gawronski, H. Gawronska, S. Drews, P. Juroszek, C. de Lucas Bueno, and R. Zanoli. 2005. Strategies of Weed Control in Organic Farming (WECOF). Final Report of the FP5 European Combined Project, online publication. Accessed April 11, 2011. http://www.wecof.uni-bonn.de.

Niemann, P. 1992. Unkrautunterdrückendes Potential von Wintergerstensorten. Zeitschrift für Pflanzenkrankheiten und Pflanzenschutz, Sonderheft XIII. pp. 149–159.

Oberforster, M., C. Krüpl, and J. Söllinger. 2003. Genotypische Unterschiede im Unkrautunterdrückungsvermögen von Winterweizen und Sommergerste – Parameter zur Bildung eines Indexwertes. 7. Wissenschaftstagung zum Ökologischen Landbau – Ökologischer Landbau der Zukunft. pp. 113–116.

O'Donovan, J.T., E.A. St. Remy, P.A. O'Sullivan, D.A. Dew, and A.K. Sharma. 1985. Influence of the relative time of emergence of wild oats (*Avena fatua*) on yield loss of barley (*Hordeum vulgare*) and wheat (*Triticum aestivum*). *Weed Science* 33: 498–503.

Ogg, A.G., and S.S. Seefeldt. 1999. Characterizing traits that enhance the competitiveness of winter wheat (*Triticum aestivum*) against jointed goat grass (*Aegilops cylindrica*). *Weed Science* 47:74–80.

Olofsdotter, M. 2001. Getting closer to breeding for competitive ability and the role of allelopathy – An example from rice (*Oryza sativa*). *Weed Technology* 15:798–806.

Olofsdotter, M., L.B. Jensen, and B. Coutois. 2002. Improving crop competitive ability using allelopathy – An example from rice. *Plant Breeding* 121:1–9.

Qingwu, X., and R.N. Stougaard. 2002. Spring wheat seed size and seeding rate affect wild oat demographics. *Weed Science* 50:312–320.

Rauber, R., and W. Böttger. 1984. Untersuchungen zur Konkurrenz zwischen Winterweizen und der Gemeinen Quecke (*Agropyron repens* [L.] Beauv.). *Zeitschrift für Acker-und Pflanzenbau* 153:246–256.

Rebetzke, G.J., and R.A. Richards. 1999. Genetic improvement of early vigour in wheat. *Australian Journal of Agricultural Research* 50:291–301.

Rebetzke, G.J., C. Lopez-Castaneda, T.L.B. Acuna, A.G. Condon, and R.A. Richards. 2008. Inheritance of coleoptile tiller appearance and size in wheat. *Australian Journal Of Agricultural Research* 59:863–873.

Richards, M.C. 1989. Crop competitiveness as an aid to weed control. *In*: Proceedings Brighton Crop Protection Conference – Weeds. pp. 573–578.

Richards, M.C., and D.H.K. Davies. 1991. Potential for reducing herbicide inputs/rates with more competitive cereal cultivars. *In*: Proceedings Brighton Crop Protection Conference – Weeds. pp. 1233–1240.

Richards, M.C., and G.P. Whytock. 1993. Varietal competitiveness with weeds. Aspects of Applied Biology 34: Physiology of Varieties. pp. 345–354.

Rooney, J.M., D.T. Clarkson, M. Highett, J.J. Hoar, and J.V. Purves. 1990. Growth of *Galium aparine* L. (cleavers) and competition with *Tritium aestivum* L. (wheat) for N. *In*: Proceedings of the European Weed Research Society Symposium, Integrated Weed Management in Cereals. pp. 271–280.

Seavers, G.P., and K.J. Wright. 1999. Crop canopy development and structure influence weed suppression. *Weed Research* 39:319–328.

Stoller, E.W., and R.A. Myers. 1989. Response of soybeans (*Glycine max*) and four broadleaf weeds to reduced irradiance. *Weed Science* 37:570–574.

Truberg, B., and M. Huhn. 2002. Contributions to the analysis of genotype x environment interactions: Comparison of different parametric and non-parametric tests for interactions with emphasis on crossover interactions. *Journal of Agronomy and Crop Science* 185:267–274.

van Elsen, T. 2000. Species diversity as a task for organic agriculture in Europe. *Agriculture, Ecosystems and Environment* 77:101–109.

Verschwele, A., and P. Niemann. 1996. Indirekte unkrautbekampfung durch soretnwahl bei weizen. 8th EWRS Symposium "Quantitative approaches in weed and herbicide research and their practical application." Braunschweig. pp. 799–806.

de Vida, F.B.P., E.A. Laca, D.J. Mackill, G.M. Fernandez, and A.J. Fischer. 2006. Relating rice traits to weed competitiveness and yield: A path analysis. *Weed Science* 54:1122–1131.

Vollmann, J., H. Wagentristl, and W. Hartl. 2010. The effects of simulated weed pressure on early maturity soybeans European *Journal of Agronomy* 32:243–248.

Watson, P.R., D.A. Derksen, and R.C. van Acker. 2006. The ability of 29 barley cultivars to compete and withstand competition. *Weed Science* 54:783–792.

Wicks, G.A., R.E. Ramsel, P.T. Nordquist, J.W. Schmidt, and R.E. Challaiah. 1986. Impact of wheat cultivars on establishment and suppression of summer annual weeds. *Agronomy Journal* 78:59–62.

Willey, R.W., and R. Holliday. 1971. Plant population shading and thinning studies in wheat. *Journal of Agricultural Science* 77:453–461.

Wilson, B.J. 1988. Shoot competition and root competition. *Journal of Applied Ecology* 25:279–296.

Wilson, B.J., and K.J. Wright. 1990. Predicting the growth and competitive effects of annual weeds in wheat. *Weed Research* 30:201–211.

Wolfe, M., E.T. Lammerts van Bueren, J.P. Baresel, D. Desclaux, I. Goldringer, S.P. Hoad, G. Kovacs, F. Löschenberger, T. Miedaner, A.M. Osman, and H. Østergård. 2008. Developments in breeding cereals for organic agriculture. *Euphytica* 163:323–346.

Wu, H., J. Pratley, D. Lemerle, and T. Haig. 2000. Evaluation of seedling allelopathy in 453 wheat (*Triticum aestivum*) accessions against annual ryegrass (*Lolium rigidum*) by the time equal compartment agar method. *Australian Journal of Agricultural Research* 51:937–944.

Wu, H.W., S.R. Walker, V.A. Osten, and G. Robinson. 2010. Competition of sorghum cultivars and densities with Japanese millet (Echinochloa esculenta). *Weed Biology and Management* 10:185–193.

Zhao, D.L., G.N. Atlin, L. Bastiaans, and J.H.J. Spiertz. 2006. Cultivar weed-competitiveness in aerobic rice: Heritability, correlated traits, and the potential for indirect selection in weed-free environments. *Crop Science* 46:372–380.

5 Breeding for Genetically Diverse Populations: Variety Mixtures and Evolutionary Populations

Julie C. Dawson and Isabelle Goldringer

Introduction

Agricultural biodiversity is essential for the improvement of sustainability in agricultural systems and for food and nutrition security (Frison et al., 2011). Among the components of agrobiodiversity, the within-species diversity level may be displayed at different scales such as within breeding programs, fields, or landscapes and thus is a key component that can be handled through a wide range of breeding strategies. Many authors have highlighted the need for increased genetic diversity in the field (Demeulenaere et al., 2008; Finckh et al., 2000; Finckh, 2008; Hajjar et al., 2008; Heal et al., 2004; Horneburg and Becker, 2008; Lammerts van Bueren et al., 2010; Murphy et al., 2005; Newton et al., 2009; Phillips and Wolfe, 2005; Wolfe et al., 2008; Zhu et al., 2000). This need is especially strong in organic and low-input systems due to the increased heterogeneity of environments in time and space (Altieri, 1999; Ceccarelli et al., 2007; Desclaux, 2005; Finckh, 2008; Østergård et al., 2009; Wolfe et al., 2008). There is great interest among organic farmers for increasing varietal diversity, and many benefits of genetic diversity at the production level are recognized in the scientific literature. Breeding for this type of sustainable system means that varieties need to adapt to environmental conditions, rather than changing the environment (through synthetic fertilizers, pesticides, irrigation, etc.) to fit crop requirements (Coffman and Smith, 1991). There is widespread agreement on the need to conserve and utilize greater levels of diversity in the face of environmental change, yet most current varieties are genetically uniform. This is due to both legal requirements for variety registration and difficulties in implementing breeding programs to generate genetically diverse varieties. Historically, crops have been grown as populations (landraces), which has allowed for the diversification of crop varieties, adaptation to contrasting environments, and use and maintenance of genetic diversity. In the twentieth century, however, pure line and F_1 hybrid breeding has become dominant in industrialized countries due to a combination of intensification of agricultural practices, which allowed farmers and breeders to homogenize growing environments, and intellectual property regimes that allowed breeders to protect and benefit from homogeneous varieties that could be officially registered and released. Especially in the second half of the twentieth century, input use has been widespread enough in conventional agriculture to significantly limit the natural environmental variability to which crops

---

Organic Crop Breeding, First Edition. Edited by Edith T. Lammerts van Bueren and James R. Myers.
© 2012 John Wiley & Sons, Inc. Published 2012 by John Wiley & Sons, Inc.

are exposed and reduce the need for genetic diversity in commercial varieties (Phillips and Wolfe, 2005). What appear to be broadly adapted varieties are, in fact, varieties that are superior across wide geographic areas but only in a narrow range of production conditions that mitigate severe stress through the use of inputs (Ceccarelli, 1996).

Box 5.1. Definitions

Variety: In agricultural contexts, "A plant grouping within a single botanical taxon of the lowest known rank, which grouping, irrespective of whether the conditions for the grant of a breeder's right are fully met, can be (1) defined by the expression of the characteristics resulting from a given genotype or combination of genotypes, (2) distinguished from any other plant grouping by the expression of at least one of the said characteristics and (3) considered as a unit with regard to its suitability for being propagated unchanged" [UPOV Convention, Article 1(vi); UPOV, 1991]. Modern varieties are generally genetically and phenotypically homogeneous and registered in national catalogs or protected through an intellectual property rights system. The term cultivar, proposed by L.H. Bailey (1923), is also used for "cultivated variety" in many situations, but we have opted for the use of variety throughout this chapter.

Varietal Mixture: A mixture of distinct varieties, grown and harvested together. Components of the mixture can be landraces, historic varieties, or modern varieties.

Landrace: (Camacho Villa et al., 2005) proposed a consensus definition which we adopt here: "A landrace is a dynamic population(s) of a cultivated crop that has a historical origin, a distinct identity and lacks formal crop improvement, as well as often being genetically diverse, locally adapted and associated with traditional farming systems."

Farmer variety: A variety developed and managed by farmers, such as a landrace or a historic variety, that has been maintained on-farm over many years; or, a new population created and selected by farmers. Farmer varieties can also include variety mixtures that are grown and harvested together over many years instead of reconstituted each year (see Bocci et al., 2010 for a detailed description).

Synthetic populations: Random mating populations formed by inter-mating a group of inbred lines (Kutka and Smith, 2007).

Composite populations: In outcrossing species, random mating populations formed by inter-mating a group of populations (Kutka and Smith, 2007).

Composite cross populations: In self-pollinating species, formed by mating a group of diverse parents, then growing and harvesting subsequent generations as a bulk population (Suneson, 1956).

Multiline: A variety made up of distinct varieties that differ only for certain traits, such as particular disease resistance alleles, but otherwise phenotypically homogeneous.

Population: Plants of a particular variety managed under the same conditions, for example, on a single farm. All the populations in a region of the same variety make up a metapopulation of that variety (Love and Spaner, 2006). A seedlot is the part of a single population used for planting the next year (Louette, 2000) or for seed exchanges.

Due to rapid turnover of varieties for crops with significant public or private-sector breeding efforts, spatial diversity has been replaced with temporal diversity in industrialized agriculture as farmers have switched from planting multiple varieties on their farm to planting single varieties but switching frequently (Jackson et al., 2007). While this may seem an appropriate strategy in some circumstances, many of these elite varieties are genetically similar, using similar resistance genes and genetic backgrounds, so the overall effect is a reduction in diversity over both time and space. Monocultures usually mean one crop species growing over a large space, but as Finckh 2008) points out, monocultures can exist at multiple levels, from the species to the variety to the gene. Many modern varieties share the same genes, so even with different varieties there can be a resistance gene monoculture for a particular pathogen (Finckh, 2008). Reliance on a narrow genetic base for continued improvement of elite varieties therefore makes breeding programs vulnerable to unpredictable environmental changes. Particularly for organic systems, there is a need to expand the genetic base of breeding programs and to keep higher levels of genetic diversity at all stages of variety development, up to and including varieties grown in farmers' fields. The purpose of this chapter is to briefly review some of the benefits of genetic diversity per se in cultivated varieties for organic systems and then to discuss in more detail possible strategies for developing such varieties.

Benefits of Genetic Diversity for Organic Agriculture

Buffering Capacity Against Biotic and Abiotic Stresses over Space and Time/Resilience

The capacity for genetically heterogeneous populations to buffer the impacts of environmental stress has been studied primarily with landraces in traditional agroecosystems and with variety mixtures for disease control. Finckh et al. (2000) and Finckh (2008) provide excellent reviews of the use of mixtures and diversity for disease control. These papers stress that combining agricultural practices that encourage biodiversity with genetic diversity of cultivated species can buffer both crop yields and quality from biotic and abiotic stresses. In addition, the use of diversity prolongs the utility of existing resistance genes and incorporates potentially useful genes into farmers' fields, so that the response to unpredictable environmental conditions takes place on-farm rather than keeping diversity in reserve for plant breeders to use when a new disease or stress appears (Finckh, 2008).

The benefits of diversity have also been studied in natural ecosystems, and this research may be relevant to organic agricultural systems, which, while managed, must use natural processes and biological self-regulatory mechanisms to succeed. In terms of maintaining critical functions of the system, genetic diversity may contribute more to the resilience of ecosystems rather than to specific functions under normal conditions (Hughes and Stachowicz, 2004). Genetic diversity within species may be able to replace species diversity in some ecosystems, making them more resilient in the face of climatic extremes (Reusch et al., 2005). Hajjar et al. (2008) argue that crop genetic diversity contributes to ecosystem services provided by agricultural land, by increasing the long-term stability of these systems and by maintaining biomass, which contributes to soil conservation, below-ground biodiversity, and above-ground biodiversity of species, such as pollinators.

Much research is needed on the mechanisms involved and the most effective way to use diverse populations in response to specific stresses in organic systems. To better understand how diversity works to increase system resilience (how to organize, what type of components, etc.), these types of populations need to be created and used by farmers, and the development of such populations will provide immediate benefits to farmers as well as making this type of research possible.

Evolution and Adaptation in Response to Environmental Variation

Predictions by climate scientists point to increasing environmental variability in the near future. For example, in Europe both temperature and precipitation are expected to become more variable (Olesen et al., 2011). Heal et al. (2004) state that current elite cultivated varieties are each adapted to a particular weather pattern. In developed countries, many farmers choose those varieties with potentially higher yields under normal weather conditions, as most have crop insurance or other support to prevent crop failures (e.g., irrigation, extra mineral nitrogen, or chemical crop protectants). While this approach might make sense for individual farmers and private breeding companies in the current context of fairly stable weather patterns and government support for conventional agriculture, it could prove extremely risky, should unpredictable weather events cause widespread crop failure. Collectively, farmers (and plant breeders) may be making poor choices with respect to the risk-yield trade-off from a societal point of view, and they may be reducing genetic diversity too much in the search for higher yields under controlled conditions (Heal, 2004).

In contrast, the use of more genetically diverse varieties maintains the evolutionary potential of these populations, as natural selection can continue to act on heterogeneous crop populations when reproduced and replanted on-farm. These varieties can develop local adaptation and evolve over time as local environmental conditions change. This type of strategy replaces broad spatial adaptation with stability over time in one location, which is of primary importance to farmers (Ceccarelli, 1989, 1994). If genetic diversity exists for key traits such as life history or fitness-related traits, evolutionary forces can act on this diversity while farmer management and selection maintains traits important to agronomic performance and quality (Almekinders and Elings, 2001; Berthaud et al., 2001; Elias et al., 2001; Louette and Smale, 2000; Smith et al., 2001). Local adaptation may reduce the diversity of a variety at any one site, but the overall diversity is maintained across the landscape as each subpopulation (such as one farmer's population of a particular variety) responds to different environmental conditions (Enjalbert et al., 1999; Goldringer et al., 2001; Lavigne et al., 2001; Porcher et al., 2004). In an experimental network of wheat populations, adapted traits such as flowering time, thousand kernel weight, and disease resistance showed significant evolution linked to geographic location over 10 generations (Goldringer et al., 2006; Paillard et al., 2000a, 2000b; Rhone et al., 2010; Rousselle, 2010). Despite significant selection for locally adapted traits, crop populations within a site can also maintain levels of diversity much higher than those found in modern varieties. The experimental wheat populations mentioned previously maintained higher than expected levels of genetic diversity assessed through "neutral" DNA markers, even those closely linked to genes putatively involved in the response to climatic conditions, and they maintained multiple alleles at candidate gene loci, including alleles not found in the parental varieties (Raquin et al., 2008; Rhone et al., 2007).

On-Farm Conservation of Useful Genetic Diversity

Landraces that have existed in situ (rather than being passed through a gene bank) are evidence that specific adaptation is not contradictory to maintaining evolutionary potential. Ex situ conservation in gene banks may lead to reduced genetic diversity, due to genetic drift in small population sizes (Horneburg and Becker, 2008; Parzies et al., 2000). Organic farmers in industrialized countries have become increasingly engaged in the conservation of genetic diversity, partially due to the lack of adapted modern varieties, and many seed-saving and seed exchange networks have emerged (Osman and Chable, 2009). In France, farmers involved in the Réseau Semences Paysannes (Farmers' Seed

Network) grow many historic and landrace varieties both out of concern about the loss of crop genetic diversity and because of the direct agricultural and quality benefits they observe. A particular historic wheat variety, 'Rouge de Bordeaux', has become well known, and was chosen for a detailed genetic study of on-farm populations in a collaborative project among farmers and researchers (Demeulenaere et al., 2008). The level of genetic diversity measured by neutral DNA markers was much higher and qualitatively different in populations that had been continuously cultivated in the Bordeaux region rather than derived from samples of the accession Rouge de Bordeaux in the national genebank. Populations showed significant differentiation among farmers, and some of this population structure can be explained by the structure of seed exchanges and development of local adaptation at each site. Even after many years of on-farm management and selection, populations maintained high levels of genetic diversity (Demeulenaere et al., 2008).

The effectiveness of in situ conservation depends on having sub-populations exposed to diverse environmental conditions, as contrasting selection pressures will preserve the greatest level of diversity at the global metapopulation level (Hodgkin et al., 2007; Goldringer et al., 2001, 2006; Lavigne et al., 2001; Louette et al., 1997; Porcher et al., 2004). On-farm cultivation of diverse varieties provides a workable method of in situ conservation where the genetic diversity in each sub-population can respond to selection pressures in the target environment. As shown by Demeulenaere et al. (2008), a relatively small number of farmers can make a large contribution to this form of conservation.

The development and on-farm use of genetic diversity can make these resources more available and relevant to breeders and farmers. Despite large ex situ collections of accessions including population varieties, landraces, historic varieties, and crop wild relatives, relatively little of this diversity (1–2%) is actually used (Wolfe et al., 2008). Breeders and farmers face difficulties in efficiently using accessions or incorporating this diversity into their programs due to the loss of diversity and adaptation to modern farming practices of previously locally adapted varieties conserved ex situ. There are also difficulties in evaluating genebank accessions and in transferring useful traits to more adapted varieties (Wolfe et al., 2008). This is why it is critical for organic breeding programs to assess the potential of existing accessions, to work with farmers currently cultivating genetically heterogeneous populations and to begin creating diverse populations that can undergo selection for local conditions and specific traits of interest to organic farmers (see fig. 5.1).

Breeding Strategies

While the production of genetically diverse varieties is not necessarily more complicated or costly than the development of genetically homogeneous varieties, there are still many practical and regulatory hurdles faced by breeding programs, which try to implement strategies for including higher levels of diversity in their varieties. Over 95% of current varieties used in organic production were bred for conventional agriculture (Lammerts van Bueren et al., 2010). While some companies are interested in developing varieties for organic agriculture, the seed market is much smaller than that of conventional agriculture, and a company relying solely on royalties from seed sales can only spend 5 to 10% of their breeding expenses on additional activities related to organic breeding (Birschitzky, 2007, cited by Löschenberger et al., 2008). This means that private-sector breeders look to combine organic and conventional breeding to the greatest extent possible and radically different variety structures for organic agriculture may not be immediately feasible for this type of selection program. Therefore it is important for public-sector researchers and farmers interested in

Figure 5.1 Farmers' selected spikes within diverse wheat varieties (Photo: FSO project, J.F. Berthellot's farm, France 2010).

plant breeding to develop methods of breeding genetically diverse varieties that meet the needs of organic agriculture that are relatively simple to implement. The following two sections deal with current work and potential strategies for creating such varieties, first through the creation of variety mixtures and second through developing multi-parental populations of outcrossing or inbreeding crop species.

Variety Mixtures/Multilines

Mixtures, multilines, and varietal associations are created from known components to provide diversity for certain traits, often disease resistance. Populations that are created and then allowed to evolve through seed saving and replanting over several years are considered in the next section on evolutionary breeding.

Successful Uses of Mixtures
As mixtures can be easily constituted before sowing, in countries where regulations permit the sale of mixtures (instead of the farmer having to buy separate cultivars and mixing on-farm), there are examples of fairly large-scale cultivation (Cowger and Weisz, 2008; Finckh et al., 2000;

Mundt et al., 1994; Wolfe, 1988). As a specific example of the effectiveness of mixtures, a particularly large experiment was conducted in China with mixtures of a high-value glutinous rice varieties and a hybrid rice variety, resulting in higher yield and dramatically reduced disease (Zhu et al., 2000).

In France, scientists and stakeholders (farmers, millers, the extension service) collaborated to evaluate yield and bread making quality of mixtures of four bread wheat varieties for two years in 19 environments under low input conditions (de Vallavieille-Pope et al., 2004). Four-way mixtures proved to be as high yielding as the pure components and more stable over space and time. Bread making quality of mixtures was also as high as the best pure stand component and more stable. It is worth noting that the miller did not raise any objections to the use of mixtures and kept on working with mixtures with the farmers after the research program ended. This was in contrast to experiments with mixtures in the Netherlands where the millers were very opposed to using mixed grains, although farmers could profit from improved baking quality combined with good yield and weed suppression in carefully composed mixtures (Osman, 2006 and personal communication).

In some countries, multilines can be released as single varieties and this strategy has been successfully used to incorporate diverse genes for disease resistance into otherwise homogeneous populations. In Washington State, USA, club wheat cultivars were released with multiple race resistance to stripe rust, one of the primary disease threats in the Pacific Northwest (Allan et al., 1993). In Iowa, Dr. J. A. Browning released 13 oat varieties composed of multilines, which prevented significant problems with crown rust for farmers that grew them, even in years of high disease pressure (Browning, 1988). In Austria, this strategy was also applied, with the company Probstdorfer Saatzucht, now Saatzucht Donau, maintaining as much heterozygosity as possible through the use of multilines for wheat varieties that had to pass testing for distinction, uniformity, and stability (DUS; Löschenberger et al., 2008).

Mixtures may be more advantageous than multilines due to the greater degree of heterogeneity possible. As multilines are generally genetically uniform except for certain resistance genes, they do not benefit from the buffering capacity of diversity for abiotic stresses or other biotic stresses (Wolfe, 1985). Browning (1988) concluded that multilines may be excessively uniform, especially related to traits that affect appearance but have little impact on crop performance, and that increased diversity would benefit crops facing multiple biotic and abiotic stressors.

Criteria to "Assemble" Components of Mixtures

When forming a mixture, multiline, or varietal association, it is important to understand the key traits for which diversity will be most beneficial and the most critical traits for which some degree of homogeneity is needed. One of the most important traits to consider in designing mixtures is the disease resistance component. While genotypes with complementary resistance genes or mechanisms should be chosen, plant-plant interactions may interfere with plant-pathogen interactions and make the choice more complicated. For instance, research in potato showed that cultivars with positive effects in mixtures are characterized by strong intra-cultivar competition in pure stands (Finckh et al., 2007). Increases in yield between 5 and 10% have been frequently recorded in potato variety mixtures, which were due to improved yield of the resistant varieties by reducing the negative effects of intra-cultivar competition together with improved yield of the susceptible varieties through reduced disease incidence due to the presence of the resistant varieties (Finckh et al., 2007).

For other traits, there are agronomic reasons for some uniformity; for example, maturity dates among the components of any mixture that will be machine harvested at one time. Some vegetable or fruit crops that are harvested by hand over a period of days or weeks may not need the same level of homogeneity for maturity, but may have other constraints, such size grading or color requirements dictated by the market. For other traits, uniformity may be more aesthetic and not of great agronomic importance. Plant height is one trait that is generally uniform in modern cereal varieties, perhaps because it is easy to spot off-types in plots of uniform height. This may not be necessary or even desirable in a more diverse population, as long as the plant heights are within a range such as to moderate competition between plants. In fact, having more diverse heights may reduce disease pressure due to increased aeration of the canopy and less humidity trapped near the leaves (Finckh, 2008). Plants of different heights may improve performance as taller genotypes improve early-season competition with weeds and shorter genotypes provide later-season support to prevent taller genotypes from lodging. Product quality is generally one of the most important factors, and components of mixtures must be selected so that the final harvested product meets market specifications for a certain use or, in the case of multiple species mixtures, can be easily separated by the farmer before sale. Farmers in organic systems that market their crops directly to consumers or in short distribution channels may have more flexibility and fewer needs for uniformity in their final products.

In addition to the choice of mixture components, the spatial deployment of these components may be important. Newton et al. (2009) state that patchiness may be important to the stability of the mixture in terms of disease resistance, and that random patches of different genotypes may delay the development of resistance in pathogen or pest populations. However, patches of multiple individuals of the same genotype may be difficult to implement in the field, and a physical mixture of the seed before planting may be the best way to ensure random distribution of genotypes. In the Chinese rice example given previously, rice varieties were planted in strips, which worked well for hand harvesting, as it was easy for farmers to separate the two rice types. While fresh market potatoes are grown in mixtures in some traditional systems, in general, however, producing different varieties in alternating rows or strips of varieties will be easier (Finckh et al., 2007). Because generalizations cannot be made about crop species and organic agricultural systems, the design of crop mixtures or multilines needs to be taken case-by-case, in consultation with farmers, processors, and end-users in specific contexts. Breeders may be surprised that farmers often have fewer requirements for homogeneity than assumed, and the primary barrier to releasing diverse varieties may be national seed regulations and varietal testing regimes.

Methods for Breeding, Specifically for Mixing Ability
In general, mixtures have been created from pure line varieties with similar genetic backgrounds but which have not been bred specifically for mixing ability or performance in heterogeneous populations (Finckh et al., 2000). This may not be optimal for mixture performance, as performance in pure stands is not necessarily a predictor of performance in mixed stands and vice versa (Phillips and Wolfe, 2005). While overall there appears to be a slight but significant advantage of cultivars mixtures over the average of their pure line components (see for instance Gallandt et al., 2001; Jensen, 1988; Smithson and Lenne, 1996), it has been suggested that mixtures could be optimized by conducting experiments to identify the exceptional mixtures that out-perform their best components (Gizlice et al., 1989; Jensen, 1988). Yet, breeding for mixing ability should go further and allow for the development of genetic resources with increased performance in mixtures assessed in terms of homeostasis (Finckh, 2008), which might depend on a greater niche exploitation or complementary resource utilization of the components as it has been shown with interspecific mixtures.

The first attempts to assess the mixing (or blending) ability of genotypes used diallels of binary mixtures, based on the analogy with the diallel crosses in quantitative genetics (e.g., Jensen and Federer, 1965; McGilchrist, 1965; Williams, 1962). Gallandt et al. (2001) used such a diallel approach to study 15 wheat binary mixtures and their six pure stand components. Following Griffing (1956), the term general mixing ability (GMA) was used to describe the average performance of a pure line in mixtures based on the performance of mixtures having that pure line as a component. Similarly, the term specific mixing ability (SMA) was used to describe the deviation in performance of a mixture from that predicted by the GMA of both parents. Across 33 environments in eastern Washington, the average yield of all mixtures was significantly higher than the average of all pure lines by 1.5% (Gallandt et al., 2001). In this study, GMA effects were significant, indicating that some pure lines promoted higher yields in mixtures than others. SMA effects were also significant, indicating that the GMA of each combination of pairs of pure lines did not account for the differences in yield observed among mixtures.

Gizlice et al. (1989) advocated that the GMA effect reflects more than the true mixing ability of genotypes; it reflects both the innate productive ability of a genotype in pure stand (called General Yielding Ability, GYA) plus its separate ability to affect blend response through competition (called True General Competitive Ability, TGCA). Partitioning GMA into GYA and TGCA can only be achieved by using the pure stand of the components and analyzing mixing response (blend response, the deviation of the mixture production from the average of the two pure stand components; Gizlice et al., 1989). The authors applied this approach to evaluate eight genetically diverse soybean genotypes and their 28 binary mixtures grown in eight environments. They found that both GMA and TGCA were highly significant and they suggested that the genotypes with large significant TGCA should be sought in breeding for blend performance.

Even if plant breeders worked only on global measures or harvest of binary mixtures, the diallel design would be too labor-intensive to screen a large set of breeding lines in early generation stages of the breeding process (six to eight components tested by Gallais, 1970; Gizlice et al., 1989; and Gallandt et al., 2001). We think an alternative strategy could take advantage of the very small plots, usually single-rows that are often used in early generations. Strong between plot competition occurs within this design, which can be modeled in terms of neighbor effects (Besag and Kempton, 1986; Foucteau et al., 2000 Goldringer et al., 1994; Kempton, 1982).

The producer-competitor model (Azais, 1987; Foucteau et al., 2000; Kempton, 1985) proved to be efficient in accounting for interactions between genotypes of adjacent plots by allowing the estimation of both the producer and the competitor effects of each genotype. But while the producer-competitor model was less efficient in bringing yield estimates closer to the pure stand values than the use of neighboring plant height covariates (Foucteau et al., 2000) we think that looking for genotypes with a high producer effect (high level of production under competition conditions) and with a high competitor effect (beneficial effects on neighbors) could allow selection for the GMA of these genotypes. Although the correlation between producer and competitor effects of genotypes tended to be negative in the study of Foucteau et al. (2000), several of the 40 genotypes (in single-row plots, with three replicates over two years) showed both moderate to high producer and competitor effects indicating possibilities for breeding. Increasing the number of replicates would improve parameter estimation by confronting each genotype with more neighbors. Another important improvement should be to assess these effects for yield components such as spike number, kernel number, and kernel weight.

Since in practice, cultivar mixtures are composed of more than two components, usually four to five components, Knott and Mundt (1990) suggested using mixing ability derived from two-way mixtures and to predict performance in higher order mixtures. Mille et al. (2006) assessed the performance of this strategy in four-way mixtures of bread wheat for disease severity, grain protein content and grain yield based on pure stands and two-way mixtures. Their results suggested that two-way mixtures should be first screened to remove unfavorable cultivar pairs before constructing more complex mixtures.

Finally, as an alternative to breeding for mixing ability by selecting among inbred lines in pedigree breeding programs, it has been suggested that selection could occur within highly diverse populations grown over time in contrasted organic or low input farming conditions (for example Finck, 2008; Goldringer et al., 2001). Allard (1961) showed that beans selected in composite crosses rather than as pure lines from early generations performed better when grown in mixed cropping systems. Allard and Adams (1969) proposed the concept of ecological combining ability. Good ecological combining ability means that the yield of all components of a mixture increase when grown in a mixture compared to when grown separately in pure lines (Finckh and Mundt, 1992). Allard and Adams (1969) found that this type of combining ability was rare in mixtures formed by varieties selected as pure lines but common in mixtures of barley lines that had been cultivated together for many generations.

Highly diverse populations are expected to become adapted through natural selection both to external biotic and abiotic pressures, and to internal competition among plants. Thus, after several years, they could constitute more appropriate sources of germplasm adapted to combining with partner plants in controlled mixtures. In that regard, there is a need for more investigation to find acceptable tradeoffs between individual competitiveness and the collective productivity of plant communities (Denison, 2010; Denison et al., 2010).

The Limits of Single Year Mixtures
Mixtures and multilines are advantageous because of their ease of adoption by farmers. Mixtures have an advantage over multilines in terms of ease of use and potential diversity, but multilines are more homogeneous if this is what the market requires. Farmers are also able to experiment on their own in composing mixtures if processors and end-users accept mixed grain lots. In addition, seed companies are able to market mixed seed because these can be approved under the Union Internationale pour la Protection des Obtentions Végétales (UPOV) and official catalog systems in most countries. However, mixtures and multilines present significant drawbacks when faced with highly heterogeneous environments or unpredictable weather patterns. Mixtures can make a significant contribution to disease and pest resistance but must be periodically recomposed. Additionally, because they are generally made up of pure lines, they cannot respond to pathogen evolution with their own evolution of improved disease resistance (Finckh, 2008).

The need for components of mixtures to pass catalog registration requirements also leads to these components being genetically homogeneous varieties selected for broad adaptation, so the use of diversity to achieve specific local adaptation in many different environments is more difficult with this strategy. Developing different mixture compositions for particular environmental conditions would be possible, but extensive experimentation is necessary to optimize variety choices and the proportion of each variety that makes up a particular mixture for a particular environment. This could be done on-farm by farmers, as is done in parts of the United States, but this does not include farmer input in the development of the component varieties that are available for mixing. Rather than such extensive testing over years, breeding methods have been proposed to develop varieties from the outset as heterogeneous populations, with simultaneous selection for traits of interest and for plant types that perform well in diverse populations (Finckh, 2000). These strategies are discussed in the following section.

Evolutionary Breeding

Suneson (1956) is the classic reference for evolutionary plant breeding. This type of selection was practiced by farmers for thousands of years; however, Suneson introduced the concept of creating complex composite crosses to generate genetic diversity within a population and then subjecting the population to natural selection over several generations to increase its evolutionary fitness. For barley in California there was a steady increase in the composite cross population (CCP) yield although these populations did not reach the level of the modern commercial cultivar. However, the CCP showed greater stability across the range of environments tested, and were particularly advantageous in poor environments (Danquah and Barrett, 2002; Soliman and Allard, 1991). According to Suneson (1965):

> "The core features of the suggested breeding method are a broadly diversified germplasm, and a prolonged subjection of the mass of the progeny to competitive natural selection in the area of contemplated use. The prospect of marked progress by competitive sorting of the fittest is challenging because a bulked hybrid population can be advanced through many generations at low cost as compared to conventional and costly early generation line testing techniques. Information available suggests that 15 generations of natural selection seem desirable. Thereafter there can be repeated recourse to three methods of breeding (1) continued natural selection with prospects for significant gains in yields to accrue throughout a working lifetime; (2) cyclic hybrid recombinations with intervening natural selection to give a kind of recurrent selection; or (3) resort to conventional selection and testing (the proportion of well adapted and high yielding lines being a partial function of generation)."

The basic concepts of this method can be adapted to a wide range of situations, from on-farm breeding by farmers to formal breeding programs seeking to commercialize varieties. Phillips and Wolfe (2005) provide a review of evolutionary breeding and point out that it is an appropriate method for decentralized selection programs, where breeding is targeted at meeting local needs. Just as farmers rely on landraces and the informal seed system to provide adapted varieties in developing countries, farmers' networks and breeding clubs have spontaneously developed within the organic agricultural context of industrialized countries. Farmers are using varietal mixtures or individual population varieties as a basis for on-farm development of locally adapted crop populations, through natural selection and simple mass selection in these populations, or through more complex methods such as gridded mass selection for maize and individual progeny testing for beans (Goldringer et al., 2010; Smith et al., 2001; Thomas et al., 2011).

Choosing the Type of Population
When starting an evolutionary breeding project, one of the most important decisions is the type of parental varieties to use (Murphy et al., 2005). If farmers have been experimenting with a diverse range of varieties, they may have a good idea of the best potential parents. Quality traits are of particular importance because, unlike agronomic traits such as yield, which can be expected to be linked to evolutionary fitness, human-defined end-use quality is not necessarily favored by natural selection (Murphy et al., 2005). Existing varieties may also lack particular traits or combinations sought by farmers or have very little variability for these traits.

When possible, varieties that already have some level of adaptation to the target environment are preferable (Murphy et al., 2005), but historic varieties and landraces sometimes have surprisingly good performance far from their region of origin (Goldringer et al., 2010). In other situations, a combination of modern varieties may be best, or crosses between modern and historic varieties, with historic varieties used for particular traits, such as nutritional value or taste (Murphy et al., 2005, 2007, 2009). Ideally, the choice of what varieties go into a diverse breeding population should be discussed with all stakeholders, including farmers, processors, and those connected to consumers (e.g., farmers who direct-market their crops, bakers, retail cooperatives). Those directly involved in breeding will then need to evaluate many varieties and accessions before choosing those that best meet the needs expressed by stakeholders and developing a strategy to combine the best traits available into populations that can undergo selection to improve those traits and eliminate undesirable aspects without losing a significant amount of genetic diversity.

Creating Initial Diversity, Building the Population
The development of a diverse population to select depends on the mating system of the crop and the available plant varieties. Outcrossing populations can be mixed together to produce inter-crossed progeny, but they must be protected from external sources of pollen. Inbreeding species are easier to manage in the field but the initial crosses must be done manually.

Murphy et al. (2005) provided potential crossing schemes and management strategies for creating populations of inbred crops. Crosses could consist of two carefully chosen parents with contrasting genetic backgrounds but possessing specific traits that farmers or breeders would want to combine. More complex crosses could be assembled by crossing parents in pairs and then bulking the progeny or by crossing each parent with all other parents (a full diallel or half-diallel) followed by bulking the progeny to form a composite cross population. Suneson (1956), Enjalbert et al. (2011), Wolfe et al. (2006), Ghaouti et al. (2008), and Kovacs et al. (2010) provide details on different experiments with evolutionary populations (see for winter wheat fig. 5.2). Additional parents will increase genetic diversity in populations and will, in general, delay the development of homogeneity within the

Figure 5.2 Winter wheat composite cross population (left) and pure line (right) at Wakelyns Farm (Suffolk, UK, 2009) in an agroforestry alley with a background of mixed hardwood trees (Photo: M.S. Wolfe, The Organic Research Centre, Elm Farm, UK).

population. However, if there are specific traits contributed by one parent, adding parents can make it more difficult to recover this trait at adequate frequencies in the resulting population and extends the length and effort required for the crossing process (Murphy et al., 2005).

In outcrossing species like maize, open-pollinated varieties often maintain high levels of genetic variation within a single population. Historic and landrace open-pollinated varieties are used directly by some farmers, especially those seeking specific quality traits. However, open-pollinated varieties often have a significant yield disadvantage, producing less than 70% of F_1 hybrids (Kutka and Smith, 2007). The use of synthetic varieties or composites might increase yields while maintaining the benefits in terms of yield stability and resistance that open-pollinated populations have. Creating these types of populations could also be the first step in a recurrent selection program (Kutka and Smith, 2007). Synthetic populations begin with a mixture of inbred parents and have exceeded open-pollinated varieties and approached F_1 hybrids for yield (Kutka and Smith, 2007). Models using Wright's (1922) equation were used to predict synthetic variety yields based on F_1 and F_2 progeny data from single crosses. Using diallel crosses with greater initial diversity, the best yield grouping was for synthetics with 5 to 12 initial lines. However, the authors point out that there are major heterotic groupings that have not yet been included in experiments designed to test synthetic populations and that using these additional sources of diversity could possibly contribute to higher relative yields (Kutka and Smith, 2007). Composite populations in outcrossing species are made by combining open-pollinated varieties and have equaled varietal hybrids and some double-cross hybrids (Kutka and Smith, 2007). Using the same model as for synthetics, Kutka and Smith found that yield benefits in composites were greatest with 3 to 4 parental populations, with a range from 2 to 7 parental populations. Most had predicted increases in yield over the best parental population, however they were still not at F_1 hybrid yield levels.

Selection Strategies, Managing Populations
Selection strategies can range from the simplest method of letting natural selection act on populations in the target environment under organic farming practices – the evolutionary breeding method (Suneson 1956) – to mass selection within segregating populations and later generations, to recurrent individual plant selection, ear-to-row methods of progeny selection, or selection followed by subsequent crossing and selection within progeny, or more complex schemes. Yield is hypothesized to be linked to evolutionary fitness, but other traits may negatively influence yield and be under stronger natural selection in heterogeneous populations, so directional selection on yield may not be achieved by natural selection alone. An example is height in cereal populations. Taller plants tend to be more competitive, and so produce more grains when they can out-compete shorter neighbors, but these plant types may not be the most productive when all the plants in the population have increased height (Goldringer et al., 2001; Raquin et al., 2008). Agronomically, increased height can be beneficial for competition with weeds, but plants that are too tall without a corresponding increase in the strength of the stem have a tendency to lodge.

An important distinction to make is that natural selection acts in terms of *individual* fitness, while agronomic value is usually accessed at the *population* level. Over thousands of years, natural selection has likely optimized many functions that increase individual plant reproductive capacity, but it has not operated at the level of the plant community, and so artificial selection may be able to make significant improvements in the performance of diverse plant communities under agricultural conditions and for agronomic traits that may not have been important in pre-agricultural plant communities (Denison et al., 2003). Individual plant resistance to pests and diseases, stress tolerance, enzyme efficiency, and other physiological mechanisms have probably undergone significant evolution through natural selection, and breeders are unlikely to improve these at the plant level

(Denison et al., 2003), although there is much progress to be made in discovering more efficient genotypes, combining these traits, and increasing their frequency in heterogeneous populations while maintaining the benefits gained from genetic diversity per se at the population level.

Whether natural selection or directed selection is the principal method used, the choice of selection environments will have a very large impact on the outcome. While this is true for all breeding programs, it is particularly critical in organic agriculture because of the diversity of farming systems and environmental conditions. Fortunately, diverse populations can continue to adapt to changing environments and for most traits they do not have to remain phenotypically uniform, so populations could be built and selected for key traits in breeding programs, then released to farmers in fairly early stages for a period of decentralized selection (Ceccarelli, 1996). This decentralized selection could also be incorporated into a cyclical process that continually generates diversity, selects the most-adapted individuals for each environment, and returns those individuals to create new populations with higher frequencies of adaptive traits. The maximum response to selection is predicted under unchanging environmental conditions, so the development of local adaptation will be greatest if each population is selected within a particular environment (Phillips and Wolfe, 2005), but where target environments are inherently heterogeneous, the best method of selection may be to expose breeding populations to environmental conditions that represent the range of year-to-year variation expected over the long term in any one particular target environment (Ceccarelli, 1996).

New breeding programs may wish to start with simpler selection strategies, especially decentralized and participatory selection programs, as these methods greatly reduce record-keeping and do not require large investments of time and equipment for farmers who are already extremely busy managing their operations. More complex strategies can be pursued if the desired selection gains are not realized. Existing breeding programs may already have systems in place and can see which strategy for introducing more diversity into finished varieties works best for their context.

Ideally, multiple strategies could be tested in parallel to add to our knowledge of the benefits and drawbacks of each method, as there are currently few examples in the literature of these methods being compared in working breeding programs. For composite crosses, there are several methods proposed to monitor progress over time. One is to compare the population to all of the parental lines, another to compare performance against a mixture of parental lines, and a third is to compare performance against that of a few check varieties (Phillips and Wolfe, 2005). Similar methods could be used for synthetics and composites of out-crossing species, with the equations used to predict synthetic and composite performance from parental data used to compare predictions and empirical data on their performance over time. Perhaps most important, multiple types of selection conducted on populations with the same initial composition could be compared to determine realized responses to selection and changes in population diversity over time under different selection or management regimes.

Natural selection. The concept proposed by Suneson (1956) of bulk breeding allowing evolutionary forces to work on diverse populations rests on the assumption that natural selection will increase the evolutionary fitness of the population and that this evolutionary fitness is highly correlated to important agronomic traits (Phillips and Wolfe, 2005). For natural selection to be efficient, population sizes must be large enough to prevent significant drift and especially to prevent a genetic bottleneck from occurring. Even after 50 generations at individual sites, the barley composite cross populations had maintained substantial genetic diversity (Allard, 1988) and were able to adapt to new environments (Danquah and Barrett, 2002). In eight to ten generations of a dynamic management experiment with wheat in France, populations diverged significantly for earliness and disease resistance along a geographic gradient in response to local selection pressures

(Paillard et al., 2000a, 2000b, 2006; Rhone et al., 2010). Lentil landraces responded positively to natural selection on organic farms in a study in Germany (Horneburg and Becker, 2008). Three generations of natural selection significantly changed landrace populations studied under the European project Farm Seed Opportunities (Goldringer et al., 2010). While some of these changes were seen as positive, others were not necessarily desirable. The effects of natural selection and the interest farmers have in evolutionary populations may greatly depend on the choice of parental material and whether desirable traits are present at a high enough frequency in the initial population and through subsequent generations of natural selection. Even if not pursued as a breeding strategy in itself, evolutionary breeding may be a good strategy to maintain diverse populations from which to pick parents for creating breeding populations or to adapt a population to a new site (such as a specific farm) after previous selection to increase yield and quality using one of the methods described further on.

Directed Selection. While natural selection may improve some traits linked to evolutionary fitness, there are other traits where directed selection is necessary in order to maintain the population or to increase the frequency of desirable genotypes. For the previous example of plant height, elimination of tall plants could be a simple form of mass selection to maintain the population at an optimum height (Goldringer et al., 2001; Phillips and Wolfe, 2005). Similarly, the effects of natural selection can be accelerated by removing diseased plants or plants with undesirable growth habit before they are able to flower or set seed. This removes them entirely from the subsequent generations rather than simply having natural selection reduce the number of viable offspring from these genotypes. Mass selection can also be conducted at harvest, where only the best plants are selected for seeding the next year, or where all the undesirable plant types are removed. For many crops, negative selection is more difficult at harvest, as it is difficult to distinguish the effects of disease once plants have senesced, and it is often more difficult to distinguish individual plants at the end of the season than earlier in the year. Positive selection is easier, however, as it is possible to evaluate the plants' performance over the whole growing season for final quality and productivity.

If individuals are selected, care must be taken not to reduce population size too drastically, unless a more homogeneous variety is the goal. More plants should be selected at harvest than will be planted the following year as some may be eliminated with post-harvest evaluations and some should be saved in cold storage as a backup, as in any breeding program. If individuals are selected, the progeny of each plant can be evaluated for post-harvest traits such as thousand kernel weight or grain protein, and progeny from the best individuals can be planted in single plots or rows the following season to evaluate progeny values. The best of these populations could then be bulked to create another population for ongoing mass selection, or crossed based on complementary traits, or continued to create several pure-lines that could be used in mixtures or synthetics. The best strategy will depend on the particular traits and species under consideration, as well as the established infrastructure for a breeding program and whether selection is being done by breeders, farmers, or both. Readers are encouraged to seek out the referenced articles for more details on particular examples.

Population Use and Release to Farmers

To be considered successful, diverse varieties selected for organic conditions need to meet the needs of farmers, and involving farmers in the selection process, either in the early stages of setting objectives, in on-farm evaluations, or in participatory selection projects is the best way to ensure that what comes out of this type of breeding program will be sought out and used by farmers.

For farmers involved in research and breeding programs, access to diverse populations is possible, and on-farm management and seed saving will let natural selection shape each population to the particular farming conditions of the farmer using it. For farmers not interested in selection or farmers not able to become involved in breeding, the sale of such varieties as seed and the possibility for farmer-to-farmer exchange become important. Seed exchanges are likewise important sources of new diversity for ongoing breeding efforts and for farmer populations, which may have lost diversity over time in one location. However, there are major regulatory hurdles in allowing farmer seed-saving and seed-exchange in many countries, particularly the European Union, because of seed regulations and legislation related to plant variety protection. The following section briefly touches on potential regulatory reforms that would allow the development and circulation of genetically diverse varieties without compromising seed quality assurance measures (phytosanitary standards) or the varietal identity protections for seed buyers. These reforms to establish an alternative, parallel system for the commercialization of heterogeneous varieties would not require changes to the existing system of varietal protection for pure-line varieties and hybrids that makes up the bulk of the commercial market. The two types of varieties and regulatory mechanisms could easily coexist and complement each other in the current agricultural context.

Regulation Consequences: New Concept for Variety Description – Traceability Plus Distinctiveness
Present laws following UPOV 1991 guidelines, such as the EU Regulation 2100/94/EC or the Plant Variety Protection Act in the United States, require protected varieties to meet standards of distinction, uniformity, and stability (DUS; see Chapter 8 for more information). This means that varieties have to be distinct in some way from other registered varieties, that they must be phenotypically homogeneous in the field and that their phenotypic character cannot change over multiple generations of testing (Lammerts van Bueren et al., 2010). Agronomic and quality tests are also required in some countries for release of a variety (Value for Cultivation and Use, VCU testing). This is an expensive procedure and for private seed companies, the expected royalties must cover the costs of testing and registration. In many cases, the organic seed market is not large enough to generate this level of royalty income, and VCU testing is often geared toward conventional agriculture, increasing the chances of rejection for varieties selected for organic systems. This is one reason why participatory approaches to breeding may be an efficient method of developing varieties for organic systems because it reduces reliance on the commercial seed sector to supply organic farmers (Lammerts van Bueren et al., 2010).

However, no mechanism exists at present for releasing the results of participatory and decentralized or evolutionary breeding programs. Two key changes could make this possible. The first would be to eliminate VCU testing for heterogeneous varieties and require the submission or online archiving of performance data under diverse environmental conditions for each generation. Detailed data could be released to the public, along with summary measures of performance over several years in major environmental zones that farmers could use to evaluate different varieties. A second change would be to replace the homogeneity and stability requirements with a traceability requirement, with the history of the population (parents, growing locations, selection stages) that is accessible to purchasers of the seed, with the requirement to update this history if they, in turn, wanted to distribute seed to others (Wolfe et al., 2008; Wolfe and Döring, 2010). Varietal descriptions would still be required, but would be more flexible, with particularly unique characteristics and quality parameters described, and with phenotypic frequencies and summary statistics and variances given for the initially released population. Updates could be required should any of these descriptions fall outside a set range, such as two standard deviations from the initial population mean.

Example of a Self-Pollinated Species: Wheat Composite Cross Populations in England

A composite cross wheat-breeding program was started at the Elm Farm Research Center, UK, in collaboration with the John Innes Center in the UK. Wheat parents that were popular in the period following their release were selected in 2002, primarily from Western Europe (Jones et al., 2010). The composite cross populations (CCP) included: Yield CCP, 10 high yielding parents (45 crosses); Quality CCP, 12 high quality parents (66 crosses); Yield/Quality CCP, intercrosses of all 22 parents (231 crosses). In the F_5 generation, population samples were distributed to farmers to be grown on larger plots on six farms to study the potential for adaptation and differentiation, and samples were also sent to researchers in France, Germany, and Hungary to estimate the potential for regional adaptation (Wolfe et al., 2006).

The populations generally performed within the range of the parental varieties but typically, the Yield CCP yielded more than the parental mean. Yield CCPs were consistently higher yielding than the quality populations, and Quality CCPs had consistently higher protein than the Yield CCPs. The Yield/Quality CCPs were closer in yield to the Yield CCPs and closer in protein content to the Quality CCPs, suggesting the potential for simultaneously improving these traits, which are generally regarded as negatively correlated (Wolfe and Döring, 2010). Importantly, the CCP populations showed higher stability over environments and time compared to parental populations grown in pure stands, and had slightly better stability than the parental lines grown as mixtures. The advantage of the heterogeneous populations was greatest in environments with lower mean yields and greater variability (Döring et al., 2010). This could be related to the greater potential diversity in the CCP populations, as the number of genotypes possible exceeded the demographic size of the CCP populations, while mixtures were limited to the initial genotypes present in the parental varieties (Wolfe and Döring, 2010).

The CCPs also showed potential to adapt to very different sites. For example, the Hungarian populations quickly developed cold hardiness to ensure winter survival and produced yields comparable to local varieties (Wolfe and Döring, 2010). With local selection and some targeted mass selection, it is hypothesized that these populations may have enough variability to respond to a wide range of uses and environments. In addition, this approach could be used to create new populations starting with parental varieties adapted to each target environment, or new varieties could be crossed into or mixed with the current composite cross varieties (Wolfe and Döring, 2010).

Example of a Cross-Pollinated Species: Maize Open-Pollinated Populations in Portugal

Moreira et al. (2008, 2009) described a participatory maize breeding project for open-pollinated populations adapted to Portuguese farming conditions. The project started in 1984 and farmer participation was essential from the outset. A regional open-pollinated variety with white grain was chosen for the project. This variety, 'Pigarro', had quality characteristics for bread making preferred by farmers. Traditional maize bread (broa) is still a market of economic importance in the region, and is one reason why some landrace populations have not been replaced by hybrid varieties. All populations were managed by farmers on their farms, and parallel selection was conducted by farmers and breeders. Farmers carried out mass selection with their own criteria, and breeders carried out both phenotypic recurrent selection and S_2 line recurrent selection (Moreira et al., 2008).

Trials were conducted to compare populations from all stages of the selection process, with the original population, populations after six cycles of mass and recurrent selection, and populations after three cycles of S_2 recurrent selection tested at three locations in Portugal. Results of mass selection showed that selection favored later maturing plants, possibly due to correlation with larger ears. Ear lengths decreased and diameters increased, with an increase in the number of kernel rows per ear. Yield had a tendency to increase but overall was not significant. With S_2 recurrent selection, populations also matured later, ear length increased and ear diameter and number of rows decreased, with kernel size also decreasing. Yield had a tendency to decrease, but again this was not significant (Moreira et al., 2008). Yield was not the factor of primary importance for farmers, as quality is the overriding concern for broa production. The authors recommended mass selection because it was simpler to implement, more economical, and more easily accessible to farmers. The S_2 recurrent selection would be appropriate if increased uniformity was necessary but would need to be monitored to ensure that the observed yield decrease did not continue and that too much diversity was not lost through the cycles of inbreeding. A follow-up study on the genetic diversity in populations that had undergone mass selection showed no significant decrease in genetic diversity based on Simple Sequence Repeat (SSR) markers (Vaz Patto et al., 2008).

Conclusion

Plant breeders and farmers have long recognized the value of diversity. Farmers developed heterogeneous landraces over hundreds of years and this wealth of agricultural heritage still provides the diversity breeders and farmers look for when seeking to adapt crops to new abiotic and biotic stresses and environmental conditions. Unfortunately, over the past century, crop genetic diversity has become increasingly threatened as modern varieties replace traditional farmers' varieties. Ironically, it is the farmers' varieties that made possible the development of these modern varieties, and the loss of these traditional varieties may endanger future plant breeding responses to environmental changes. There have been many efforts to encourage in situ and ex situ conservation of genetic resources. However, the goal is usually to protect genetic diversity that can be tapped for specific traits or alleles by plant breeding programs that release pure-line cultivars or F_1 hybrids, so the initial diversity never reaches farmers' fields. This has two major consequences: Farmers do not receive the benefits of having genetic diversity, such as improved buffering of environmental stress, and the existing diversity is not exposed to evolutionary forces that can shape and create new populations adapted to current environmental conditions in a specific geographic location.

The growth of organic agriculture provides an opportunity to incorporate genetic diversity in varieties to capture the benefits of diversity and evolution in farmers' fields. Organic farmers are particularly interested in having more diversity because of higher levels of heterogeneity in their environmental conditions and a marketing context that does not require overly standardized products. Many are also concerned about the loss of genetic diversity as a common good and are willing to make extraordinary efforts to preserve it. But the creation of this type of variety is not simple, and breeders and farmers need to work together on strategies for breeding and cultivating heterogeneous varieties. We have attempted in this chapter to provide an overview of existing efforts, and look forward to the continuing development and assessment of methods to breed crops for genetic diversity.

References

Allan, R.E., C.J. Peterson, R.F. Line, G.L. Rubenthaler, and C.F. Morris. 1993. Registration of 'Rely' wheat multiline. *Crop Science* 33:213–214.

Allard, R.W. 1961. Relationship between genetic diversity and consistency of performance in different environments. *Crop Science* 1:127–133.

———. 1988. Genetic changes associated with the evolution of adaptedness in cultivated plants and their wild progenitors. *Journal of Heredity* 79:225–238.

Allard, R.W., and J. Adams. 1969. Populations studies in predominantly self-pollinating species: 13. Intergenotypic competition and population structure in barley and wheat. *American Naturalist* 103:621–645.

Almekinders, C.J.M., and A. Elings. 2001. Collaboration of farmers and breeders: Participatory crop improvement in perspective. *Euphytica* 122:425–438.

Altieri, M.A. 1999. The ecological role of biodiversity in agroecosystems. *Agriculture, Ecosystems and Environment* 74:19–31.

Azais, J.M. 1987. Design of experiments for studying intergenotypic competition. *Journal of the Royal Society Series* B49: 334–345.

Bailey, L.H. 1923. Various cultigens, and transfers in nomenclature. *Gentes Herbarum* 1:113–136.

Berthaud, J., J.C. Clément, L. Emperaire, D. Louette, F. Pinton, J. Sanou, and G. Second. 2001. The role of local-level geneflow in enhancing and maintaining genetic diversity. *In*: H.D. Cooper, C. Spillane and T. Hodgkin (eds.), Broadening the genetic base of crop production. p. 452. CABI Publishing.

Besag, J., and R. Kempton. 1986. Statistical analysis of field experiments using neighboring plots. *Biometrics* 42:231–251.

Bocci, R., V. Chable, G. Kastler, and N.P. Louwaars. 2010. Set of recommendations on farm conservation strategy, the role of innovative market mechanisms, legislative framework for landraces, conservation varieties and amateur varieties in Europe. European Commission - Sixth Framework Programme.

Browning, J.A. 1988. Current thinking on the use of diversity to buffer small grains against highly endemic and variable foliar pathogens: Problems and future prospects. *In*: Breeding strategies for resistance to the rusts of wheat. pp. 76–90, 968th edition. Mexico: D.F. CIMMYT.

Camacho Villa, T., N. Maxted, M. Scholten, and B. Ford-Lloyd. 2005. Defining and identifying crop landraces. *Plant genetic resources: Characterization and utilization* 3:373–384.

Ceccarelli, S. 1989. Wide adaptation - How wide? *Euphytica* 40:197–205.

———. 1994. Specific adaptation and breeding for marginal conditions. *Euphytica* 77:205–219.

———. 1996. Adaptation to low high input cultivation. *Euphytica* 92:203–214.

Ceccarelli, S., S. Grando, and M. Baum. 2007. Participatory plant breeding in water-limited environments. *Experimental Agriculture* 43:411–435.

Coffman, W.R., and M.E. Smith. 1991. Role of Public, Industry, and International Research Center Breeding Programs in Developing Germplasm for Sustainable Agriculture. *In*: D.A. Sleeper, T.C. Barker, and E.J. Bramel-Cox (eds.) Plant breeding and sustainable agriculture, Considerations for Objectives and Methods, no. 18:1–9. Madison, WI: CSSA Special Publication.

Cowger, C., and Weisz, R. (2008). Winter wheat blends (mixtures) produce a yield advantage in North Carolina. *Agronomy Journal* 100:169–177. doi:10.2134/agronj2007.0128

Danquah, E.Y., and J.A. Barrett. 2002. Grain yield in composite cross five of barley: Effects of natural selection. *Journal of Agricultural Science* 138:171–176.

Demeulenaere, E., C. Bonneuil, F. Balfourier, A. Basson, J.F. Berthellot, V. Chesneau, H. Ferté et al. 2008. Étude des complémentarités entre gestion dynamique à la ferme et gestion statique en collection: Cas de la variété de blé Rouge de Bordeaux. *Les Actes du BRG* 7:117–138.

Denison, R.F. 2010. Past evolutionary tradeoffs represent opportunities for crop genetic improvement and increased human lifespan. *Evolutionary Applications* 4:216–224.

Denison, R.F., J.M. Fedders, and B.L. Harter. 2010. Individual fitness versus whole-crop photosynthesis: Solar tracking tradeoffs in alfalfa. *Evolutionary Applications* 3:466–472.

Denison, R.F., E.T. Kiers, and S.A. West. 2003. Darwinian agriculture: When can humans find solutions beyond the reach of natural selection? *Quaterly Review of Biology* 78:145–168.

Desclaux, D. 2005. Participatory plant breeding methods for organic cereals. *In*: E.T. Lammerts van Bueren, I. Goldringer, H. Østergård, COST SUSVAR/ECO-PB Workshop on Organic Plant Breeding Strategies and the Use of Molecular Markers. pp. 17–23. Driebergen: The Netherlands.

Döring, T.S., M.S. Wolfe, H. Jones, H. Pearce, and J. Zhan. 2010. Breeding for resilience in wheat – Nature's choice. *In*: I. Goldringer, J.C. Dawson, A. Vettoretti, F. Rey (eds.) Breeding for resilience: A strategy for organic and low-input farming systems? Proceedings of the Eucarpia conference Section on Organic and Low-input Agriculture, December 1–3, 2010, Paris. pp. 47–50. France: INRA and ITAB.

Elias, M., D. McKey, O. Panaud, M. Anstett, and T. Robert. 2001. Traditional management of cassava morphological and genetic diversity by the Makushi Amerindians (Guyana, South America): Perspectives for on-farm conservation of crop genetic resources. *Euphytica* 120:143–157.

Enjalbert, J., J.C. Dawson, S. Paillard, B. Rhoné, Y. Rousselle, and I. Goldringer. 2011. Dynamic management of crop diversity: From an experimental approach to on-farm conservation. *CR Biologies*.

Enjalbert, J., I. Goldringer, S. Paillard, and P. Brabant. 1999. Molecular markers to study genetic drift and selection in wheat populations. *Journal of Experimental Botany* 50:283–290.

Finckh, M.R., M.S. Wolfe, and E.T. Lammerts van Bueren. 2007. The canon of potato science: 32. Variety mixtures and diversification strategies. *Potato Research* 50:335–339.

Finckh, M.R. 2008. Integration of breeding and technology into diversification strategies for disease control in modern agriculture. *European Journal of Plant Pathology* 121:399–409.

Finckh M.R., E.S. Gacek, H. Goyeau, C. Lannou, U. Merz, C.C. Mundt, L. Munk et al. 2000. Cereal variety and species mixtures in practice, with emphasis on disease resistance. *Agronomie* 20:813–835.

Finckh, M.R., and C.C. Mundt. 1992. Stripe rust, yield, and plant competition in wheat cultivar mixtures. *Phytopathology* 82:905–913.

Foucteau, V., P. Brabant, H. Monod, O. David, and I. Goldringer. 2000. Correction models for intergenotypic competition in winter wheat. *Agronomie* 20:943–953.

Frison, E.A., J. Cherfas, and T. Hodgkin. 2011. Agricultural biodiversity is essential for a sustainable improvement in food and nutrition security. *Sustainability* 3:238–253.

Gallais, A. 1970. Modèle pour l'analyse des relations d'associations binaires. *Biometrie-Praximetrie XI* 2–3:51–80.

Gallandt, E.R., S.M. Dofing, P.E. Reisenauer, and E. Donaldson. 2001. Diallel analysis of cultivar mixtures in winter wheat. *Crop Science* 41:792–796.

Ghaouti, L., W. Vogt-Kaute, and W. Link. 2008. Development of locally-adapted faba bean cultivars for organic conditions in Germany through a participatory breeding approach. *Euphytica* 162:257–268.

Gizlice, Z., T.E. Carter, J.W. Burton, and T.H. Emigh. 1989. Partitioning of blending ability using 2-way blends and components lines of soybean. *Crop Science* 29:885–889.

Goldringer, I., P. Brabant, and R.A. Kempton. 1994. Adjustement for competition between genotypes in single-row-plot trials of winter-wheat (*Triticum aestivum*). *Plant Breeding* 112:294–300.

Goldringer, I., J. Enjalbert, A.L. Raquin, and P. Brabant. 2001. Strong selection in wheat populations during ten generations of dynamic management. *Genetic Selection and Evolution* 33:S441–S463.

Goldringer, I., C. Prouin, M. Rousset, N. Galic, and I. Bonnin. 2006. Rapid differentiation of experimental populations of wheat for heading time in response to local climatic conditions. *Annals of Botany* 98:805–817.

Goldringer, I., J.C. Dawson, E. Serpolay, N. Schermann, S. Giuliano, V. Chable, E. T. Lammerts van Bueren et al. 2010. Analysis of the bottlenecks and challenges identified for on-farm maintenance and breeding in European agricultural conditions. Opportunities for farm seed conservation, breeding and production. Deliverable D2.3. European Commission within the Sixth Framework Programme Specific Targeted Research Project Thematic Priority 8.1. France: INRA.

Goldringer, I., J. Enjalbert, J. David, S. Paillard, J.L. Jean-Louis Pham, and P. Brabant. 2001. Dynamic management of genetic resources: A 13-year experiment on wheat. *In:* H.D. Cooper, C. Spillane, and T. Hodgkin (eds.) Broadening the genetic base of crop production. Oxford: CABI Publishing.

Griffing, B. 1956. Concept of general and specific combining ability in relation to diallel crossing systems. *Australian Journal of Biological Sciences* 9:463–493.

Hajjar, R., D.I. Jarvis, and B. Gemmill-Herren. 2008. The utility of crop genetic diversity in maintaining ecosystem services. *Agriculture, Ecosystems & Environment* 123:261–270.

Heal, G., B. Walker, S. Levin, K. Arrow, P. Dasgupta, G. Daily, P. Ehrlich et al. 2004. Genetic diversity and interdependent crop choices in agriculture. *Resource and Energy Economics* 26:175–184.

Hodgkin, T., J. Rana, J. Tuxill, D. Balma, A. Subedi, I. Mar, D. Karamura et al. 2007. Seed systems and crop genetic diversity in agroecosystems. *In:* D. Jarvis, C. Padoch, and H.D. Cooper (eds.), Managing biodiversity in agricultural ecosystems. pp. 77–116. New York: Columbia University Press.

Horneburg, B., and H.C. Becker. 2008. Crop adaptation in on-farm management by natural and conscious selection: A case study with lentil. *Crop Science* 48:203–212.

Hughes, A.R., and J.J. Stachowicz. 2004. Genetic diversity enhances the resistance of a seagrass ecosystem to disturbance. *Proceedings of the National Academy of Sciences of the United States of America* 101:8998–9002.

Jackson, L.E., U. Pascual, and T. Hodgkin. 2007. Utilizing and conserving agrobiodiversity in agricultural landscapes. *Agriculture, Ecosystems & Environment* 121:196–210.

Jensen, N.F. 1988. Plant breeding methodology. New York: John Wiley & Sons.

Jensen, N.F., and W.T. Federer. 1965. Competing ability in wheat. *Crop Science* 5:449–452.

Jones, H., S. Clarke, Z. Haigh, H. Pearce, and M. Wolfe. 2010. The effect of the year of wheat variety release on productivity and stability of performance on two organic and two non-organic farms. *The Journal of Agricultural Science* 148:303–317.

Kempton, R.A. 1982. Adjustment for competition between varieties in plant breeding trials. *Journal of Agricultural Science* 98:599–611.

———. 1985. Statistical models for interplot competition. *Aspects of Applied Biology* 10:110–120.

Knott, E.A., and C.C. Mundt. 1990. Mixing ability analysis of wheat cultivar mixtures under diseased and non-diseased conditions. *Theoretical and Applied Genetics* 80:313–320.

Kovacs, G., P. Miko, and M. Megyeri. 2010. Evolutionary breeding of cereals under organic conditions. *In*: I. Goldringer, J.C. Dawson, A. Vettoretti, F. Rey (eds.) Breeding for resilience: A strategy for organic and low-input farming systems? Proceedings of the EUCARPIA conference Section on Organic and Low-Input Agriculture, December 1–3, 2010, Paris, France. pp. 17–19. Paris, France: INRA and ITAB.

Kutka, F.J., and M.E. Smith. 2007. How many parents give the highest yield in predicted synthetic and composite populations of maize? *Crop Science* 47:1905–1913.

Lammerts van Bueren, E.T., S.S. Jones, L. Tamm, K.M. Murphy, J.R. Myers, C. Leifert, and M.M. Messmer. 2010. The need to breed crop varieties suitable for organic farming, using wheat, tomato and broccoli as examples: A review. *NJAS Wageningen Journal of Life Sciences* 58:193–205.

Lavigne, C., X. Reboud, M. Lefranc, E. Porcher, F. Roux, I. Olivieri, and B. Godelle. 2001. Evolution of genetic diversity in metapopulations: *Arabidopsis thaliana* as an experimental model. *Genetic Selection and Evolution* 33:S399–S423.

Löschenberger, F., A. Fleck, H. Grausgruber, H. Hetzendorfer, G. Hof, J. Lafferty, M. Marn, A. Neumayer, G. Pfaffinger, and J. Birschitzky. 2008. Breeding for organic agriculture: The example of winter wheat in Austria. *Euphytica* 163:469–480.

Louette, D., and M. Smale. 2000. Farmers' seed selection practices and traditional maize varieties in Cuzalapa, Mexico. *Euphytica* 113:25–41.

Louette, D. 2000. Traditional management of seed and genetic diversity: What is a landrace? *In*: S. Brush (ed.) Genes in the field: On-farm conservation of crop diversity. pp. 109–142. Rome, Italy: Bioversity International.

Louette, D., A. Charrier, and J. Berthaud. 1997. *In Situ* conservation of maize in Mexico: Genetic diversity and Maize seed management in a traditional community. *Economic Botany* 51:20–38.

Love, B.E., and D. Spaner. 2006. Review of agrobiodiversity: Its value, measurement, and conservation (*In-Situ* and *Ex-Situ*) in the context of sustainable agriculture. *Sustainable Agriculture* 31:53–82.

McGilchrist, C.A. 1965. Analysis of competition experiments. *Biometrics* 21:975–985.

Mille, B., M.B. Fraj, H. Monod, and C. de Vallavieille-Pope. 2006. Assessing four-way mixtures of winter wheat cultivars from the performances of their two-way and individual components. *European Journal of Plant Pathology* 114:163–173.

Moreira, P., S. Pêgo, C. Vaz Patto, and A. Hallauer. 2008. Comparison of selection methods on 'Pigarro,' a Portuguese improved maize population with fasciation expression. *Euphytica* 163:481–499.

Moreira, P.M.M., M.C. Vaz Patto, M. Mota, J. Mendes-Moreira, J.P.N. Santos, J.P.P. Santos, E. Andrade, A.R. Hallauer, and S.E. Pego. 2009. ' Fandango': Long term adaptation of exotic germplasm to a Portuguese on-farm conservation and breeding project. *Maydica* 54:269–285.

Mundt, C.C., Hyes, P.M., Schon, C.C. 1994. Influence of barley variety mixtures on severity of scald and net blotch and on yield. *Plant Pathology* 43:356–361.

Murphy, K., D. Lammer, S. Lyon, B. Carter, and S.S. Jones. 2005. Breeding for organic and low-input farming systems: An evolutionary-participatory breeding method for inbred cereal grains. *Renewable Agriculture and Food Systems* 20:48–55.

Murphy, K.M., K.G. Campbell, S.R. Lyon, and S.S. Jones. 2007. Evidence of varietal adaptation to organic farming systems. *Field Crops Research* 102:172–177.

Murphy, K.M., L.A. Hoagland, P.G. Reeves, B.K. Baik, and S.S. Jones. 2009. Nutritional and quality characteristics expressed in 31 perennial wheat breeding lines. *Renewable Agriculture and Food Systems* 24:285–292.

Newton, A.C., G.S. Begg, and J.S. Swanston. 2009. Deployment of diversity for enhanced crop function. *Annals of Applied Biology* 154:309–322.

Olesen, J.E., M. Trnka, K.C. Kersebaum, A.O. Skjelvåg, B. Seguin, P. Peltonen-Sainio, F. Rossi, J. Kozyra, and F. Micale. 2011. Impacts and adaptation of European crop production systems to climate change. *European Journal of Agronomy* 34: 96–112.

Osman, A., and V. Chable. 2009. Inventory of initiatives on seeds of landraces in Europe. *Journal of Agriculture and Environment for International Development* 103:95–130.

Osman, A. 2006. The effect of growing cultivar mixtures on baking quality of organic spring wheat. *In*: H. Østergård and L. Fontaine (eds.) COST SUSVAR workshop on Cereal Crop Diversity: Implications for Production and Products. June 13–14, 2006, Domaine de La Besse, France. pp. 17–22. Paris, France: ITAB.

Østergård, H., M.R. Finckh, L. Fontaine, I. Goldringer, S.P. Hoad, K. Kristensen, E.T. Lammerts van Bueren, F. Mascher, L. Munk, and M.S. Wolfe. 2009. Time for a shift in crop production: Embracing complexity through diversity at all levels. *Journal of Agricultural & Food Information* 89:1439–1445.

Paillard, S., I. Goldringer, J. Enjalbert, M. Trottet, J. David, C. de Vallavieille-Pope, and P. Brabant. 2000. Evolution of resistance against powdery mildew in winter wheat populations conducted under dynamic management. II. Adult plant resistance. *Theoretical and Applied Genetics* 101:457–462.

Paillard, S., I. Goldringer, J. Enjalbert, G. Doussinault, C. de Vallavieille-Pope, and P. Brabant. 2000. Evolution of resistance against powdery mildew in winter wheat populations conducted under dynamic management. I – Is specific seedling resistance selected? *Theoretical and Applied Genetics* 101:449–456.

Parzies, H.K., W. Spoor, and R.A. Ennos. 2000. Genetic diversity of barley landrace accessions (*Hordeum vulgare* ssp vulgare) conserved for different lengths of time in *ex situ* gene banks. *Heredity* 84:476–486.

Phillips, S.L., and M.S. Wolfe. 2005. Evolutionary plant breeding for low input systems. *Journal of Agricultural Science* 143:245–254.

Porcher, E., T. Giraud, I. Goldringer, and C. Lavigne. 2004. Experimental demonstration of a causal relationship between heterogeneity of selection and genetic differentiation in quantitative traits. *Evolution* 58:1434–1445.

Raquin, A. L., P. Brabant, B. Rhoné, F. Balfourier, P. Leroy, and I. Goldringer. 2008. Soft selective sweep near a gene that increases plant height in wheat. *Molecular Ecology* 17:741–756.

Reusch, T.B.H., A. Ehlers, A. Hammerli, and B. Worm. 2005. Ecosystem recovery after climatic extremes enhanced by genotypic diversity. *Proceedings of the National Academy of Sciences of the United States of America* 102:2826–2831.

Rhone, B., R. Vitalis, I. Goldringer, and I. Bonnin. 2010. Evolution of flowering time in experimental wheat populations: A comprehensive approach to detect genetic signatures of natural selection. *Evolution* 64:2110–2125.

Rhone, B., A-L. Raquin, and I. Goldringer. 2007. Strong linkage disequilibrium near the selected Yr17 resistance gene in a wheat experimental population. *Theoretical and Applied Genetics* 114:787–802.

Rousselle, Y. 2010. Rôle de la migration dans la gestion dynamique des ressources génétiques végétales. Thèse de Docteur de l'Université Paris-Sud 11, France.

Smith, M.E., F. Castillo, and F. Gomez. 2001. Participatory plant breeding with maize in Mexico and Honduras. *Euphytica* 122:551–565.

Smithson, J.B., and J.M. Lenne. 1996. Varietal mixtures: A viable strategy for sustainable productivity in subsistence agriculture. *Annals of Applied Biology* 128:127–158.

Soliman, K.M., and R.W. Allard. 1991. Grain yield of composite cross populations of barley - Effect of natural selection. *Crop Science* 31:705–708.

Suneson, C.A. 1956. An evolutionary plant breeding method. *Agronomy Journal* 48:188–191.

Thomas, M., J.C. Dawson, I. Goldringer, and C. Bonneuil. 2011. Seed exchanges, a key to analyze crop diversity dynamics in farmer-led on-farm conservation. *Genetic Resources and Crop Evolution* 58:321–338.

UPOV. 1991. International convention for the protection of new varieties of plants. Accessed April 12, 2011. http://www.upov.int/en/publications/conventions/1991/act1991.htm

de Vallavieille-Pope, C., M. Belhaj Fraj, B. Mille, and J.M. Meynard. 2004. Les associations de variétés: Accroître la biodiversité pour mieux maîtriser les maladies. *Dossiers de l'environnement de l'INRA* 30:101–109.

Vaz Patto, M., P. Moreira, N. Almeida, Z. Satovic, and S. Pego. 2008. Genetic diversity evolution through participatory maize breeding in Portugal. *Euphytica* 161:283–291.

Williams, E.J. 1962. Analysis of competition experiments. *Australian Journal of Biological Sciences* 15:509–525.

Wolfe, M.S. 1985. The current status and prospects of multiline cultivars and variety mixtures for disease resistance. *Ann. Rev. Phytopathol* 23:251–73.

Wolfe, M.S., J.P. Baresel, D. Desclaux, I. Goldringer, S. Hoad, G. Kovacs, F. Löschenberger, T. Miedaner, H. Østergård, and E.T. Lammerts van Bueren. 2008. Developments in breeding cereals for organic agriculture. *Euphytica* 163:323–346.

Wolfe, M.S., and T.S. Döring. 2010. Steps towards an ecological future. *In*: I. Goldringer, J.C. Dawson, A. Vettoretti, F. Rey (eds.), Breeding for resilience: A strategy for organic and low-input farming systems? Proceedings of the EUCARPIA conference, section on Organic and Low Input Agriculture, December 1–3, 2010, Paris. pp. 39–42. Paris, France: INRA and ITAB.

Wolfe, M.S., K.E. Hinchsliffe, S.M. Clarke, H. Jones, and Z. Haigh. 2006. Evolutionary breeding of healthy wheat: From plot to farm. *In*: Atkinson, C., Ball, B., Davies, D.H.K., Rees, R., Russell, G., Stockdale, E.A., Watson, C.A., Walker, R., and Younie, D. (eds.) Aspects of Applied Biology 79:47–50.

Wright, S. 1922. The effects of inbreeding and crossbreeding on guinea pigs. United States Department of Agriculture Bulletin 1121. Washington D.C.: U.S. Gov. Print. Office.

Zhu, Y.Y., H.R. Chen, J.H. Fan, Y.Y. Wang, Y. Li, J.B. Chen, J.X. Fan, S.S. Yang, L.P. Hu, and H. Leung. 2000. Genetic diversity and disease control in rice. *Nature* 406:718–722.

6 Centralized or Decentralized Breeding: The Potentials of Participatory Approaches for Low-Input and Organic Agriculture

Dominique Desclaux, Salvatore Ceccarelli, John Navazio, Micaela Coley, Gilles Trouche, Silvio Aguirre, Eva Weltzien, and Jacques Lançon

Introduction

Organic or low input production in northern countries can be characterized by:

1. *Heterogeneous environments* that cannot be standardized by input supply and also by greater diversity in farming systems (e.g., from high organic input level to low input, from strict observance of organic guidelines to a new approach to holistic system);
2. *Broad diversity of farmers' needs* that can be determined also by the targeted outlets (e.g., self-consumption, direct marketing or via middleman, hand-made or industrialized products);
3. *Lack of varieties,* adapted to environments, to outlets or coming from breeding methods accepted by organic sector (i.e., cauliflower hybrids produced by cytoplasmic male sterility and not compatible with principles of the International Federation of Organic Agriculture Movements (IFOAM));
4. *Unresponsiveness* of formal seed sector, which considers the organic sector a niche market, and not sufficiently profitable.

Considering these four points, organic production shows similarities with production in marginal areas in developing countries. In such countries, the poor adoption of modern varieties raises questions about applying formal breeding systems to these environments.

Formal breeding programs can be briefly described as a centralized sequential process in which breeders collect germplasm, evaluate it at carefully controlled and well-managed experiment stations, and cross among superior "materials." The created genetic variability is then drastically

---

Organic Crop Breeding, First Edition. Edited by Edith T. Lammerts van Bueren and James R. Myers.
© 2012 John Wiley & Sons, Inc. Published 2012 by John Wiley & Sons, Inc.

reduced through selection and few elite lines are tested in farmers' fields. For economic reasons the focus is on "broad adaptability" – the capacity of a plant to produce a high average yield over a wide range of growing environments and years. Therefore, candidate varieties yielding well in one zone but less in another are quickly eliminated (Ceccarelli and Grando, 1997). Yet, "specific adaptability" may be exactly what organic farmers require (Vernooy, 2003).

How to better grasp farmers' criteria? How to valorize Genotype × Environment (G×E) interactions and favor specific or locally adapted cultivars? How to select adapted varieties? Two concepts are important to answer these questions: The location (centralized or decentralized) and the organization (participatory or not) of breeding.

Centralized and Decentralized Breeding: Definitions

A classic plant breeding program may be partitioned into five main stages: (1) setting objectives, (2) creating variability, (3) selection, (4) evaluation, and (5) diffusion (Sperling et al., 2001).

Centralized breeding is defined as breeding programs entirely conducted on one or more research stations except for the evaluation and diffusion stages. Therefore, we restrict the use of the term "decentralized breeding" to mean decentralized selection, as opposed to decentralized evaluation, the last stage of any breeding program (Ceccarelli, 2009).

Reasons for centralization are numerous: To benefit from a perennial location, with a high traceability of past events (databases available on rotation, crop management); to control or predict several factors concerning biophysical environment or crop management and therefore to reduce environmental variance; to share a breeding program among staff belonging to a common institution and interact daily; to get the infrastructure (equipment to plant, harvest, and handle seed) when needed; to have the resources and knowledge to evaluate small seed quantities; to facilitate organizational aspects such as land allocation, organization of human resources, data capture, storage, and analysis. Centralized breeding is particularly efficient in cropping systems similar to those of the experiment stations but is unsuitable when G×E interactions are strong (Ceccarelli et al., 2009).

Usually, the term "environment" (E) covers biophysical environment and crop management. But centralized breeding is much more unsuitable when considering the term E in a broad sense. Looking for a variety adapted to organic conditions means looking for a variety adapted to targeted (1) biophysical environment, (2) crop management, (3) uses, market, and outlets, (4) regulations linked with seed sector (variety structures, specifications, contracts, environmental measures, etc.), (5) roles, skills, responsibilities, and resources of breeding actors, and (6) more generally, societal dynamics (Desclaux et al., 2008). Considering these different components of E, the main question is which component should be centralized and which should be decentralized – and at what cost?

What Can Be Decentralized in Breeding and Why?

Centralizing or Decentralizing Biophysical Environment and Crop Management?

The efficiency of a trial location is determined by (1) the genotypic correlation between the performance of a cultivar at this location and the performance obtained at the target location, and (2) the precision (involved in broad-sense heritability, proportion of variance explained, and power of discrimination) with which the performances of the cultivar are measured at the same location.

A centralized breeding program enables heritability to be maximized, while decentralized breeding enables correlation to be maximized (Atlin et al., 2001). The question of centralizing or decentralizing can also be seen as a compromise between bias and variance. The research station location is powerful for discriminating among varieties by minimizing variances, but the more the selected trait depends on environment, the greater the bias and vice versa for the farm location.

It is possible to achieve optimal results by increasing the number of locations when bias increases. When resources are limited, a plant breeder has to juggle the number of locations and the number of repetitions at each location. The search for a strong $G \times E$ interaction involves many trials with a limited number of repetitions and acceptance of the risk of unjustified expenditure if the interaction turns out to be weak (Gozé, 1992).

The potential gap between the breeding location and the target agro-ecological environment should be represented by the frequency and intensity deviations of the limiting factors. If this cannot be simulated at the experimental station, then trials must be decentralized in farmers' fields.

Crop management depends on the farmers' objectives as a function of their constraints and is defined by farmers' decision rules. It is difficult for an experimenter to represent the farmers' choices and for this reason decentralization also appears to be useful for breeding. However, if decision rules are clear, it is possible to apply them at the research station and thus control experimental bias. In the biophysical conditions of a breeding station and with a judicious choice of crop management parameters, the diversity of target agro-ecological environments may be represented by identifying the limiting factors and predicting $G \times E$ (Desclaux et al., 2008).

Centralizing or Decentralizing Actors?

Actors usually considered in PPB (farmer, breeder, processor, NGO, etc.), may be involved directly either at the research station or on-farm. Their participation can also take the form of individual interviews, working groups, debates, etc. In all cases, specific principles and procedures are needed to facilitate critical participation – and not merely passive attendance (Friedberg, 1988). These principles, based on empathy, include transparency, democracy, open doors, and respect of the individual (Desroches, 1976), with rules of conduct explicitly negotiated. Selection by farmers, on their own farms, facilitates this type of critical participation by enabling the participants to recognize the constraints and limits of plant improvement (Chiffoleau and Desclaux, 2006). Experimental varieties and plots can be used as "intermediary objects" (Vinck, 1999) to link the actors in the network and to facilitate the expression and sharing of knowledge and agreements about what constitutes a feasible innovation and for whom.

Centralizing and Decentralizing Outlets, Regulations, and Society

Two stages can be envisaged to deal with target markets and outlets and with the economic, political, and territorial coordinating structures. In situ audits and joint information gathering should enable listing and weighting criteria that can subsequently be used as entry variables in economic simulations. Moreover, some regulations such as variety registration must be relaxed to enable creation and use of heterogeneous genotypes needed to meet the diversity of needs (Desclaux et al., 2008).

Concerning societal dynamics, insofar as "one cannot relocate society" (Cauderon, 2003), the challenge is to consider the individuals who make up society both as actors and as owners of diverse values able to constructing relevant rules for both global and territorial contexts (Pecqueur, 2007) and of making "the cultural effort to take part in the debate." Societal dynamics thus promote a technical and proximate democracy that facilitates not only genetic progress but also ethical and social progress.

We define decentralized breeding as a program in which selection and evaluation are conducted in the target environment. A decentralizing program is relevant to address localized environmental stresses, but without farmer participation, it may fail in its aim if farmers' preferences are ignored.

Participatory Approaches

Participatory Plant Breeding (PPB) is described as an approach involving all the actors not only in establishing breeding objectives, but also in managing the breeding process and the creation of varieties (Gallais, 2006). It aims to respond to systemic issues and demands for which classic breeding appears to be unsuited (see table 6.1; Almekinders and Hardon 2006; Ceccarelli et al., 2001; Witcombe et al., 2003). PPB that involves farmers as well as other partners such as extension staff, seed producers, traders, processors, and NGOs in the development of a new variety is expected to produce varieties that are targeted (focused on the appropriate farmers), relevant (responding to real needs, concerns, and preferences), and appropriate (able to produce varieties that can be adopted; Bellon, 2006). The science behind participatory and conventional plant breeding is the same; the main difference is that conventional plant breeding is a process where priorities, objectives, and methodologies all are decided by one or more scientists with little or no participation of farmers, while PPB gives equal weight to the opinions of farmers (and others actors) and scientists. PPB must be distinguished from farmers' breeding, which can be defined as the complex of various breeding activities that farmers conduct on their own without any participation by researchers.

PPB projects have been initiated by international research institutes to improve the adoption of cultivars by small farmers in developing countries (Almekinders and Hardon, 2006). More recently, they emerged in European and North American countries as both socio-political and scientific projects, conducted by farmers' associations promoting sustainable agriculture, and by researchers preoccupied with biodiversity preservation. These actors refer to southern PPB initiatives to justify their projects of "transfer" to the north. However, the huge diversity in PPB programs must be assessed to determine which conditions and along which modalities PPB may be relevant for low input and organic agriculture.

PPB: A Single Term Yielding Different Approaches

Participatory plant breeding (PPB) is a relatively recent concept, but already it refers to a large set of approaches and breeding methods. All these approaches could be integrated into an n-dimension matrix where the following items would be integrated.

Objectives

PPB usually mixes Functional and Process approaches as defined by Thro and Spillane (2000). Functional approaches consist of obtaining better-adapted crop varieties (i.e., more closely tailored to small-scale farmers' needs), whereas, process approaches aim to empower farmers to develop

Problem	Examples	References	Description	Causes	Proposed Solution
Erosion of local genetic resources	Food crops, Crops with cultural value, Locally domesticated crops	Vaksmann et al., 2005 Caillon et al., 2006 Barnaud et al., 2008	The crop is regressing, Conservation practices are vanishing	Other crops or species are more competitive, Loss of indigenous knowledge	Improving competitiveness of the species (plant breeding), Valuing the crop (new markets), Valuing the genetic resources (seed fair), Valuing in situ conservation
Lack of formal breeding program	Orphan crops, Niche crops, Organic agriculture (Europe & US)	Sperling et al., 1993 Chable et al., 2006 Desclaux, 2009 (a, b) Desclaux et al., 2003 Chiffoleau and Desclaux, 2006	Farmers cannot get adapted varieties	Insufficient or poorly identified markets	Train farmers to breed and produce their own varieties
Lack of competition among breeding programs	Contract crops, Genetically Modified Crops (developing countries)	Lançon and Mucchielli, 2006 Hofs et al., 2006	Farmers have no variety choice	Captive market (monopolistic strategy) – Agreement market	Allowing farmers to access or to produce other adapted varieties
	Niche crop	Sthapit et al., 1996 Zimmermann, 1996 Joshi and Witcombe, 1996			
Low efficacy of breeding programs	Food crops, Cultural crops, Locally domesticated crops	Sperling et al., 1993 Weltzien et al., 1998 Thiele et al., 1997 Sthapit and Witcombe, 1996 Zimmermann, 1996 Trouche et al., 2011	Supply is disconnected from demand and a minority of actors enjoy adapted varieties	Actors' needs and know-how are not fully identified by the breeder	Associating farmers with genetic resources evaluation or management
	High altitude crops, Crops under limiting factors (water, photoperiod, mineral content, etc.), Crops on salty soil, Crops under infested E (diseases or insects),	Sthapit et al., 1996 Ceccarelli et al., 2001 Machado and Fernandes, 2001 Zimmermann, 1996	The varieties bred on station are not suited to farmers	Important GxE interactions (where E = biophysical environment or crop management)	Decentralizing part of the breeding process
	Industrial or commercial crops grown in highly variable environments	Lançon et al., 2004			

(*continued*)

Table 6.1 (Continued)

Problem	Examples	References	Description	Causes	Proposed Solution
Low efficiency of breeding programs	Crops assigned to a small and low-cost market		The unit of genetic progress is too costly to produce and disseminate	Resources are too limited to provide the actors with a sustainable service	Dividing up breeding operations and costs
Lack or ineffectiveness of seed system*	Subsistence or self-consumption crop	Jones and Wopereis-Pura, 2004; Thiele, 1999	Actors have no access to seeds of adapted varieties	Farmers cannot adopt the adapted varieties	Associating farmers to strategic phases of the breeding process
	Crops exchanged through barter	Rubyogo, 2005; Weltzien et al., 2005; vom Brocke et al., 2003		Collective stakes are too limited to justify the implementation of an organized formal seed system	Building up a seed production and dissemination system with the beneficiaries

*The seed system is made-up of all the components (actors, organizations, norms) that influence seed dissemination. It may be formal when institutionally organized or informal when controlled by unwritten traditional rules.

their skills as plant breeders. In a truly participatory process, these two approaches cannot be separated (Ceccarelli, 2009).

From the literature, the main contemporary PPB objectives are obtaining adapted varieties; accelerating cultivar transfer and adoption; identifying, verifying, and testing specific selection criteria (Weltzien et al., 2000); improving local adaptation; promoting genetic diversity; recognizing the potential social and economic value of local traditional varieties that have long been excluded from the formal seed system (Imaizumi and Hisano, 2010); and empowering farmers (including women) that traditionally have been left out of the development process (McGuire et al., 1999; see also the section "Participatory Barley Breeding in Syria").

Institutional Context

It is useful to differentiate a formal-led PPB program that is initiated by researchers inviting farmers to join breeding research from a farmer-led PPB program, where scientists seek to support farmer's own systems of breeding, varietal selection, and seed multiplication and dissemination. Based on the work of Franzel et al. (2001), a more elaborate differentiation involves identifying leaders of breeding process designs and those of management (McGuire, 2008). A Formal-led PPB can gradually evolve into farmer-led PPB, and can eventually be entirely transferred to farmers (Ceccarelli, 2009).

Forms of Interaction between Actors

The various modes of participation can be thought of as points along a continuum representing different levels of interaction. Each mode of participation can be characterized in terms of how farmers and plant breeders interact to set objectives, make decisions, and share responsibility for decision making, implementation, and product generation (Morris and Bellon, 2004). Three kinds of participation are usually distinguished: Consultative (information sharing), collaborative (task sharing), and collegial (sharing responsibility, decision making, and accountability; Sperling et al., 2001). In practice, field experience indicates that as farmers become progressively more empowered, PPB continuously evolves.

Location of Selection

As previously discussed, decentralized selection emphasizes favorable $G \times E$ interactions, but it can miss its objectives if it does not utilize farmers' knowledge of the crop and the environment. PPB can also be conducted in centralized research stations. Farmers are invited to visit and select among lines grown at experimental stations.

Stage of Selection

In many cases, farmers' participation is limited to the final breeding stages: Evaluating and commenting on a few near-finished or advanced varieties just prior to official release. This is known as participatory varietal selection (PVS), while PPB includes participatory selection within unfinished or segregating "breeding material" (i.e., with a high degree of genetic variability; Witcombe et al., 1996). Both terms are included in the participatory crop improvement concept.

PVS can be useful before beginning a PPB process because it helps to identify parents and important target traits. As suggested by Witcombe and Virk (2001), a PPB program may use only a few crosses from which large populations were produced. Because few parents are employed, their choice is crucial to obtaining relevant genetic diversity. Many of the varieties reaching on-farm trials would have been eliminated from testing years earlier if farmers had been given the chance to critically assess them (Toomey, 1999).

This review emphasizes the great diversity of PPB approaches. However, all of them have in common the aim of shifting the focus of plant genetic improvement research toward the local level by directly involving the end user in the breeding process (Morris and Bellon, 2004).

Some Examples of PPB for Organic and Low Input Agriculture in Southern Countries

Interest of PPB in Southern Countries

Considering the low impact of agricultural research in developing countries and on poor farmers, international agronomic institutes showed an increasing interest in participatory research and in PPB in particular. Indeed, about 2 billion people still lack reliable access to safe, nutritious food, and 800 million of these are chronically malnourished (Reynolds and Borlaug, 2006).

Agricultural research is organized (1) by a research agenda usually decided unilaterally by the scientists and not discussed with farmers, and (2) into compartmentalized disciplines and/or commodities in contrast to the integration existing at the farm level. There is an imbalance between the large number of innovations (such as varieties) generated by the agricultural sciences and the relatively small number actually adopted and used by the farmers. Even if the new varieties are acceptable to farmers, the seed is often either not available or too expensive.

Participatory Barley Breeding in Syria

The International Center for Agricultural Research in the Dry Areas (ICARDA) conducts PPB programs in 24 villages across Syria. Most of them are located in marginal areas affected by droughts and subsequent crop losses. The collaboration of research station staff located in the provinces facilitates access to the farmers. Farmers and researchers together decided that developing improved barley varieties adapted to the climate and to breed in farmers' fields with farmers' participation should be the main priorities.

Initial Phase of the Project

The main objectives were to build human relationships, understand farmers' preferences, measure farmers' selection efficiency, developed a scoring methodology, and enhance farmers' skills. The exploratory work included the selection of farmers and test sites (see table 6.2; Ceccarelli et al., 2000).

The initial barley experiment lasted three cropping seasons and included 200 new barley types representing a wide range of variability for plant height, flowering and maturity date, leaf color, row type (two vs. six rows), seed color (white, black, grey), stem diameter and associated lodging resistance, and straw palatability. Eight farmer-selected cultivars were also included. The 208 barley types could be sorted into 100 fixed lines derived from modern germplasm and 108 segregating landrace populations.

Table 6.2 Characteristics of the test sites used for the participatory barley breeding in Syria

Sites			
Name	No.	Climatic Conditions	Other Characteristics
Villages	9	From wet to dry (200–400 mm)	Variable farm size (5–160 ha) and levels of on-farm and off-farm income; cropping choices limited and barley is important
ICARDA Research station	2	Breda rainfall: 243 mm	Low input and limited choice of crops
		Tel Hadya rainfall: 343 mm	High input with large choice of crops

Agronomic management of trials was left to the host farmer. The trials were conducted under rainfed conditions both in the farmers' fields and on the research stations. Each participating farmer was given a field book in which they recorded daily rainfall and their plot observations. Farmers used a mix of quantitative scores for some traits and qualitative descriptors for others. Illiterate farmers were assisted in recording their scores by other farmers or by the scientists. Four different types of selection processes were performed as shown in table 6.3.

Local landraces and improved varieties (chosen by the farmers) were used as systematic checks. The process of evaluation and selection conducted the first year was repeated the following years. The results of the first three years indicated that (Ceccarelli et al., 2000):

– Farmers were able to handle large numbers of entries, make many observations during the cropping season, and develop their own scoring methods.
– Farmers select for specific adaptive traits, and in some cases, such as modern germplasm versus landraces, selection was mostly driven by environmental effects.
– There was more diversity among farmers' selections in their own fields than among farmers' selections on research stations and among breeder's selections irrespective of the selection location.
– The selection criteria used by the farmers were nearly identical to those used by the breeders.
– In their own fields, farmers were slightly more efficient than the breeders in identifying the highest yielding entries. Breeders were more efficient in selecting at a research station located in a high rainfall area, but less efficient than the farmers at the research station located in a low rainfall area.

Table 6.3 Different types of selection performed during the first phase of the PPB project in Syria

Type of Selection	Conducted by:	Location of Selection
Centralized – Nonparticipatory	Breeder	Research stations (2)
Centralized – Participatory	Farmers	Research stations (2)
Decentralized – Nonparticipatory	Breeder	Farmers' fields (9)
Decentralized – Participatory	Farmers	Farmers' fields (9)

The first phase of the participatory barley breeding project in Syria led to an increased awareness among farmers of what plant breeding is and what it can offer. Requests to extend the project to other crops besides barley revealed a substantial interest among farmers for PPB.

The pilot project was essential to move from the linear process used in the first phase to a cyclic process and a truly participatory program. This meant clarifying with the farmers that the project would not be short-term, but an ongoing and evolving activity.

The Second Phase of the PPB Project

During the second phase, research stations only provided support for seed multiplication. The number of villages taking part in the project increased from nine to eleven the first year and to 24 two years after, the number of farmers directly involved in the project also increased. Village-based seed production was initiated in some villages. The role of scientists was to cross lines (landraces × improved cultivars, landraces × wild relatives), grow the first two generations on research stations, measure traits the farmers defined as important, and analyze and preserve electronic copies of all data. The farmers evaluated and scored breeding material, deciding what to select and what to discard, adopted and named varieties, and produced and distributed seed of the adopted varieties.

The evaluation process was composed of four stages; initial, advanced, elite and large scale trials. The initial yield trials in Syria included 165 entries. Due to the large diversity in the crop, and to different farmers' preferences, more than 400 genetically different entries were grown.

The advanced and elite trials, comprising the entries selected during the initial and advanced trials of the previous year, were replicated trials with two replications. The data were analyzed with spatial models in GenStat: The analysis produced the best linear unbiased predictors of genotypic values and several variables, including heritability. These PPB trials generated the same quantity and quality of data as the data generated by the Multi Environment Trials in a conventional breeding program. In addition, these trials provided information on farmers' preferences, data that are usually not available through conventional trials. The resulting varieties qualified for official variety release. The best-selected lines were used as parents in a new cycle of recombination and selection, exactly as in a conventional breeding program. The difference was that these lines were selected by farmers and potentially varied from location to location. This cyclic aspect had an enormously empowering effect for farmers: The use of their best materials by breeders for the next cycle of breeding created a strong sense of ownership. Particular care was taken to design a scientifically robust model for two reasons. First, the farmers could be provided with scientifically correct information (the same information a breeder usually has) on which to base their decisions. Second, PPB programs are often criticized, and sometimes rightly so, for not using rigorous experimental design and statistical analysis, and the model used protected the project against such criticisms (Ceccarelli and Grando, 2007).

Benefits of the PPB Project

Increased Crop Biodiversity. The number of different entries at the end of a breeding cycle in farmers' fields was higher than the number of lines the Syrian National Program evaluated at the beginning of its on-farm testing, which usually ends with one or two recommended varieties for the entire country. The PPB facilitated increase in biodiversity took place not only in space (because different villages select different lines) but also in time. The cyclic nature of the process ensured a rapid turnover of varieties within the same location and therefore an increase of biodiversity both in time and space (Ceccarelli et al., 2007).

As the PPB project progressed, the farmers also contributed suggestions for modification of methodology, such as the use of mixtures. Given that farmers in Syria do not generally like heterogeneous plots, it was surprising when a farmer decided to mix two very different barley varieties.

He learned about the characteristics of the two varieties by conducing PPB trials, and he thought that mixing these two varieties could be a good strategy to stabilize the yields. During three years of production, farmers reported that this mixture produced better yields than all the other varieties. This example of evolutionary-participatory plant breeding showed that handling of mixtures or complex populations is very simple, since all that is needed is to cultivate them in locations affected by either abiotic or biotic stresses and let natural selection slowly increase the frequency of the best-adapted genotypes. With experience and skills developed through PPB, farmers and breeders can superimpose artificial selection for traits that are important in specific locations.

Increased Number of Farmers Involved in the PPB Program. Farmer participation increased from 5 to 10 per village during the selection phase and from 10 to 15 per village for the data discussion step. As a result, between 200 and 400 farmers were directly involved in two of the most important decision-making events each cropping season. In addition, in some villages as many as 60 farmers bought seeds of the varieties developed through the PPB program. On average, more than 1,000 farmers benefited from the program each cycle.

Adapted and Yielded Varieties. Data from several years, including a very dry year, showed that PPB lines outperformed both the commonly used landraces and conventionally bred modern varieties (see table 6.4).

Economic Considerations. A number of farmers have started to produce seed from the resulting varieties. This seed was either provided for free or at a slightly higher price than the seed of "improved" varieties. Because they were buying seed of a variety they had seen grown in the field by another farmer with agronomic practices similar to their own, the farmers were sometimes willing to pay more than what they would for other varieties.

Analysis of the farm-level benefits and costs of barley production show that the participation of farmers in the breeding program did not mean higher costs of production (Ceccarelli et al., 2011). However, farmers adopting varieties bred through PPB projects would likely pay higher input costs, but gain higher net returns. In addition to the economic benefits, participating farmers appreciated other benefits such as increased knowledge of barley production, variety selection, and collaboration with scientists and other farmers.

When assuming a 50% adoption rate and 50% of the yield gain obtained for the varieties of the two programs, the analysis showed that the benefit-cost ratio, as well as the internal rate of return,

Table 6.4 Location, climatic data, and number and performance of barley lines developed through PPB in local villages in Syria

Village	Rainfall (mm)		Barley Lines	Yield	Check
	Mean	Range	(no. & description)	(% local check)	
Kherbit El Dieb	189	139–206	4 (PPB lines)	112–123	Local black-seeded landrace
Om El Amad	249	183–328	2 (PPB lines)	111–119	Local white-seeded landrace
			1 (modern bred variety)	105–113	
Bari Sharky	204	130–238	2 (PPB lines from crosses with wild barley progenitor)	133	Local landraces
Suran	277	198–300	2 (PPB sister lines from landrace crosses	115–125	Local landrace
			1 bred modern variety	118–127	

would be higher for PPB. This finding still stands with an adoption rate of 10% and 33% yield gain. It is important to ensure the availability of seeds to farmers, since most of the interested farmers operate in marginal areas where drought is a major risk factor.

The Gender Dimension. A study revealed that women farmers in Syria were interested in PPB but were not informed about the possibility or assumed they could not participate. A female researcher has supported the integration of Syrian women farmers into the PPB project by combining gender analysis with action research (Ceccarelli et al., 2011). Multi-criteria mapping was used to discuss women's expectations of the program, their views on the validity of the current PPB process, and suggestions for improvement. PPB activities were organized to facilitate the involvement of women farmers (coordinating the events directly with the women, collaborating with local institutions, and by creating women-only spaces). The project tried to respect local sensitivities, particularly with regard to the participation of young female farmers in public events, and created arenas for discussion where women may interact more easily with male strangers. Because women, on average, are more illiterate than men and have less access to technology, the reports from the PPB project were produced both in digital and hard copy and included visual and oral material. The PPB project enlarged its research portfolio beyond barley to reflect women's priority crops – such as chickpea and cumin – and included priority traits for selection suggested by women, such as spike hardness necessary for hand harvesting and palatability.

Knowledge and Empowerment. The knowledge that farmers gained through participation in the program improved their decision-making abilities regarding variety testing, evaluation, and selection. Almost all participating farmers said that even if the PPB project ends they will continue to practice, and they will maintain the seeds of the new varieties and keep looking for good varieties with other farmers. Many farmers feel that their participation in the PPB project improved their knowledge of barley production, as well as agriculture in general. They say that their experience increased as a result of their interaction with other farmers, indicating that an increase in social capital has taken place. The majority of the farmers believe that the benefits of selected varieties should be shared and distributed at a community level. They view the local plant genetic resources as their common heritage and not something only a few should benefit from (Halewood et al., 2007).

Regulation Considerations. The legislation in Syria that regulates variety release and seed multiplication and distribution is often seen as an obstacle to PPB projects due to their limiting the amount of seed that can be produced and distributed. However, the only legislation existing on this subject is a Ministerial Decree from 1975 (available only in Arabic), and it contains no specific restrictions. The legislative situation in Syria has been unclear and is a barrier to scaling up seed production. As of 2011, the Ministry of Agriculture and Agrarian Reform was drafting a seed law. This law may bring legal certainty to the field, but depending on restrictions, it might also end up being detrimental to farmers' rights.

Conclusion on the PPB Program in Syria
The PPB project on barley in Syria inspired other countries in the region (Jordan, Egypt, Eritrea, Ethiopia, Yemen, Tunisia, Morocco, Algeria, and Iran) to conduct similar efforts. To ensure the success of such projects, especially in reaching out to substantive numbers of farmers, seed laws must allow sufficient seed multiplication and distribution. The PPB project strengthened seed systems by improving the production, allowing selection, maintaining diversity, and providing access to seeds. It helped to increase food security of all involved villages.

Participatory Breeding of Sorghum for Matching the Needs of Resource-Poor Farmers in Nicaragua

In the dry hillsides of Nicaragua, sorghum has become an important staple crop for resource-poor farmers. Because they considered it as evidence of their extreme poverty, farmers of these regions up until the 1980s did not reveal to foreigners that they consumed sorghum grain.

Farmer-Led Changes in the Sorghum Cropping Systems and Cultivars

In the northern region of Nicaragua, increasing numbers of farmers were growing sorghum (white-grain early maturing cultivars, sorgo tortillero, or more traditional late photoperiod-sensitive cultivars, sorgo millón), instead of maize. Older farmers related the change in their cropping systems to climatic changes (the first major drought was experienced in 1972) and reported that changes mainly occurred without intervention of the government's extension services or NGO programs. Most national development programs focused on the production of maize and bean, but farmers felt that sorghum was a major crop for ensuring their food security and expressed interest in improving their cultivars.

Participatory Variety Selection Phase

A research team from the French Agricultural research center for development (CIRAD) met the national agricultural research institute (INTA), local NGOs, and local organized farmer groups and proposed to develop a decentralized participatory sorghum breeding program. This program focused on diversifying and improving the sorghum varieties for matching the needs of resource-poor farmers in the dry hillsides areas. The project team, breeders, farmers, and extensions workers first introduced improved inbred lines or cultivars from Africa. These represented a wide genetic diversity. Agronomic performances were evaluated in farmer's fields.

As the result of this program, farmers adopted and now grow several new varieties of tortillero and millón types (Trouche et al., 2009). 'Blanco Tortillero' was officially registered at the national level by a smallholder's cooperative involved in this research, and it is being disseminated in the dry marginal areas of Northern, Central, and Western departments of Nicaragua. This variety showed excellent adaptation to the low-input cropping systems of this area, earliness, stable yields, and good grain appearance and quality. Another new variety highly praised by farmers and rapidly adopted is the millón cultivar Coludo Nevado. It adapts very well to the maize-sorghum intercropped systems in the hillsides areas and showed high tolerance to drought and an excellent grain quality. At least four other lines derived from this phase were also locally disseminated by the main NGO partner, CIPRES, in the same areas.

Participatory Breeding for Developing New Ideotypes for Coping with Specific Cropping Systems

As a first strategy defined in collaboration with some farmers, the best local cultivars and Africa lines presenting complementary traits were crossed to create new genetic variability for the tortillero and millón breeding programs managed in situ with farmer breeder groups (Trouche et al., 2008). Later, a second strategy was to create broader genetic base synthetic populations from intercrossing between six to eight elite cultivars and lines (Trouche et al., 2011). The final goal of these programs was to develop new diversified lines that better satisfied farmers' needs and preferences for specific cropping systems, such as bean-sorghum intercropped, dual purpose sorghum, and ratooning sorghum.

A key result of these programs was the formation of a core group of skilled and motivated farmer-breeders involved in plant selection phases (see fig. 6.1) and/or subsequent final lines evaluations (see fig. 6.2), and making decisions in dialog with scientists and NGO agents. A large number

Figure 6.1 The female farmer-breeder Cleotilde Soto Vargas selecting plants in segregating tortillero sorghum populations at Palacagüina, Nicaragua (photo courtesy of Gilles Trouche).

Figure 6.2 Farmers of the Unile locality evaluating new lines derived from a PPB program focused on sorghum-bean intercropped systems, Nicaragua (photo courtesy of Gilles Trouche).

of men and women, young and old, were involved at different steps in these participatory breeding activities. Many of them were later involved in developing or evaluating new progenies of maize or beans adapted to their environments. About 30 sorghum lines derived from these programs were advanced to multilocal yield trials in order to identify those that might be more suitable to be released. The first line, renamed 'Crema Nacional,' was released by the CIPRES and INTA partners. More than two years after the departure of the professional sorghum breeder from the CIRAD-CIAT project, the breeding programs were continued with sole support of CIPRES, and two new regions of Nicaragua were included in this program.

Farmers' Empowerment
This PPB program strengthened individual and collective skills and capacities of involved resource-poor farmers. Resulting new knowledge of the plant and crop should be highlighted and includes:

– Knowledge obtained of the breeding process, which is applicable to other crops.
– Increased social status for young men and women farmer-breeders (FBs) involved in these actions.
– The FBs have improved communication skills that are beneficial in solving other problems in their communities.
– Strengthened community capacity.
– Integration of involved farmers into cooperatives and a federation of cooperatives. This federation has become recognized by local authorities and NGOs.
– Recognition of these PPB programs and the know-how acquired by FBs groups by the National Research Institute.

Some Examples of PPB for Organic and Low Input Agriculture in Northern Countries

Organic production in northern countries shows great similarities with production in marginal areas in developing countries. Crops produced with low input organic farming practices are frequently exposed to a greater set of environmental challenges including variable nutrient availability, drought conditions, and weed or pest pressure than crops grown under high input conventional systems. Developing crop genetic resources adapted to low input, diverse organic cropping systems is essential for the continued improvement of organic agriculture.

A North American Participatory Approach to Breeding a Genetically Resilient Zucchini Variety

Introduction
Many organic farms in the North are not located on prime agricultural land in large-scale, centralized production areas. These farms face similar challenges to many in the South in that they are exposed to a greater environmental variability both within field and between locations (Grube, 2007). Three key lessons learned from PPB projects are (1) strong farmer involvement, (2) breeding under the challenges of the targets environment and market, and (3) maintaining greater genetic variability in any resultant varieties than would normally be retained in centralized breeding programs (Ceccarelli and Grando, 2007). These core concepts offer many valuable parallels when applied to breeding crops for the diverse, low external input, organic farms of the North.

The success of a new, superior dark green zucchini squash (*Cucurbita pepo*) bred under organic, low input conditions in coastal Northern California, USA, demonstrates the potential of participatory plant breeding to fulfill the needs of commercial organic vegetable farmers in developed countries. A professional plant breeder of Organic Seed Alliance (OSA) and a farmer-breeder of Eel River Produce cooperated in breeding a zucchini for the organic wholesale market of San Francisco. The farmer initiated this project on his own before the collaborative effort with OSA to address his need for a reliable seed source of a zucchini that fit both the target market and was appropriate to his low input organic cultural practices.

OSA interacts with many experienced organic farmers frustrated by the lack of commercial varieties suited to their climate, organic cropping systems, or unique markets. Consolidation in the seed industry and a focus on breeding for areas of large-scale agricultural production has resulted in fewer varietal choices and "less than appropriate" varieties for these decentralized organic producers. Many share the opinion that the older standard varieties were more resilient or "field tough" and frequently better adapted to the challenges of their low input farming systems. Some are highly motivated to participate in on-farm breeding to remedy this situation.

In all OSA PPB projects, recurrent selection breeding schemes are used to retain substantial genetic variation so the varieties produced will remain more resilient for a number of important agronomic traits (Rattunde et al., 2009). OSA recognizes that the organic market is accustomed to modern varieties bred for uniformity in key phenotypic traits. OSA has encouraged their farmer-breeder partners to select for a relatively high degree of uniformity in important market traits, while leaving phenotypic variation intact in agronomic traits where uniformity is not essential for their production practices (Dawson et al., 2008).

The challenge is to balance the genetic heterogeneity required for varietal resiliency with a high level of uniformity in the necessary traits. One of the strengths of PPB is that growers with first-hand experience in their production practices and the necessary market parameters for their crop can best determine which traits require uniformity and to what degree. The vegetable grower that markets directly or is in close contact with the distributor is also well suited to challenge the market standards and influence the perceptions of people who are selling, preparing, and eating the crops.

Objectives
The ideotype is a dark green, cylindrical zucchini, 15 to 19 cm long with faceted sides and a length to diameter ratio of approximately 3.5:1. The plant must have an open, compact determinate habit with relatively few leaves for ease of harvest. Plants must produce an early, steady harvest and have petioles with small spines to minimize scratching of the fruit as it develops. On the Eel River Produce farm a zucchini must quickly establish a strong, vigorous taproot as the farmer does not irrigate and practices a dryland farming technique where the crop is seeded with spring precipitation and must reach the water table before the weather turns to the seasonal dry summer climate.

The beginning of this project traces to farmers' inability to acquire adequate quantities of seed of the preferred dark green zucchini variety, 'Raven' F_1, over several seasons. He then augmented his limited Raven planting with 'Black Beauty' (BB), an older, open-pollinated (OP) zucchini variety. While BB is a dark fruited type, he found it to have much phenotypic variation, both in its leafy, closed plant canopy and with fruit that is often bulbous or curved. Only 30 to 40% of the plants had marketable fruit. However, BB presented vigorous growth and an extended harvest period, several weeks longer than Raven.

The farmer was motivated to harvest seed from the best BB plants having (1) a reliable source of seed of a variety adapted to his system, and (2) a large quantity of seed so that he could drill extra seed when planting to compensate for losses from wireworms (*Limonius* spp.) and common ravens (*Corvus corax*) feeding on the freshly sown seed. The cost of hybrid Raven seed had made this practice prohibitive. The initial selection of BB was open-pollinated and it also crossed with Raven, which contributed good quality traits to this population. The farmer-breeder repeated mass selection (M) for five seasons, selecting both agronomic and pertinent market traits. Each season, he would drill the seed at ~6 times the normal rate and select in stages for the most vigorous, robust plants that would reach the water table the fastest. This initial breeding work resulted in a promising M_5 population with many plants containing multiple traits of the ideotype. OSA was then contacted to advise on the breeding of a commercially viable zucchini variety from this population.

Methods
The participatory breeding in this project began with a series of self-pollinations among selected plants of the farmer-breeder's BB × Raven M_5 population in YR 1 (see table 6.5.) From these, a series of S_1 families were evaluated and four half-sib families (S_1HS_1) were created in YR 2 for further testing and breeding. In YR 3 one of the four half-sib families (Family #41 S_1HS_1) exhibited surprising uniformity with a preponderance of plants with key traits found in Raven. Selected plants were allowed to intermate to form $\#41S_1HS_1M_1$. From YR 4 through YR 6 this population was massed with stringent selection to the ideotype for farmer-breeders' market and agronomic traits for his production system, resulting in $\#41S_1HS_1M_4$ in YR 6.

Results
This PPB project has been a success in three regards. First, the farmer-breeders' original selection of superior breeding populations, his intensity of selection, and the numerous cycles of selection and recombination before the first self-pollinations generated were instrumental in breaking undesirable linkages in BB and increasing the favorable recombinants in the population (Allard, 1960). Second, the resultant population variety by YR 5 was being used for commercial production both in the target area of farmer-breeders' farm and for winter organic production in Mexico. Third, the interaction between the farmer-breeder and the formal breeder has been well balanced and successful. The farmer has taken the lead on the choice of germplasm, the selection of traits, and selection of individuals for the breeding population. The formal breeder determined the breeding methodology and overall strategy.

One of the primary objectives was to retain adequate levels of allelic variability to insure genetic elasticity in the final variety. While this objective seems to have been achieved in the primary population, $\#41S_1HS_1M_4$, the farmer-breeder has put most of his energy into further developing the $\#13S_1M_3$ sub-population for the last three seasons. This tendency to rely on the more narrowly selected and most refined version of the ideotype by a breeder, without considering the full scope of its adaptive capacity, is what usually happens in most centralized breeding programs. All breeders tend toward the most uniform products of a breeding program, but they need to ask the question, "Which type of variety will give the greatest adaptive advantage over time?" In a recent very wet, cool summer the more diverse primary population proved to have more vigor and perform more reliably than the derived $\#13S_1M_3$ in field trials.

The primary zucchini population from this work, $\#41S_1HS_1M_4$, exhibits a high degree of uniformity in fruit qualities and plant architecture, yet contains a degree of variability in several phenotypic traits. By maintaining intra-varietal heterogeneity it is possible to exploit residual heterosis, which

Table 6.5 Methods used in the North American participatory approach to breeding a genetically resilient zucchini variety

Year	Field Layout	Breeding Method	Breeding Population	Resulting Progeny
YR 1	Commercial planting of farmer-breeders' BB X Raven M_5 population	Series of self-pollinations attempted in Reynolds' BB X Raven M_5 population	26 successful self-pollinated fruits produced	26 first cycle S_1 families
YR 2	Planted 26 S_1 family progeny rows (~60 plants in each)	Progeny selection; allowed best plants of the 4 superior S_1 progeny rows to inter-mate	Open-pollinated (OP) mass of 4 superior S_1 families	Seed harvested separately as bulk from each of 4 half-sib families (S_1HS_1)
YR 3	2 replicates of 4 S_1HS_1 bulk families (~200 plants in each) planted in RCBD	Evaluated the 4 S_1HS_1 families and eliminated 3 most undesirable families	OP mass of selected plants of single S_1HS_1 progeny row family (#41)	Single half-sib family bulk population; #41$S_1HS_1M_1$
YR 4	Planted 0.4 ha plot of #41 $S_1HS_1M_1$	1) Mass selection of #41$S_1HS_1M_1$ plants 2) Second cycle of selfing initiated	1) OP mass of selected #41$S_1HS_1M_1$ plants 2) 16 successful self-pollinated fruits produced	1) Bulk of selected OP #41$S_1HS_1M_2$ 2) 16 second cycle S_1 families
YR 5	Planted 0.4 ha plot of #41$S_1HS_1M_2$ 2) Plant 16 S_1 family progeny rows (~60 plants in each)	1) Mass selection of #41$S_1HS_1M_2$ plants 2) Evaluated 16 S_1 families and eliminate 15 inferior families	1) OP mass of selected #41$S_1HS_1M_2$ plants 2) OP bulk of selected plants of single S_1HS_1 progeny row family (#13)	1) Bulk of selected OP #41$S_1HS_1M_3$ 2) Single half-sib family bulk pop.; #13S_1M_1
YR 6	1) Planted 0.25 ha plot of #41$S_1HS_1M_3$ 2) Planted 0.25 ha plot of #13S_1M_1	1) Mass selection of #41$S_1HS_1M_3$ plants 2) Mass selection of #13S_1M_1 plants	1) OP mass of selected #41$S_1HS_1M_3$ plants 2) OP mass of selected #13S_1M_1 plants	1) Bulk of selected OP #41$S_1HS_1M_4$ pop. 2) Bulk of selected OP #13S_1M_2 pop.
YR 7	Planted 0.25 ha plot of #13S_1M_2	Mass selection of #13S_1M_2 plants	OP mass of selected #13S_1M_2 plants	Bulk of selected OP #13S_1M_3 pop.
YR 8	1) Planted 0.25 ha plot of #41$S_1HS_1M_4$ 2) Planted 0.25 ha plot of #13S_1M_3	1) Series of selfing of #41$S_1HS_1M_4$ pop 2) Series of selfing of #13S_1M_3	1) 10 successful selfs of #41$S_1HS_1M_4$ produced 2) 12 successful selfs of #13S_1M_3 produced	A total of 22 third cycle S_1 families from #41$S_1HS_1M_4$ and #13S_1M_3 populations

is expressed when multiple hybrid matings occur between distinct individuals in the population (Shull, 1908).

Farmer-breeders' zucchini is currently produced on his farm and for winter markets on 12 ha in Baja California, Mexico. The farmer plans to produce it on 20 ha in Mexico. This zucchini variety (1) produces high quality, marketable fruit, (2) appears to resist downy mildew (*Pseudoperonospora cubensis*), (3) produces fruit for a longer harvest period than Raven, (4) has an open, compact habit that facilitates harvest, and (5) has petioles with negligible spines to damage fruit. He also reports that the plants of his variety are stockier, with a thicker, stronger main stem than other commercial varieties, which is particularly important for production in Baja California where winter winds can result in scratched fruits from the spines and branches scraping against young fruit. Notably, the increased resistance to downy mildew and the thick, strong central stem of this type of variety were not traits that were selected for directly, but were secondary characters that were possibly unconsciously selected.

There is no formal variety registry system in the United States or Canada that would limit use or commercialization of this variety. This project may continue by the possible future varietal distribution through participatory seed diffusion (Rios, 2009) via a farmers' seed cooperative in North America.

French Examples of Organic Durum Wheat PPB

Introduction: Program Genesis

Organic durum wheat produced in France does not match the requirements of pasta processors. Organic farmers asked geneticists from INRA (National Institute for Agronomic Research) for access to older genetic resources. A PPB program was implemented with farmers in two areas located in southern France: Pays Cathare and Camargue. The objective was to build a research-action, involving not only the farmers but also all the actors of the organic durum wheat chain (producers, collectors, processors, consumers) to take into account their diverse objectives, constraints, and skills.

The First Step

The first step was to collectively evaluate the demands of the whole supply chain. During the first four years, the partners conducted a participatory diagnosis and developed guidelines to evaluate durum wheat lines according to the needs of the classic market (semolina and pasta industrialist producers). Participatory diagnosis revealed lock in at the genetic level (no breeding for organic agriculture in France), economical level (organic is considered as a niche market, not economically viable for private breeders), and legislative level (registration criteria in the national catalog are not suitable for OA). Actors of the project first conducted an agronomic evaluation of pure lines, mixtures, and populations offered by geneticists, at both INRA and on organic farms. Field visits on a regular basis and especially at key crop stages allowed informal exchanges on varietal diversity. Farmers, researchers, and processors communicated opinions orally and in tabular form. The novel genetic variability on display at the research station created new debates about previously unanticipated ideotypes that might have application under organic production. The actors collectively established the objectives and main breeding criteria with broad identification and understanding of the needs of each actor. Subjective traits such as taste, aroma, appearance, texture, and other criteria important to organic farmers and not usually integrated in conventional program (tilling vigor, weeds competition) as well as those that are negatively associated with OA (e.g., stem height) were considered. A broad inquiry conducted by local producers' organizations integrated questions on agroecosystems and farmers' preferences and permitted formulation of an "ideotype." In Camargue, the high soil saline levels lead to "rice–durum wheat" rotations that induce nitrogen deficiency. Producers wanted varieties able to remobilize nitrogen from vegetative parts to the grain. In Pays Cathare, the main limiting factor was weed competition. Weed-competitive varieties must have vigorous root systems that are able to withstand cultivation. It became apparent that additional effort was required to refine guidelines and ideotypes adapted to different situations. Sociologists and agronomists were also introduced to the project. Some exchanges with researchers and farmers from Africa and working in a PPB project contributed to defining the objectives and modalities of the organic durum wheat PPB program.

The Second Step

The second step was breeding for a diversification of cropping systems, products, and markets. During the four following years, the goal was to evolve from participatory evaluation to PPB. Two important events permitted evolution of the project: (1) a seminar where agronomists, geneticists, statisticians, economists, sociologists, anthropologists, historians, and legislative representatives exchanged opinions on the role of PPB in the evaluation and interpretation of G×E interactions, and (2) a study travel to Sardinia where farmers and researchers examined the diversity of durum wheat uses and the importance of artisan production chains.

The goal of breeding was associated with the goal of diversification of cropping system and markets. This evolution allowed (1) a direct contribution of farmers to the creation of adapted varieties, (2) market diversification to build equity between actors and to develop territories, and (3) exchanges and co-learning among actors of different cultural practices. New actors were involved including researchers in ecophysiology and technology, artisanal processors, cooperatives, as well as consumers. New objectives accounted for diverse outlets (fresh pasta, durum wheat bread, and cookies), different market scales, and new territories.

Each actor agreed that genetic progress must be an aim as well as social or ethical progress. Several analyses were conducted including an agronomic evaluation of biophysical environments and evaluation of farmers cultural practices based on the observation of genotypes able to reveal the frequency and the nature of the main limiting factors (Nolot, 1994; Desclaux et al., 2008).

Two distinct and complementary chains were identified. One, the "long" chain, mobilized a network of organic farms and experimental stations to evaluate the quality of varieties to produce organic dry pasta. A pure line (tall stem, long spikes, and large seeds) was selected and proposed for registration. Farmers from diverse territories were federated into an organic producer's organization to oversee maintenance and multiplication of this line. The other ("short") chain involved farmers at the first steps of the breeding scheme. Heterogeneous populations of durum wheat were adapted to diverse farm environments and the first fresh pasta was realized by pasta farmers and marketed via direct sales.

The Third Step
The third step was to act to recognize ethical systems of innovation and commercialization.

Legislative Frame for Organic Variety. French legislation prevents the production and dissemination of varieties that are not registered in the catalog. The registration implies the evaluation of the variety for DUS (Distinctness, Uniformity, stability) and VCU (Value for Cultivation and Uses) into a national network. The U of DUS prevents the registration of heterogeneous populations that can, however, be very relevant for organic conditions (Wolfe, 2008). Until 2009, VCU criteria were evaluated only under conventional conditions and targeted only the demands of dominant market. Varieties intended for organic production and for alternative markets (direct selling, hand made production, etc.) could not be registered. Facing this situation, the actors asked the French Government for testing of the PPB-derived durum wheat line in a national organic network. After three years of demand, the government allowed a special network of four "low input" locations and one organic location.

Ultimately, the line was not registered, but the process led to:

– An awareness of the need to allow evolution in the registration system for innovation to occur.
– Establishment of new criteria for variety evaluation.
– The implementation of an organic registration network and integration of new environmental characteristics.

Legislative Frame for PPB Varieties. The actors of the project envision a legislative framework for collective breeding and question the notion of intellectual property around varieties. Two points were tackled: Who can breed and what is the regulatory framework for the recognition of the involved actors?

A farmer may innovate to fulfill his needs, but problems arise when a decision about monetary compensation for his innovation or his variety must be made. We confronted questions not only about regulations (e.g., certificate of vegetal creation; whether to consider variety as a common good), but also about economic feasibility (who will multiply the seeds, maintain the stock, and disseminate the variety, and with what financial means?).

Partners Until the Final Steps (Multiplication and Dissemination of Variety). The need for diversity is not limited today to only new varieties but also to the integration of the process of evaluating for a diversity of functions and traits. Multifunctional varieties are needed for landscape enhancement (via color), health contribution (via nutriments), and equilibrium of agroecological system (e.g., via mycorrhizal infection, competition, soil remediation) and socio-economic benefits (via family seed companies). The PPB program has been constantly evolving, such as actors from fair breeding working jointly with others local actors (e.g., consumers) within the Northern countries (Chiffoleau and Desclaux, 2008).

Outcomes of the French PPB Durum Wheat Program
By being decentralized and pluri-participatory, this PPB program accounted for the diversity and complexity of organic and low input farming systems, by creating varieties specifically adapted not only to the biophysical environment (climate, soil) but also to the agronomical, economical, and social components of the environment. The difficulties: a close multidisciplinary collaboration is fundamental to an in-depth analysis and management of the PPB program. Sometimes, requested actors were unavailable, especially over time. It was also destabilizing to be working at the limits of one's disciplinary field. It was also difficult to integrate the diversity of individual knowledge into a collective and durable dynamic. It takes time to confront new knowledge and understand its nuances, to harmonize that knowledge collectively, to evolve the methods, and to collectively share knowledge. Learning takes time, because PPB supports an evolution from researcher to farmer driven activities.

General Conclusions and Limits of PPB Approaches in Organic Farming

There is a much richer history of established programs utilizing decentralized participatory approaches to breeding for low-input, heterogeneous systems in the South than in the North. The South's success is partly due to the cultural integration of seed management as a key aspect of subsistence farming systems and the lack of financial interest to fulfill the diverse needs of marginal agricultural environments. It is only in recent years that a similar approach has been adopted in North countries where the established seed industry failed to meet the requirements of low-input organic systems. Much is to be learned from the success of models developed in the South that may be applied to parallel economic and environmental challenges in the North. However, difference in the cultural, economic, and physical environments must also be taken into consideration in developing an effective program in the North. Three keys to success in PPB projects include a strong farmer involvement, breeding in the target environment and for market constraints, and maintaining or even enhancing genetic diversity and variability in traits that are not essential for uniformity in production and markets.

One of the most important lessons learned from PPB projects is the valuable role the farmer may play in evaluating and selecting individuals from segregating type of varieties grown under a particular region or farm with low external inputs. An experienced farmer's knowledge of how the crop interacts with the challenges of the environment is invaluable in the selection process. Experienced farmers can often interpret the subtleties of the G×E interaction of a crop under the fluctuations of the environment. To some degree this enables the farmer to be able to distinguish between phenotypes in environments where traits would normally have very low heritability.

Organic farming systems, like low-input marginal systems in the South, have a limited number of crop management input options to mitigate pest and fertility challenges. These farms, when outside the centralized production areas, are also characterized by a high degree of heterogeneity in the environment. For these reasons, the productive capacity of these diverse, low-input organic systems needs resilient crop and yield stability. Both may come from varieties bred under the target environment, maintaining a greater degree of heterogeneity than those bred in conventional, centralized systems.

Variability of organic farming systems is so high that developing a variety fitting all situations is not conceivable. Because they are aware of the breeding cost necessary to meet several objectives and also to develop several locally adapted varieties, private breeding companies do not want to enter the organic seeds market. But considering an approach like PPB, we can imagine without additional costs developing varieties adapted to an area, a region, or a specific environment and why not at the farmer-field scale? For these reasons, PPB appears to be a more suitable solution for organic conditions than formal breeding.

Moreover, compared to conventional breeding, PPB seems to be the best alternative to fit the principle aims of organic agriculture for production and processing prescribed by IFOAM, and especially: "(1) to maintain and conserve genetic diversity through attention to on-farm management of genetic resources, (2) to recognize the importance of, and protect and learn from, indigenous knowledge and traditional farming systems."

Indeed, because breeding for organic conditions means breeding for sustainability, the process of breeding is as important as the results. Therefore, breeding process must comply with the three following criteria for organic production: Minimized off-farm inputs, natural self-regulation, and agro-biodiversity.

According to Lammerts van Bueren et al. (1999), equivalent criteria at the socio-economic level are: Close interaction between farmers, trade, industry, and breeders, and regulations geared to organic agriculture and cultural diversity.

PPB is an interesting alternative to answer the new exigencies of agriculture aiming for high economical and environmental performance, and it contributes further by integrating the social component.

References

Allard, R.W. 1960. Principles of plant breeding. New York: John Wiley & Sons.

Almekinders, C., and J. Hardon (eds.). 2006. Bringing farmers back into breeding. Experiences with participatory plant breeding and challenges for institutionalisation. Agromisa Special 5. Wageningen: Agromisa.

Atlin, G.N., M. Cooper, and A. Bjornstad. 2001. A comparison of formal and participatory breeding approaches using selection theory. *Euphytica* 122:463–475.

Barnaud, A., H. Joly, D. McKey, C. Khasah, S. Monné, and E. Garine. 2008. *In situ* management of sorghum (Sorghum bicolor ssp. bicolor) genetic resources among Duupa farmers in northern Cameroon: Selection and seed exchange. *Cahiers Agricultures* 17:178–182.

Bellon, M.R. 2006. Crop research to benefit poor farmers in marginal areas of the developing world: A review of technical challenges and tools. *CAB Reviews: Perspectives in Agriculture, Veterinary Science, Nutrition and Natural Resources*. Online, ISSN 1749–8848.

Caillon, S., J. Quero-Garcia, J.P. Lescure, and V. Lebot. 2006. Nature of taro (*Colocasia esculenta* L. Schott) genetic diversity prevalent in a Pacific island, Vanua Lava, Vanuatu. *Genetic Resources and Crop Evolution* 53:1273–1289.

Cauderon, A. 2003. Un cas d'école dans l'accueil d'une innovation: les OGM. *Semences et Progrès* 115:15–18.

Ceccarelli, S. 2009. Selection methods. Part 1: Organizational aspects of a plant breeding programme. *In:* Ceccarelli, S., E.P. Guimaraes, and E. Weltzien (eds.) Plant breeding and farmer participation. pp. 63–74. Rome: FAO.

Ceccarelli, S., and S. Grando. 1997. Increasing the efficiency of breeding through farmer participation. *In*: Ethics and equity in conservation and use of genetic resources for sustainable food security, pp. 116–121. Proceedings of a workshop to develop guidelines for the CGIAR, April 21–25, 1997, Foz de Iguacu, Brazil. Rome, Italy: IPGRI.

Ceccarelli, S., S. Grando, R. Tutwiler, J. Baha, A.M. Martini, H. Salahieh, A. Goodchild, and M. Michael. 2000. A Methodological Study on Participatory Barley Breeding. I. Selection Phase. *Euphytica* 111:91–104.

Ceccarelli S., S. Grando, E. Bailey, A. Amri, M. El Felah, F. Nassif, S. Rezgui, and A. Yahyaoui. 2001. Farmer participation in Barley breeding in Syria, Morocco and Tunisia. *Euphytica* 122:521–536.

Ceccarelli, S., and S. Grando. 2007. Decentralized-participatory plant breeding: An example of demand driven research. *Euphytica* 155:349–360.

Ceccarelli, S., S. Grando, and M. Baum. 2007. Participatory plant breeding in water-limited environment. *Experimental Agriculture* 43:411–435.

Ceccarelli, S. 2009. Selection methods. Part 1: Organizational aspects of a plant breeding programme. *In:* S. Ceccarelli, E.P. Guimarães, and E. Weltzien (eds.) Plant breeding and farmer participation. pp. 259–273. Rome: FAO.

Ceccarelli, S., A. Galié, Y. Mustafa, and S. Grando. 2011. Chapter 6. Syria: Participatory barley breeding — farmers' input becomes everyone's gain. *In*: Manuel Ruiz and Ronnie Vernooy (eds.) The custodians of biodiversity: Sharing access and benefit sharing of genetic resources. London and Sterling: Earthscan.

Chable, V., and J.F. Berthellot. 2006. La sélection participative en France: Présentation des expériences en cours pour les agricultures biologiques et paysannes. *Le courrier de l'environnement* 30. pp. 129–138. INRA.

Chiffoleau, Y., and D. Desclaux. 2006. Participatory plant breeding: The best way to breed for sustainable agriculture? *International Journal of Sustainable Agriculture* 4:119–130.

Chiffoleau, Y., and D. Desclaux. 2008. La sélection participative pour un commerce éthique en agriculture: L'exemple de la filière blé dur biologique dans le sud de la France. 3eme colloque international sur le commerce équitable, Montpellier, May 14–16, 2008, France.

Dawson, J.C., K.M. Murphy, and S.S. Jones. 2008. Decentralized selection and participatory approaches in plant breeding for low-input systems. *Euphytica* 160:143–154.

Desclaux, D. 2009a. Diversité et complémentarité des approches de la sélection pour l'AB Séminaire Techniques de sélection végétale, compatibilité avec l'AB et perspectives-ITAB-Paris, 28–29 Avril 2009.

Desclaux, D. 2009b. La sélection participative pour élaborer des variétés de blé dur pour l'agriculture biologique. Colloque CIAG – Variétés innovantes et modes de culture adaptés pour une agriculture durable'. 5 Nov 2009 Angers.

Desclaux, D., Y. Chiffoleau, F. Dreyfus, and J.C. Mouret. 2003. Cereal cultivar innovations adapted to organic production: A new challenge. *In*: Lammerts van Bueren, E.T. and K.P. Wilbois (eds.) Organic seed production and plant breeding – strategies, problems and perspectives. Proceedings of ECO-PB 1st international symposium on organic seed production and plant breeding in Berlin, Germany, November 21–22, 2002, European Consortium for Organic Plant Breeding, Driebergen, Frankfurt.

Desclaux, D., J.M. Nolot, Y. Chiffoleau, E. Gozé, and C. Leclerc. 2008. Changes in the concept of genotype × environment interactions to fit agriculture diversification and decentralized participatory plant breeding: Pluridisciplinary point of view. *Euphytica* 163:533–546.

Desroches, H. 1976. Le Projet coopératif. Son utopie et sa pratique, Ses appareils et ses réseaux. Ses espérances et ses déconvenues, Éd. Ouvrières.

Franzel, S.R., R. Coe, F. Cooper, F. Place and S.J. Scherr. 2001. Assessing the adoption potential of agroforestry practices in sub-Saharan Africa. *Agricultural Systems* 69:37–62.

Friedberg, E. 1988. L'analyse stratégique des organisations. Pour, numéro spécial, Cahier 28.

Gallais, A. 2006. Préface. In: Lançon, Floquet, Weltzien (eds.) Partenaires pour construire des projets de sélection participative. Actes de l'atelier. Cotonou, Bénin, March 14–18, 2005. Cirad; INRAB.

Gozé, E. 1992. Détermination de la dimension des réseaux d'essais. *Cotton et fibres Tropicales* 47:81–94.

Grube, R. 2007. Breeding for organic and sustainable systems: One size does not fit all. *HortScience* 42:813.

Halewood, M., P. Deupmann, B. Sthapit, R. Vernooy, and S. Ceccarelli. 2007. Participatory plant breeding to promote Farmers' Rights. Rome, Italy: Biodiversity International.

Hofs, J.L., M. Fok, M. Gouse, and J.F. Kirsten. 2006. Diffusion du coton génétiquement modifié en Afrique du Sud: Des leçons pour l'Afrique zone franc. *Revue tiers monde* 43:799–823.

Imaizumi, A., and S. Hisano. 2010. Farmers' seed system and the institutionalisation of genetic resources use and management in agriculture: a case of seed supply for local traditional vegetables in Japan. ISDA 2010, Montpellier, France. Accessed April 20, 2011. http://hal.archives-ouvertes.fr/hal-00522971/en/.

Jones, M., and M. Wopereis-Pura. 2004. Development of agriculture and food production in Africa: WARDA's success in technology development and dissemination. *In:* Sperling, L., J. Lançon, and M. Loosvelt (eds.) Participatory breeding and participatory plant genetic resources enhancement. An Africa-wide exchange. Proceedings of the Symposium Participatory Research and Gender Analysis Program (PRGA), Mbé, Cote d'Ivoire, Cali, Colombia, May 7–10, 2001, Cirad. Africa Rice Center (WARDA).

Joshi, K.D., and J.R. Witcombe. 1996. Farmer participatory crop improvement. II. Participatory varietal selection: A case study in India. *Experimental Agriculture* 32:461–477.

Lammerts van Bueren, E.T., M. Tiemens-Hulscher, M. Haring, J. Jongerden, J.D. van Mansvelt, A.P.M den Nijs and G.T.P. Ruivenkamp. 1999. Sustainable organic plant breeding. Final report: a vision, choices, consequences and steps. Driebergen, The Netherlands: Louis Bolk Institute.

Lançon, J., M. Djaboutou, S. Lewicki, and E. Sêkloka. 2004. Decentralised and participatory cotton breeding in Benin: Farmer-breeders' results are promising. *Experimental Agriculture* 40:419–431.

Lançon, J., and A. Mucchielli. 2006. The panoramic table: A contribution to strategic and prospective analysis of a plant breeding situation. *Experimental Agriculture* 42:229–250.

Machado, A.T., and M.S. Fernandes. 2001. Participatory maize breeding for low nitrogen tolerance. *Euphytica* 122:567–573.

McGuire, S.J, G. Manicad, and L. Sperling. 1999. Technical and institutional issues in participatory plant breeding – done from a perspective of farmer plant breeding. CGIAR Systemwide Programme on participatory research and gender analysis for technology development and institutional innovation working Doc. 2. CGIAR.

McGuire, S.J. 2008. Path-dependency in plant breeding: Challenges facing participatory reforms in the Ethiopian sorghum improvement program. *Agricultural Systems* 96:139–149.

Morris, M.L., and M.R. Bellon. 2004. Participatory plant breeding research: Opportunities and challenges for the international crop improvement system. *Euphytica* 136:21–35.

Nolot, J.M. 1994. Parcours d'élaboration du rendement. *In:* INRA (ed.) CR Réunion Sci.gpe céréales, Dijon, March 1994. France: INRA.

Pecqueur, B. 2007. L'économie territoriale: Une autre analyse de la globalisation. *L'économie politique* 33:41–52.

Rattunde, F., K. vom Brocke, E. Weltzien, and B.I.G. Haussmann. 2009. Selection methods. Part 4: Developing open-pollinated varieties using recurrent selection methods. *In:* S. Ceccarelli, E.P. Guimaraes, and E. Weltzien (eds.) Plant breeding and farmer participation. pp. 63–74. Rome: FAO.

Reynolds, M.P., and N.E. Borlaug. 2006. Applying innovations and new technologies for international collaborative wheat improvement. *Journal of Agricultural Science* 144:95–110.

Rios, H. 2009. Participatory seed diffusion: Experiences from the field. *In:* S. Ceccarelli, E.P. Guimaraes, and E. Weltzien (eds.), Plant breeding and farmer participation. pp. 589–612. Rome: FAO.

Rubyogo, J.C. 2005. Renforcer les capacités des systèmes semenciers pour accélérer l'adoption des variétés améliorées de haricot en Afrique de l'Est, centrale et australe. *In:* J. Lançon, A. Floquet, and E. Weltzien (eds.), Partenaires pour construire des projets de sélection participative. pp. 167–170. Actes de l'atelier. Cotonou, Bénin, March 14–18, 2005. Cirad; INRAB.

Shull, G.H. 1908. The composition of a field of maize. *American Breeders Association Report* 4:296–301.

Sperling, L., M.E. Loevinsohn, and B. Ntabomvura. 1993. Rethinking the farmer's role in plant breeding, local bean experts and on-station selection in Rwanda. *Experimental Agriculture* 29:509–519.

Sperling, L., J.A. Ashby, M.E. Smith, E. Weltzien, and S. McGuire. 2001. A framework for analyzing participatory plant breeding approaches and results. *Euphytica* 122:439–450.

Sthapit, B.R., K.D. Joshi, and J.R. Witcombe. 1996. Farmer participatory crop improvement. III. Participatory plant breeding: A case study for rice in Nepal. *Experimental Agriculture* 32:479–496.

Thiele, G. 1999. Informal potato seed systems in the Andes: Why they are important and what we should do with them. *World Development* 27:83–99.

Thiele, G., G. Gardner, R. Torrez, and J. Gabriel. 1997. Farmer involvement in selecting new varieties: Potatoes in Bolivia. *Experimental Agriculture* 33:275–290.

Thro, A.M., and C. Spillane. 2000. Biotechnology-assisted participatory plant breeding: Complement or contradiction? Working document CGIAR, 4. April 2000. CGIAR.

Toomey, G. 1999. Farmers as researchers: The rise of participatory plant breeding. IDRC Project number 950019. Canada: IDRC.

Trouche, G., K. vom Brocke, S. Aguirre, and Z. Chow. 2009. Giving new sorghum variety options to resource-poor farmers in Nicaragua through participatory varietal selection. *Experimental Agriculture* 45:451–467.

Trouche, G., S. Aguirre Acuña, H. Hocdé, R. Obando Solís, N. Gutiérrez Palacios, and Z. Chow Wong. 2008. Valorisation de la diversité génétique du sorgho par des approches de sélection participative au Nicaragua. *Cahiers Agricultures* 17:154–159.

Trouche, G., S. Aguirre Acuña, B. Castro Briones, N. Gutiérrez Palacics, and J. Lançon. 2011. Comparing decentralized participatory breeding with on-station conventional sorghum breeding in Nicaragua: I. Agronomic performance. *Field Crops Research* 121:19–28.

Vaksman, M., M. Kouressy, A. Toure, M. Coulibaly. 2005. Troisième cas: Valorisation de la diversité génétique des sorghos en zone cotonnière du Mali grâce à la sélection décentralisée et participative. In: Lançon, J., A. Floquet et E. Weltzien (eds.) Partenaires pour construire des projets de sélection participative. pp. 39–48. Actes de l'atelier, Cotonou, March 14–18, 2005. Cirad; INRAB.

Vernooy, R. 2003. Seeds that give: Participatory plant breeding. Canada: IDRC.

Vinck, D. 1999. Les objets intermédiaires dans les réseaux de coopération scientifique. Contribution à la prise en compte des objets dans les dynamiques sociales>>. *Revue Française de Sociologie* vol. XL(2):385–414.

vom Brocke, K., A. Christinck, E. Weltzien, T. Presterl, and H.H. Geiger. 2003. Farmers' seed systems and management practices determine pearl millet genetic diversity patterns in semiarid regions of India. *Crop Science* 43:1680–1689.

Weltzien R.E., M.L. Whitaker, H.F.W. Rattunde, M. Dhamotharan, and M.M. Anders. 1998. Participatory approach in pearl millet breeding. *In*: Witcombe, J.R., D.S. Virk, and J. Farrington (eds.) Seeds of choice: Making the most of new varieties for small farmers. pp. 143–170. Oxford and IBH Publishing Co.

Weltzien, E., M.E. Smith, L. Meitzner, and L. Sperling. 2000. Technical and institutional issues in participatory plant breeding – from the perspective of formal plant breeding. A global analysis of issues, results, and current experience. Working Document No. 3. CGIAR Systemwide Program on Participatory Research and Gender Analysis for Technology Development and Institutional innovation.

Weltzien, E., A. Christinck, M. Ag Hamada, A. Toure, and H.F. Rattunde. 2005. Quatrième cas: Améliorer l'accès des paysans maliens aux variétés de sorgho grâce à la sélection participative. In: Lançon, J., A. Floquet, and E. Weltzien (eds.) Partenaires pour construire des projets de sélection participative, pp. 43–54. Actes de l'atelier. Cotonou, Bénin, March 14–18, 2005. Cirad; INRAB.

Witcombe, J.R., A. Joshi, K.D. Joshi, and B.R. Sthapit. 1996. Farmer participatory crop improvement I: Varietal selection and breeding methods and their impact on biodiversity. *Experimental Agriculture* 32:445–460.

Witcombe, J.R., and D.S. Virk. 2001. Number of crosses and population size for participatory and classical plant breeding. *Euphytica* 122:451–462.

Witcombe, J.R., A. Joshi, and S.N. Goyal. 2003. Participatory plant breeding in maize: A case study from Gujurat, India. *Euphytica* 130:413–422.

Wolfe, M.S., J.P. Baresel, D. Desclaux, I. Goldringer, S. Hoad, G. Kovacs, F. Löschenberger, T. Miedaner, H. Østergård, and E.T. Lammerts Van Bueren. 2008. Developments in breeding cereals for organic agriculture. *Euphytica* 163:323–346.

Zimmermann, M.J. de O. 1996. Breeding for marginal/drought prone areas in Northeastern Brazil. *In*: P. Eyzaguirre and M. Iwanaga (eds.) Participatory plant breeding, Proceedings of a workshop, pp. 117–122. Wageningen, July 26–29, 1995. IPGRI.

7 Values and Principles in Organic Farming and Consequences for Breeding Approaches and Techniques

Klaus P. Wilbois, Brian Baker, Maaike Raaijmakers, and Edith T. Lammerts van Bueren

Introduction

Organic agriculture is a movement continuously adapting to developments in agriculture and society. Plant breeding and seed production are among the latest areas being discussed and developed in the organic sector. Plant breeding programs to enhance disease resistance and increase yield were considered to benefit conventional and organic farming systems alike. Since the 1970s, organic farmers and researchers have organized variety trials to identify the best existing varieties suitable for organic farming systems from an ecological point of view. In these trials, traits such as yield stability, weed competitiveness, disease and pest resistance, and storage without post-harvest chemicals were evaluated (Lammerts van Bueren et al., 2002). A next step was to move away from chemical seed treatments that were applied prior to planting. The first step was accomplished by requesting seed companies to provide non-treated seeds, with the eventual goal of stimulating organic seed production. With the introduction of genetic engineering (GE) in the 1990s, organic farmers realized that some breeding techniques were inconsistent with the principles of organic agriculture. At the end of the twentieth century, the organic movement decisively rejected GE with unprecedented global solidarity (IFOAM, 2002a). As a result, the use of GE and its products and derivatives in organic farming and food processing was officially banned – first by the EU in 1999 with an amendment to EC 2092/91, followed by the United States in 2000 by the NOP final rule (7 CFR 205.105(e)) (Anonymous, 2000). The organic sector realized it had to more clearly define its future direction with respect to variety development to be compliant with the basic principles of organic agriculture (Lammerts van Bueren et al., 2003). International Federation for Organic Agricultural Movements (IFOAM) published its first draft standards for plant breeding in 2002 (IFOAM, 2002b).

This chapter describes why genetic engineering and other technologies are considered incompatible with organic principles. In this context, the future perspectives for the organic sector with respect to the development of breeding will be discussed.

Organic Crop Breeding, First Edition. Edited by Edith T. Lammerts van Bueren and James R. Myers.
© 2012 John Wiley & Sons, Inc. Published 2012 by John Wiley & Sons, Inc.

Arguments Against Genetic Engineering

The organic movement initiated debate on whether to accept or ban GE in the 1990s, as the first transgenic crops were developed (Youngberg, 1986). By the mid-1990s, a consensus emerged that GE should be banned (Bullard et al., 1994). Space does not permit an exhaustive review of the available literature, but the case against GE in organic farming can be grouped into three categories (Verhoog, 2007):

- Unpredictable health and environmental risks,
- socio-economic impacts, and
- incompatibility with basic values in organic agriculture.

Health and Environmental Risks

The first critical category concerns the risks posed by biotechnology for human health, food safety, and the ecosystem (Pimentel et al., 1989). The organic sector raised concerns that human health consequences, such as allergenicity, could not be predicted or adequately assessed (FAO/WHO, 2001). Even more difficult to predict were risks to the ecosystem in case of unintended gene transfer to wild plants (Tiedje et al., 1989). Another example of a concern in the organic sector was the use of a gene construct based on the *Bacillus thuringiensis* (Bt) gene, derived from soil bacteria in many crops: E.g., to manage the corn borer. Bt is foliar-applied in organic cropping systems as a natural pesticide against Lepidopteran and Coleopteran pests. Organic farmers in temperate zones apply Bt only a few times during the growing season and rely primarily on non-biocidal methods, while insects are exposed to the Bt toxin expressed by transgenic plants throughout the growing season. Organic farmers are concerned that constant exposure will sooner or later lead to accelerated resistance by target pests to Bt, thus increasing reliance on more toxic pesticides in the long run (Altieri, 2000; Shelton et al., 2002). Verhoog (2007) argues that from the organic sector's holistic point of view, the unpredictable environmental and human health risks of gene technology are inherent to the reductionist approach of genetic engineering. So even when scientific analysis would show low levels of risks or when science would make progress to reduce risks, the organic sector would still reject the use of GE based on the precautionary principle.

Socio-economic Aspects

The second category of critical arguments against GE is based on the fact that GE is used as a tool by industry to concentrate control over agriculture and the food supply. In many GE-producing countries consumers have limited access to GE-free food due to the dominant role of the food industry tied to biotech companies and the risks of contamination in the food chain. Farmers in these countries experience reduced availability (on the market) of conventional, non-GM seed (OSA, 2011). The organic sector is also concerned about the use of patents that eliminate breeders' rights, which have been introduced in the breeding sector as a consequence of GE techniques. Plant breeders' rights laid down in the international UPOV Convention and the European plant protection right include a so-called breeders' exemption (UPOV, 1991), which allows breeders to use protected varieties as parent lines in breeding programs to develop new varieties, and to use and exploit those new varieties without permission from the original holder of the breeders rights (Plantum NL, 2009). Patent law does not recognize this kind of breeder's exemption and thus inhibits the open exchange of germplasm and open innovation in plant breeding. The monopoly status given to a patent is a factor that increases the consolidation and power of breeding companies.

Incompatibility with Basic Values of Organic Agriculture

The first two categories of arguments are focused on the products and the consequences of applying GE. The technology itself is often believed to be value-free. The approach is very much consequential (utilitarian), i.e., focused on the usefulness and potential risks of the product, but the approach does not take the inherent worth of organisms into account (Verhoog et al., 2003; Verhoog, 2007). However, organic agriculture is a process-based standard and does not only consider the consequences (the product) but also has inherent concerns about the process of gene technology itself, based on the basic principles of organic agriculture (Lammerts van Bueren et al., 2007). Thus, the third category of arguments makes the case that GE as part of the breeding process is incompatible with the principles of organic agriculture.

By this reasoning, GE is considered a reductionist rather than holistic technology. When applied to plant breeding, plants are viewed merely as composed of genes that can be altered according the preference of mankind. As a tool, GE operates directly at the DNA level – and thus below the cellular level – at the very lowest entity of self-organizing life. By inserting synthetically produced gene constructs in the genome, the integrity of organisms is violated (Lammerts van Bueren et al., 2003, see Box 7.1). GE conflicts with the values and basic principles of organic agriculture, and this may be the most durable defense against GE, as health and environmental risks may be reduced as science and technology improves over time (Verhoog, 2007).

Organic Basic Principles

Basic Attitude toward Nature

The organic sector's critique of GE can be best understood in light of its view of nature (Verhoog et al., 2003). Kockelkoren (2005) distinguishes four basic attitudes toward nature: Ruler, stewardship, partner, and participant. From the ruler attitude, and even to a certain extent from a dominion or stewardship philosophy, humans are placed apart from and superior to nature. Alternatively, many organic farmers consider themselves as partners of nature and respect nature not only for its extrinsic value to produce food and other products, but also for its intrinsic value. According to bioethics, such a partner attitude leads toward a normative bio-centric framework of action instead of an anthropocentric framework that considers the ethical value only of humans, or a zoocentric framework that extends ethical value only to higher animals. A bio-centric framework values all living entities, including plants. When applied to agriculture a bio-centric framework accounts for both the extrinsic and intrinsic values of nature. Organic farming thus aims to manage ecosystems by cooperating with nature and intensifying ecological relationships rather then excluding or controlling nature (Verhoog et al., 2007). By contrast, GE is considered to violate the integrity of life, because it interferes with the self-regulative ability and autonomy of plants and does not respect the autonomous nature of living organisms.

> **Box 7.1. Concepts on intrinsic value and integrity of plants (adapted from Lammerts van Bueren et al., 2003)**
>
> *Intrinsic values of plants*
> Acknowledging the intrinsic value of plants indicates that one considers plants ethically relevant, and thus plants have a value of their own, independent of the instrumental or extrinsic values for man. Ethically relevant means that one takes the intrinsic value into account in one's decisions for actions, out of respect for the intrinsic value.
>
> It is a personal choice to assess not only humans or animals as ethically relevant but also plants. The organic agriculture is based on a partner attitude toward nature and departs from a bio-centric framework of action. In this context, cultivated and wild plants have an intrinsic value based on respect for their integrity and autonomy.
>
> *Integrity of plants*
> The integrity of plants takes into account the characteristic nature or way of being of living entities, their wholeness, completeness, their species-specific characteristics, and their being in balance with the species-specific environment. Taking integrity of plants seriously does not imply that one cannot interfere with the plant object. In organic agriculture, including plant breeding, one interferes in such a way that one interacts with nature in an ecological way and within the boundaries of life. Organic plant breeding will respect and not violate the autonomy and self-regulatory ability of plants, their abilities to reproduce and interact with the environment, and their reproductive barriers. Crossing techniques should allow pollination, fertilization, embryo growth, and seed formation on the (whole) plant. Selection should focus on plant types that can maintain themselves and also potentially complete their life cycle in an organic farming system.
>
> Breeding techniques beyond the cellular level destroy the natural context of a cell as lowest entity of life. However, the organic sector considers also techniques at cellular level to be outside the scope of the principles of the soil-bound organic production methods.

IFOAM Basic Principles

The organic sector has formulated its values in the following four basic principles: Health, ecology, fairness, and care (IFOAM, 2005). Each principle has implications for breeding and seed production in organic farming systems (Lammerts van Bueren, 2010).

Principle of Health
The principle of health in organic agriculture is framed in a positive, holistic way, taking the interrelationship of soil, plant, animal, and human health into account. Rather than view "health" as the absence of disease, strategies for organic plant breeding emphasize the resilience of crops and abilities to adapt to various, unpredictable, and unfavorable conditions without losing their yielding ability.

Applying the principle of health in organic breeding aims at respecting the wholeness and integrity of life including plants, supporting the immunity, autonomy, resilience of the agro-ecosystem, and its ability to regenerate and sustain. Organic farming is soil-bound, and organic growers require varieties that are robust and adapted to organic farming systems. Varieties should also strengthen the resilience of the farming system, and robust and adaptable varieties bred for horizontal disease resistance are preferable. While organic management can reduce disease pressure, selected plants need not be pathogen-free. In such cases, low levels of infection can be tolerable and horizontal resistance is desirable as long as the plants are able to produce a marketable yield (Robinson, 2007, see also Chapter 3). Ideally, continuous improvement and selective adaptation involves farmers in the reproduction of varieties under their own farming conditions. Varieties should thus be reproducible and not made male sterile resulting in a "dead end branch of a tree" and leading to a reduction in genetic diversity.

Principle of Care
At the heart of the principle of care is the precautionary principle (Wingspread Conference, 1988):

> The rationale behind the precautionary principle is that in organic farming the interaction between Nature and Man is an important ingredient of the philosophy.
> Organic farming builds on the concept that Nature is an integrated whole that people have a moral duty to respect, both for its intrinsic value and because, by using its regulatory mechanisms, one can establish a more self-sustaining agro-ecosystem. Nature is a very complex, coherent system, of which Man has often little understanding to appreciate the consequences of specific actions. Damage to Nature and the environment will ultimately damage Man. (DARCOF, 2000)

Holism in organic farming integrates scientific, ethical and cultural aspects, and places man in the role as part of, partner with, and participant in nature. Therefore, man may have the power to manipulate nature for economic objectives, but in organic farming this power must respect borders of intervention where intrinsic value of nature may be violated. From this point of view, GE transgresses such a border. Another violation of the precautionary principle is the asymmetry between how risks and potential profits in the biotech industry are distributed. Companies that own intellectual property rights (IPRs), especially patents, stand to reap the rewards of the technology but shield themselves from the external risks.

Indigenous knowledge is an integral part of the principle of care (Luttikholt, 2007). With plant breeding, there is a recognition that heritage breeds and heirloom varieties are the result of thousands of years of selection and breeding by farmers, and that modern breeding techniques are built upon this prior art. In situ/on farm conservation of genetic resources requires that these traditional varieties be grown out in quantities that can be planted back and shared among farmers and also with breeders.

Principle of Ecology
The principle of ecology aims to preserve and apply biodiversity to support optimal functioning of an assortment of site-specific ecological production systems. Based on an agro-ecological approach, humans are regarded as part of nature. Agriculture is a human activity that interacts in a holistic way with the self-organizing capacity of nature. Farmers who adopt an ecosystem approach are expected to think in more or less closed nutrient cycles, work in harmony with nature, seek ecological balance of their production with non-farm biological systems, and strive for resilience of their system.

With the increasing specialization and technological development within the breeding industry, plant breeding has shifted from farmers' fields to trial plots to greenhouses and laboratories. As a consequence, regionally determined plant by environment interaction that plays a crucial role in organic farming has diminished. Evolutionary processes allowed (e.g., cereal heterogeneous populations) to evolve on-farm over time are essential for organic farming systems and have to be applied in plant breeding to adapt to the continuous change in the agro-ecosystem (see also Chapter 5).

Biodiversity is one of the key themes of the principle of ecology. Three levels of biodiversity are involved: Intraspecific (diversity within a population/variety); interspecific (diversity among crop species), aiming at wide rotations with many crops species; and ecosystem-wide diversity. Intraspecific biodiversity is threatened by the loss of indigenous varieties in their centers of origin. Monocultures, the decline of seed saving, and increased use of IPRs to control access to germplasm have all resulted in a decline in available varieties – particularly open-pollinated varieties. Modern varieties rely on uniformity, which is the antithesis of biodiversity. Ecosystem biodiversity provides an assortment of communities of plants, animals, and soil organisms within a given zone (Mäder et al., 2002). Selection and breeding of plants that provide ecosystem services, like habitat for pollinators, predators, and parasitoids of pests is in its infancy (Suso et al., 2010). Genetic enhancement of beneficial habitat offers a promising approach for breeders to improve the ecology of organic farming systems.

Principle of Fairness
The fairness principle encourages equity, respect, justice and stewardship for the shared world, including food sovereignty. People have the right to determine their own food systems and food values, and the right to produce their own food (Luttikholt, 2007). From food sovereignty, it follows that farmers have the right to save, exchange, and sell their own seeds (FAO, 2010). At the same time, breeders' work needs to be appropriately protected and fairly compensated, while also maintaining breeders' exemptions to allow for use of each other's varieties in breeding programs to speed up progress and contribute to food security. The balance of plant breeders' rights (PBRs) and farmers' rights (FRs) poses one of the greater challenges of reconciling plant breeding with the organic principle of fairness.

Toward Organic Breeding

Organic plant breeding is a relatively new enterprize. While organic seed production and selection of varieties suitable for organic farming goes back to the 1970s in the United States and Europe, organic producers were not obliged to use organic propagated seed, let alone organically bred varieties. Although organic regulations from the start stated that organically produced seeds should be used when available, only since 2004 has the EU regulation made it more difficult to receive exceptions for use of conventional seed and the use and production of organic seed is now actively promoted in most EU member states.

As a consequence of the investment in organic seed production, several seed companies have also started to develop an interest in breeding for organic and other low-input farmers.

The bio-dynamic community – a well-established subgroup within the organic sector – has a longer history in stimulating breeding as they consider plant breeding as an indispensable tool in the whole concept of bio-dynamic agriculture, and several farmers have incorporated crop improvement as part of their farming system (http://www.kultursaat.org).

However, to facilitate discussion and awareness among the broader organic community, and to be able to communicate with breeders about future directions, clear criteria are needed to evaluate the existing breeding tools.

From Values to Criteria: Evaluation of Breeding Techniques

Following the ban on GE in organic production, farmers, breeders, researchers, and policymakers in Europe gathered in several workshops to discuss criteria by which to evaluate all current breeding techniques at three levels: Whole plant, cellular or in vitro level, and the DNA level (Lammerts van Bueren et al., 1998). The basis of the evaluation was key criteria derived from the principles of organic agriculture aiming at closing the production cycles and supporting the autonomous self-regulative ability or resilience of the farm agro-ecosystem. The aim was not only to apply the principle of diversity but also to produce agro-biodiversity (IFOAM, 2005; see table 7.1). Plant breeding should therefore support such farming systems by focusing on crops that can evolve within the organic farming system and are able to adapt to unpredictable or unfavorable growing conditions by exploiting and creating genetic diversity. Besides that, plant breeding is not merely a technical activity but also a socio-economic construction. Plant breeding should therefore include participatory approaches and should support regional diversity of breeding programs, rather than facing the current consolidation, to ensure maintenance of a broad genetic base.

The final report presented a comprehensive evaluation of acceptable plant breeding techniques with respect to organic farming (Lammerts van Bueren et al., 1999). Different breeding techniques were subsequently summarized and categorized in a separate review by Wyss et al. (2001). The conclusions of these two sources pointed out that organic agriculture needs a breeding system that complies with the ecological principles to account for the complexity and biodiversity of agro-ecological systems and which works at high levels of plant and farm organization. Organic researchers choose to conduct plant breeding on whole plant level, instead of cellular or DNA level. This supports the overall principle of organic agriculture that aims to support to life processes. Working on whole plant level allows the breeder to work within the context of life of a plant in connection to the soil.

Techniques at cellular level might still be considered to operate within the context of life, as cells are the lowest entity of self-organizing life. However, with in vitro techniques, the plants or part of the plants (e.g., embryos) are grown in the laboratory on artificial media and not in soil and are therefore considered as inappropriate and not complying with the organic principles.

DNA techniques violate the integrity of life as it operates below cellular level by dissolving the cell wall to be able to fuse cells for somatic hybridization, e.g., with cell or protoplast fusion. Or DNA is extracted from cells, multiplying genes synthetically and forcing them (at random) into the genome of the host plant, by-passing natural crossing barriers.

Organic farmers also want to respect the socio-economic principles, which are similar to ecological criteria, of organic agriculture in breeding activities. To achieve this, breeders and

Table 7.1 Key criteria derived from the organic farming system level and transformed to the crop/plant and socio-economic level (adapted from Lammerts van Bueren et al., 1999)

Farming system	Crop/plant	Socio-economic
Closed production cycles	Self-reproductive ability	Close interaction with farmers and other stakeholders
Autonomous self-regulative ability	Ability to adapt to variable growing conditions	Regulations adapted to organic principles
(Functional) agro-biodiversity	Enhancing genetic diversity	Cultural diversity represented by a diversity of breeders and breeding programs

farmers should maintain close ties and ideally work together toward the development of regional breeding systems.

With regard to suitable plant breeding techniques both publications concluded that:

- Crossing and selection techniques at the whole plant and crop level are most appropriate for organic plant breeding;
- Crossing techniques are appropriate in an organic plant breeding system, provided that pollination and seed-formation occur on the plant growing in soil;
- F_1 hybrids may have a role in organic farming, provided that the F_1 is fertile and that the parent lines can be maintained under organic growing conditions;
- Techniques at the cell level (in vitro techniques) are generally considered inappropriate for organic breeding, as they operate beyond the context of organic production systems;
- Genetic engineering is not allowed in organic sector. It operates below cellular level and no longer in the direct context of life, and therefore violates the integrity of plants;
- DNA diagnostic methods such as molecular marker-assisted selection can, in principle, supplement other selection methods used in organic plant breeding, such as selection based on morphological traits, or disease screening tests with inoculation of the pathogen, or by protein detection in baking quality tests.

Table 7.2 summarizes the available plant breeding techniques at three different levels (whole plant, cellular or in vitro, and DNA level), which are proposed to be or not to be consistent with organic principles.

Although it is clear that GE-based varieties are excluded, there are still other non-preferred techniques such as in vitro techniques and cell or protoplast fusion applied in breeding, and such varieties are not easy to recognize in the market. In order to find ways to deal with such varieties, three questions need to be addressed:

1. How are varieties currently on the market distinguished from those that might be developed by non-compliant techniques?
2. What are the appropriate standards to breed new organic varieties?
3. How can organically bred varieties be identified in the market?

How to Deal with Varieties Bred with Non-compliant Techniques?

Varieties Bred with Techniques at Cellular Level

With respect to the question of how to distinguish current varieties on the market that might be developed by non-compliant techniques, the organic sector realizes that varieties produced with techniques at the cellular level have been introduced since the 1980s and are widely distributed. For example, some disease resistances in tomato varieties are derived from wild relatives introduced by embryo rescue. This technique increases the efficiency of the crosses compared with natural fertilization but does not deal with breaking natural crossing barriers as in the cases that cell fusion is applied. An immediate ban on such varieties would exclude most modern tomato varieties and would set back the organic tomato production to a large extent. The sector currently accepts existing varieties due to lack of alternatives, but it would not allow such breeding techniques in breeding programs that aim to be certified as organic.

Table 7.2 Suitability of breeding techniques at different levels of plant organization (adapted after Wyss et al., 2001)

Methods	Crop level	Whole plant level	Whole plant level + intervention	Cell level	DNA level
Suitable for organic breeding?	Yes	Yes	No	No, with exceptions	No
Selection	Mass selection, Pedigree selection, Site-determined selection, Change in environment, Change in sowing time, Ear bed method, Test crosses, Indirect selection	Yes, with exceptions		In-vitro selection	Yes, with exceptions Diagnostics, Indirect selection
Variation and recombination		Combination breeding, Crossing varieties, Bridging crosses, Backcrossing, Hybridization with fertile F1	Temperature treatment, Grafting style, Cutting style, Untreated mentor pollen	Irradiated mentor pollen, Colchicine, Mutation breeding, Hybrids with sterile F1	Polyploidization, Anther culture, Microspore culture, In-vitro pollination, Ovary and embryo culture	Cell and protoplast fusion, "Natural" gene transfer, Agrobacterium tumefaciens, Viral vectors, Direct gene transfer, PEG, Electrofusion, Micro-injection, Particle gun, Antisense DNA
Propagation	Generative propagation, Vegetative propagation			Micro-propagation, Meristem culture, Somatic embryogenes	

Varieties Bred with Cell or Protoplast Fusion

More urgent in the short term is the problem with cell fusion, a technique within the definition of GE and therefore incompatible with the principles of organic agriculture.

Cell or protoplast fusion (somatic hybridization) is often used to develop cytoplasmic male sterility (CMS) in *Brassica* species such as cabbage, cauliflower, and broccoli. CMS occurs naturally in only a few species such as carrot and onion (Rey, 2009), but not in *Brassica* species, and can in such cases only be incorporated from non-related species with cell or protoplast fusion. A CMS mother line is of great advantage when it comes to the production of F_1 hybrids, since targeted pollination by the respective father-inbred line is needed to achieve pure hybrids. Mother plants with CMS do not produce fertile pollen, which ensures no accidental self-pollination. An alternative strategy in *Brassica* species is to employ self-incompatibility lines. This alternative is less effective than CMS because it still allows a certain percentage of self-fertilization.

Cell fusion "by means of methods that do not occur naturally" (see Anonymous, 2001), is a breeding technique under the EU, NOP, Demeter, and IFOAM definitions of genetic engineering. At the same time, cell fusion is excluded from the scope of the EU directive when it is a fusion "of plant cells of organisms which can exchange genetic material through traditional breeding methods" (Annex 1B). The European Commission (EC) proposed a very broad interpretation of "traditional breeding methods." This means in practice that all kinds of cell fusion are excluded from the European GMO regulation and varieties originated from it are not labelled as a "GMO." This makes it hard for (organic) growers to identify and avoid these varieties. As a consequence, CMS hybrids based on cell fusion – e.g., from cabbage crops – are now widely used by organic farmers. This makes it difficult, if not impossible, to implement a ban (Wilbois et al., 2010).

Many organic farmers and breeders regard cell fusion as a technique of genetic engineering that is not indispensable for plant breeding. Private trading organizations in Europe, like Demeter, Bioland and Naturland (Germany), and Carrefour (France), have already forbidden their farmers to use CMS-varieties based on cell or protoplast fusion (Rey, 2009). These organizations provide lists of banned varieties to ensure growers comply with their standards. In the United States, the status of CMS varieties in organics is less clear. A more effective way to prevent cell fusion techniques from being used in organic farming is to maintain enough CMS-free varieties in the market (with competitive quality traits) and inform farmers of their value. Therefore the breeding of CMS-free (organic) varieties should be stimulated (see Chable et al., 2008 for organic brassica breeding programs).

How to Deal with Novel Breeding Techniques?

Currently, several novel breeding techniques are being developed such as cisgenesis, reverse breeding, tilling, site-directed mutagenesis, and gene silencing (COGEM, 2006; Haring and Lammerts van Bueren, 2008). Some of these techniques, such as reverse breeding, lead to end products that do not contain recombinant DNA although in the process GE is applied. According to the European legislation for genetic engineering (Anonymous, 2001), such are still considered a GMO as the European regulation is based both on process and product, in contrast to U.S. regulations that evaluate only the end product.

The discussion currently being conducted in Europe with respect to novel techniques is whether all products technically based on GE need to be considered and labelled as GE products. Schouten et al. (2006) question whether a variety produced by cisgenesis needs a full risk assessment to be commercialized in the European market. They argue that products of cisgenesis that contain DNA sequences could be achieved by traditional breeding and are similar to products derived by traditional breeding. However, they ignore the fact that in traditional breeding, in contrast with GE, the "desired" gene is imbedded in its chromosomal context (De Cock Buning et al., 2006). For instance, a resistance gene introduced by gene transfer technology will rarely integrate close to its original position – as usually occurs in traditional breeding – and its expression may therefore be influenced by the genomic position (De Cock Buning et al., 2006).

Exempting cisgenesis varieties from the EU's GE labeling requirements would make a ban on such techniques difficult for the organic sector to enforce. From a process point of view and applying the precautionary principle, cisgenesis would not be compatible as a breeding technique for organic varieties (Lammerts van Bueren et al., 2008). Although no natural barriers are crossed when inserting a gene from crossable species, isolating a gene from its natural genomic context and its random insertion could be considered as a violation of the genotypic integrity.

As additional new breeding techniques arise in future, a framework with which to evaluate such technologies for compatibility with the principles, values, and standards of organic agriculture will be a work in progress. Several research and dissemination projects are running with the objective to provide the organic sector with expert information to evaluate existing and new breeding methods and techniques (e.g., Haring and Lammerts van Bueren, 2008; Lammerts van Bueren et al., 2010; Oehen and Stolze, 2009).

Toward Appropriate Standards to Promote Organic Plant Breeding

IFOAM approved an organic plant breeding draft standard based on the list with compliant breeding techniques presented from the studies cited above (IFOAM, 2002b). IFOAM usually promotes standards from draft to full status after a period of time for the organic sector to become familiar with the new requirements and make appropriate amendments. Until now, the draft standards for organic plant breeding have not been made full standards. Two important reasons behind the long transition phase are: a) plant breeding in the organic sector is relatively new and plant breeding standards are complex, and b) the draft standards created concern that a positive list that limits permitted techniques might exclude new breeding techniques suitable for organic. Therefore the European Consortium for Organic Plant Breeding (ECO-PB) discussed a new approach in the process of creating Basic Standards for certifying organic plant breeding activities at the 1st International IFOAM Conference on Organic Animal and Plant Breeding held in Santa Fe/New Mexico in August 2009 (IFOAM, 2010). Instead of including a list of acceptable breeding methods, only those methods and techniques that fell under the GE definition given in the IFOAM Norms for Organic Production and Processing were definitely excluded. The IFOAM's GE definition includes cell fusion techniques, mutagenesis, cis-genesis, and so on. In order to allow for an evaluation on whether other applied breeding methods are in line with the principles of organic agriculture, the breeder shall be obliged to document the breeding process together with all relevant information on parental material, breeding methods, etc., and make the breeding process subject to an independent third party inspection. In order to be recognized or marketed as an organically bred variety, it must have been bred and selected for at least three generations from seed to seed in an organically managed and certified environment.

As of this writing, the discussion of this proposal has not been settled within IFOAM. But due to this proposal a new dynamic has emerged to develop appropriate standards for organic plant breeding.

Discussion and Challenges for Organic Plant Breeding

The organic sector requires a diversity of varieties and needs techniques to deliver varieties that are productive without the use of chemical fertilizers and pesticides. While there is a strong consensus that GE is not compatible, drawing the line with regard to novel breeding techniques is not always easy. The debate on specific techniques and products will likely become more complex as new techniques, varieties, and applications are commercially introduced. Against this background it is clear that the market segmentation will further progress in the future, namely into a segment that produces varieties also useable for the organic sector and another part with varieties that are off-limits in organic production systems. The demand for low-input and identity preserved (IP) varieties as alternatives to GE offers an opportunity as well as a challenge for the organic market. While the organic market is growing, productive and high-quality low-input varieties remain limited. Low-input and IP varieties create an opportunity to expand the non-GE market, but might also dilute the incentives to breed for organic farming conditions.

The organic seed industry is already challenged and has a way to go to close the loop of the organic production cycle by requiring use of organically produced seed if available (Baker, 2008). The current variety market segmentation may limit options for organic farmers, reducing their intra-specific biodiversity. There is a consensus that a reasonable transition period is needed to allow the market to adjust to producing an appropriate assortment of new varieties. In the meantime, several product types will co-exist for the organic farmers: (1) organically produced seed from organically bred varieties, (2) organically produced seed from conventionally bred varieties, and (3) in case the first two categories are not available exceptions can be granted for conventionally produced, non-GE seed.

Organic seed production and breeding programs are interrelated but distinct activities. The IFOAM standards distinguish between these two activities. As the sector increases production and use of organically propagated seeds, organic plant breeding also becomes more important. As plant breeders become more aware of the needs for organic farming, breeding activities within the organic sector continue to expand. It is important to find breeders, public institutions, and private companies that are willing to provide the organic sector with suitable propagation material to ensure that the organic market can continue to grow, thrive, and prosper. The organic sector needs ways to stimulate organic breeding programs and a way to identify varieties from organic breeding programs in the market. A certification system will provide an incentive to breed organic varieties.

References

Altieri, M.A. 2000. Economic, ecological, food safety, and social consequences of the deployment of Bt transgenic plants. *Ecosystem Health* 6:13–23.

Anonymous. 2000. Organic Products Act of 1990, Title 7, Part 205 of the Code of Federal Regulations. Accessed April 20, 2011. http://www.ams.usda.gov/AMSv1.0/nop.

Anonymous. 2001. Directive 2001/18/EC of the European Parliament and of the Council of 12 March 2001 on the deliberate release into the environment of genetically modified organisms and repealing Council Directive 90/220/EEC May 31, 2001. Accessed April 20, 2001. http://europa.eu.int/eur-lex/en/lif/dat/2001/en_301L0018.html.

Anonymous. 2007. Council Regulation (EC) No 834/2007 of June 28, 2007 on organic production and labelling of organic products and repealing Regulation (EEC) No 2092/91. Accessed March 29, 2011. http://ec.europa.eu/agriculture/organic/eu-policy/legislation_en.

Baker, B. 2008. Organic seed in the United States: challenges and opportunities. IFOAM General Assembly June 2008. Accessed March 29, 2011. http://orgprints.org/view/projects/conference.html.

Bullard, L., B. Geier, F. Koechlin, D. Leskin, I. Meister, and N. Verhagen. 1994. Organic agriculture and genetic engineering. *Ecology and Farming* 8:25–26.

Chable, V., M. Conseil, E. Serpolay, F. Le Lagadec. 2009. Organic varieties for caulifowers and cabbages in Brittany: From genetic resources to participatory plant breeding. *Euphytica* 164:521–529.

COGEM. 2006. New techniques in plant biotechnology. COGEM report CGM/061024-02. Commission on Genetic Modification, Bilthoven, The Netherlands.

DARCOF. 2000. Principles of Organic Farming. Discussion document prepared for the DARCOF Users Committee. Danish Research Centre for Organic Farming, Copenhagen, Denmark. Accessed April 20, 2011. http://www.darcof.dk/organic/Princip.pdf.

De Cock Buning, T., E.T. Lammerts Van Bueren, M.A. Haring, H.C. De Vriend, and P.C. Struik. 2006. " Cisgenic" as a product designation. *Nature Biotechnology* 24:1329–1331.

FAO. 2010. See Article 9 regarding Farmers' Rights of the International Treaty on Plant Genetic Resources for Food and Agriculture (ITPGRFA). Accessed April 20, 2011. http://www.unutki.org/default.php?doc_id=54.

FAO/WHO. 2001. Allergenicity of genetically modified foods, a joint FAO/WHO consultation on foods derived from biotechnology, Rome, Italy, January 22–25, 2001. Accessed April 16, 2011. http://www.who.int/foodsafety/publications/biotech/ec_jan2001/en/index.html.

Haring, M.A., and E.T. Lammerts van Bueren. 2008. Genetic modification hidden in novel breeding techniques. *In*: Proceedings of the 16th IFOAM Organic World Congress Cultivate the future, June 18–20, 2008 in Modena, Italy. Accessed April 20, 2011. http://orgprints.org/view/projects/conference.html.

IFOAM. 2002a. Basic standards for organic production and processing. Bonn, Germany: International Federation of Organic Agriculture Movements (IFOAM).

IFOAM. 2002b. Position on genetic engineering and genetically modified organisms. Bonn, Germany: IFOAM.

IFOAM. 2005. Principles of organic agriculture. http://www.ifoam.org/about_ifoam/principles/.

IFOAM. 2008. Definition of organic agriculture. Accessed April 20, 2011. http://www.ifoam.org/growing_organic/definitions/sdhw/pdf/DOA_German.pdf.

IFOAM. 2010. Report on the discussion workshop on international organic plant breeding standards. *In*: 1st IFOAM International Conference on Organic Animal and Plant Breeding "Breeding Diversity," August 25–28, 2009, Santa Fe, New Mexico. Bonn, Germany: IFOAM.

Kockelkoren, P.J.M. 2005. Ethical aspects of plant biotechnology – report for the Dutch Government Commission on Ethical Aspects in Plants. Appendix 1 in: Agriculture and Spirituality, essays from the crossroads Conference at Wageningen Agricultural University. Utrecht, The Netherlands: International Books.

Lammerts van Bueren, E.T. 2010. Ethics of plant breeding: The IFOAM basic principles as a guide for the evolution of organic plant breeding. *Ecology & Farming* 47:7–12.

Lammerts van Bueren, E.T., M. Hulscher, J. Jongerden, M. Haring, J. Hoogendoorn, J.-D. van Mansvelt, and G.T.P. Ruivenkamp. 1998. Sustainable organic plant breeding, Subproject 1. Discussion paper: Defining a vision and assessing breeding methods. Driebergen, The Netherlands: Louis Bolk Institute.

Lammerts van Bueren, E.T., M. Hulscher, J. Jongerden, G.T.P. Ruivenkamp, M. Haring, J.-D. van Mansvelt, and A.M.P. den Nijs. 1999. Sustainable organic plant breeding. Final report: a vision, choices, consequences and steps. Driebergen, The Netherlands: Louis Bolk Institute.

Lammerts van Bueren, E.T., P.C. Struik, and E. Jacobsen. 2002. Ecological aspects in organic farming and its consequences for an organic crop ideotype. *Netherlands Journal of Agricultural Science* 50:1–26.

Lammerts van Bueren, E.T., P.C. Struik, M. Tiemens-Hulscher, and E. Jacobsen. 2003. The concepts of intrinsic value and integrity of plants in organic plant breeding and propagation. *Crop Science* 43:1922–1929.

Lammerts van Bueren, E.T., H. Verhoog, M. Tiemens-Hulscher, P.C. Struik, and M.A. Haring. 2007. Organic agriculture requires process rather than product evaluation of novel breeding techniques. *NJAS Wageningen Journal of Life Sciences* 54:401–412.

Lammerts van Bueren, E.T., M. Tiemens-Hulscher, P.C. Struik. 2008. Cisgenesis does not solve the late blight problem of organic potato production: Alternative breeding strategies. *Potato Research* 51:89–99.

Lammerts van Bueren, E.T., H. Østergård, H. de Vriend, and G. Backes. 2010. Role of molecular markers and marker assisted selection in breeding for organic and low-input agriculture. *Euphytica* 175:51–64.

Lukkiholt, L.W.M. 2007. Principles of organic agriculture as formulated by the International Federation of Organic Agriculture Movements. *NJAS Wageningen Journal of Life Sciences* 54:347–360.

Mäder, P., A. Fliessbach, D. Dubois, L. Gunst, P. Fried, and U. Niggli. 2002. Soil fertility and biodiversity in organic farming. *Science* 296:1694–1697.

Oehen, B., and M. Stolze. 2009. Die Kosten der Koexistenz von gentechnisch veränderten und biologischen Kulturen: Fallbeispiele aus Frankreich und der Grenzregion. Paper at: 10. Wissenschaftstagung Ökologischer Landbau, Zürich, February 11–13, 2009. http://orgprints.org/14755/1/WS_Z%C3%BCchtung_14755.pdf.

OSA. 2011. State of organic seed: Advancing the viability and integrity of organic seed systems. Accessed April 26, 2011. http://www.seedalliance.org/Publications/.

Pimentel, D., M.S. Hunter, J.A. LaGro, R.A. Efroymson, J.C. Landers, F.T. Mervis, C.A. McCarthy, and A.E. Boyd. 1989. Benefits and risks of genetic engineering in agriculture. *BioScience* 39:606–614.

Plantum NL. 2009. Position on patents and plant breeder rights 2009. Accessed April 20, 2011. http://www.plantum.nl/plantum/documenten/Standpunt%20Octrooi%20en%20Kwekersrecht%20volledig%20ENG.pdf.

Rey, F. (ed.). 2009. Strategies for a future without cell fusion techniques in varieties applied in organic farming. *In*: Proceedings of the ECO-PB workshop in Paris, France, April 27–28, 2009. ITAB/ECO-PB. Accessed March 29, 2011. http://www.ecopb.org/09/proceedings_Paris_090427.

Robinson, R. 2007. Return to Resistance: Breeding Crops to Reduce Pesticide Dependence. Accessed April 16, 2011. http://www.sharebooks.ca/system/files/Return-to-Resistance.pdf.

Schouten, H.J., F.A. Krens, and E. Jacobsen, 2006a. Do cisgenic plants warrant less stringent oversight? *Nature Biotechnology* 24:753.

Shelton, A.M., J.-Z. Zhao, and R.T. Roush. 2002. Economic, ecological, food safety, and social consequences of the deployment of Bt transgenic plants. *Annual Review of Entomology* 47:845–881.

Suso, M.J., S. Nadal, and R.G. Palmer. 2010. Potential power of the plant-pollinator relationship as a tool to enhance both environmental and production services of grain legumes in the context of low-input agriculture: what do we do? *In*: Goldringer, I., J. Dawson, F. Rey, and A. Vettoretti. Breeding for resilience: A strategy for organic and low-input farming systems? pp. 24–27. Proceedings of EUCARPIA conference Section Organic and Low-input Agriculture, December 1–3, 2010, Paris. France: INRA and ITAB.

Tiedje, J.M., R.K. Colwell, Y.L. Grossman, R.E. Hodson, R.E. Lenski, R.N. Mack, and P.J. Regal. 1989. The planned introduction of genetically engineered organisms: Ecological considerations and recommendations. *Ecology* 70:298–315.

UPOV. 1991. The International Union for the Protection of New Varieties of Plants. Accessed April 4, 2011. http://www.upov.int/index_en.html.

Verhoog, H., M. Matze, E.T. Lammerts van Bueren, and T. Baars. 2003. The role of the concept of the natural (naturalness) in organic farming. *Journal of Agricultural and Environmental Ethics* 16:29–49.

Verhoog, H. 2007. Organic agriculture versus genetic engineering. *NJAS Wageningen Journal of Life Sciences* 54:387–400.

Verhoog, H., E.T. Lammerts van Bueren, M. Matze, and T. Baars. 2007. The value of "naturalness" in organic agriculture. *NJAS Wageningen Journal of Life Sciences* 54:333–345.

Wilbois, K.P., A. Spiegel, and M. Raaijmakers. 2010. Why and how we should ban cell fusion techniques from organic farming? *Ecology and Farming* 47:21–24.

Wingspread Conference. 1988. Statement on the Precautionary Principle. Accessed April 15, 2011. http://www.sehn.org/wing.html.

Wyss, E., E.T. Lammerts van Bueren, M. Hulscher, and M. Haring. 2001. Plant breeding techniques – An evaluation for organic plant breeding. FiBL Dossier No. 2. Frick, Switzerland: FiBL.

Youngberg, G. (ed.). 1986. Biotechnology in agriculture: Implications for sustainability. *In*: Proceedings of the Institute for Alternative Agriculture, Third Annual Scientific Symposium, Washington, D.C. Greenbelt, MD: Institute for Alternative Agriculture.

8 Plant Breeding, Variety Release, and Seed Commercialization: Laws and Policies Applied to the Organic Sector

Véronique Chable, Niels Louwaars, Kristina Hubbard, Brian Baker, and Riccardo Bocci

Introduction

Seed regulations emerged with the commercialization and industrialization of what has been predominantly conventional agriculture. Conventional seed regulations may not optimally encompass the needs of organic crop production. This chapter first describes the historic origins of formal seed production and modern plant breeding in Europe and the United States, illustrating different approaches (section 1). We then analyze how the legal systems regulate variety registration based on assessment of their value for cultivation and use and the distinctiveness and morphological uniformity and stability (section 2), followed by the seed certification and quality-control rules that regulate the seed markets (section 3).

We then focus on the special needs of organic production with regard to varieties and seeds and the special rules (Jaffee and Howard, 2010) that apply to the organic seed market (section 4), and we analyze recently introduced provisions on "conservation varieties" that could prove useful to overcome some of the European regulatory limitations (section 5).

Since intellectual property rights (IPRs) have an increasing impact on the seed sector, we introduce the historical developments of IPRs on seeds in section 6 and some implications for plant breeding. Finally, we compare the United States and the European Union in their approaches to seed regulation in general and their roles in organic seed markets.

The Developments of Plant Breeding and the Emergence of Seed Laws

Before commercial seed production was established, there were no seed or variety registration laws. Because farmers traditionally obtained seed from saved stocks through exchange with neighbors and relatives, or through local markets, issues related to "consumer protection" and intellectual property that developed later did not merit regulation.

The Development of Plant Breeding in Europe

Seed production as a specialized business emerged in mid-nineteenth century Europe along with selection of preferred traits. Plant breeding received a boost after the discovery of the laws of heredity. Plant breeding in Europe thus originated in the private sector in the nineteenth century – e.g., in the Netherlands and in France through family-owned enterprizes. One of them, Vilmorin-Andrieux, was founded in 1774 by Philippe-Victoire De Vilmorin. Initially, varieties were selected through introductions and mass selection, followed by some crossing. This family-owned company has remained independent through six generations (Bonneuil, 2006), but since 1975 has been a subsidiary of the Limagrain trust. In Sweden, advanced methods of plant breeding developed through Svalöf, a non-profit co-operative association established in 1886 by progressive larger farmers to breed varieties and produce seed. Svalöf lost its public service mission at the end of the twentieth century, when integrated into the central holding company of the Swedish farmers' cooperatives (Einarsson, 1992).

Emerging seed companies sought to "add value" by selling distinct varieties with useful traits. These efforts initially built upon mass selection methods to select particular traits in their crops. In the latter part of the nineteenth century, however, plants were crossed and pedigree selection was developed, permitting quicker advances. Plant breeders aimed to maximize the positive alleles in the population — creating higher yields and more uniform varieties. The results of breeding could only be verified when the varieties could be distinguished from one another. An account of the agricultural fair in Malmö, Sweden, in 1898 stated that the breeders had created "hundreds of distinct varieties" (Hjalmar-Nilsson, 1898).

Because farmers cannot visually distinguish most qualities of good seed, seed testing stations were established beginning in the 1860s to provide independent quality assessment with respect to germination, weed seeds, and other variables. These were commonly public laboratories.

The public sector got involved in breeding in most European countries only after the rediscovery of Mendel's laws of inheritance by Hugo de Vries and Carl Correns in 1900, which laid the basis for their recognition by the scientific community. The Agricultural School in Wageningen, The Netherlands (later Wageningen University), established its Institute for Plant Breeding in 1912. Mendelian genetics, and later quantitative genetics, were gradually applied by the private breeders, thereby increasing the importance of statistical analysis. However, many breeders kept a pragmatic approach by persisting in relying on the "breeder's eye" (Bonneuil, 2006), and the saying that plant breeding is both a science and an art is still widespread. It became clear that many agriculturally important characteristics depend of a number of genes. A deeper understanding of gene control emerged only after the discovery of DNA.

The Development of Plant Breeding in the United States

In the United States – in contrast to Europe – plant breeding became dominant in the public sector. Pre-Columbian Native Americans in North America had domesticated and improved various plants, including sunflower (*Helianthus annuus*) and squash (*Cucurbita pepo;* Smith and Yarnell, 2009).

European colonization of the Americas introduced exotic Old World species of food crops and techniques for production and breeding. Crops from the New World were also introduced into Europe, some of which found their way back to new regions of the Americas. Introduced varieties were adapted to the climate and ecology of the New World through trial and error. Farmers were the main agents of plant breeding and crop improvement organized and funded by the government (Fowler, 1994; Kloppenberg, 1998, 2005). Between 1760 and 1830, wheat varieties introduced from Europe and Asia were selected and hybridized by farmers for resistance to the Hessian fly (Gilstrap, 1961).

In 1838, the U.S. Patent Office requested the funds to distribute seeds "of the choicest varieties" for free via the U.S. Postal Service (Gilstrap, 1961). Congress authorized that request the following year and by 1855, the Federal government distributed over one million packets of seed to U.S. farmers. When the U.S. Department of Agriculture (USDA) was established in 1862, it took over the Patent Office's seed distribution program, which it continued until 1923. Congress directed the USDA to "collect new and valuable seeds and plants; to test, by cultivation the value of such of them as may require tests; to propagate such as may be worthy of propagations; and to distribute them among agriculturalists." The National Plant Germplasm System had its foundation from this directive. The Morrill Act established the Land Grant Colleges in 1862, followed in 1887 by the Hatch Agricultural Experiment Stations Act. Among other tasks, these stations were initially to select imported varieties from Europe and elsewhere for their adaptation to the farming conditions in the various states.

The rediscovery of Mendelian genetics in Europe and its early application to the science of plant breeding in the United States evolved into the selection and breeding of public varieties, given out to farmers and specialized seed producers for grow out or multiplication. Some basic research in plant genetics and plant breeding took place in the private sector at institutions such as the Station for Experimental Evolution at Cold Spring Harbor and through individuals such as Luther Burbank and Henry A. Wallace. However, most basic research was conducted by the USDA and the Land Grant Universities.

The identification of hybrid vigor in maize spurred the development of private breeding in maize from the 1920s when Henry A. Wallace established the Pioneer Hi-Bred Company in 1929 (Culver and Hyde, 2000), but the breeding of many cereals and legumes remained a public task. The uniformity of varieties and hybrids became more important with the gradual mechanization of agriculture throughout North America and Europe from the 1930s onward.

Driven by the advantages uniformity conferred to standardization of the agricultural production, social and ecological changes occurred in the European and American countrysides. Duvick's description of the development of F1 hybrid maize illustrates how innovation can be at odds with organic agriculture principles based on natural processes and traditional social practices:

"When hybrid maize was invented and presented to U.S. farmers in the first decades of the twentieth century, it was based on two new operations, one biological and the other socio-economic. First, strange manipulations (forced inbreeding and controlled hybridization) produced biological products that had never before existed in nature. Second, farmers gave up their time-honored practice of saving their own varieties of seed in favor of annual purchases of hybrid maize seed" (Duvick, 2001).

Seed regulations can be seen as an outcome of the modernization of agriculture and the emergence of plant breeding. Therefore challenging the paradigm of agricultural modernization based on chemical inputs – as does organic agriculture – involves a deep revision of various policies and laws in order to avoid conflicts with organic principles.

Variety Registration

The Origins of Variety Registration

As breeders developed new varieties, cultivars also became the subject of government protection. Variety registration was initiated in order to create transparency in the market. Before variety registration, different seed suppliers used the same names for different varieties. When a variety became popular, it was tempting to sell any seed under that popular name. The opposite also happened where to create a brand name for the seed company different names for the same popular variety (perhaps bred by a competitor) were used. This created confusion in the marketplace. Farmers' associations called upon governments to develop a system that would protect them from misrepresentation. In order to create a variety list, it was necessary to be able to identify the different varieties and to validate the claims of the seed producers with respect to their value for farmers. This led to coordinated variety trials for agronomic value organized by farmers' associations (e.g., The Netherlands) or by the government (e.g., France), combined with some form of morphological description.

Variety registration thus served an additional objective. It provided an independent source of agronomic information for farmers. Variety testing gradually became a specialized procedure, separated from plant-breeding activities. The first official catalogs of recommended varieties appeared in the 1920s in different European markets.

Moreover, Distinctness, Uniformity, and Stability (DUS) norms were established in the 1940s, which served to distinguish "a variety" from any cultivated population. For most crop species since then, varieties have to pass a thorough battery of analyses and field trials before being put on the market (Bonneuil and Thomas, 2009).

The main intent of U.S. seed regulations is truth-in-labeling. The assumption is that farmers should have the freedom to choose which varieties are best suited for their conditions and market. Public institutions may conduct variety trials, but with only a few notable exceptions, there are no restrictions that govern what varieties a seed company can sell or a farmer can plant. Various states regulated seed quality and labeling prior to the establishment of federal jurisdiction. Connecticut passed the first law in the United States regulating seed purity in 1821, which banned seed sold with certain noxious weeds (Rollin and Johnston, 1961). Laws on the proper identification of vegetable seeds were passed by Florida in 1889 and California in 1891. Eventually, all states would have seed labeling and testing laws. Enforcement is primarily carried out at the state level by members of the Association of American Seed Control Officials (AASCO, 2010).

The first federal seed law in the United States was passed in 1905. It concentrated entirely on farmer protections by introducing compulsory labeling. The law regulates label information such as name and address of the supplier, variety name, and seed quality parameters such as germination percentage. Farmers have to be sure that when they find a registered name on the label of the seed bag, the seed will indeed be of that variety. The supplier is responsible for determining these qualities. States must uphold the minimum federal requirements. Additional rules often promulgated by states with accredited private or state seed certification agencies provide an independent check on the seed qualities with standardized sampling and testing methodologies. However, seed certification remains largely voluntary and it is up to the seed producer to determine whether the added value of the label outweighs the cost of certification.

Historic Overview

The compulsory registration and release systems in Europe and Canada, which provide protection to farmers against purchasing poor quality seed of dubious varieties, actually outlaw the marketing of seeds from unregistered varieties. A broad interpretation of the term "marketing" also means that farm-saved seed may not be exchanged, even among farmers. European compulsory variety registration systems require variety testing for DUS for most crops. Such tests link the variety to their name. Second, varieties of field crop plant species are also tested to assess their value for cultivation and use (VCU) before their seeds can be marketed. Both tests have an impact on the number and the types of varieties in the market.

Variety Registration

Variety registration thus initially developed as part of an attempt to create transparency in naming of varieties in the market. Varieties needed to be identifiable and distinct from one another, and had to be stable because the name had to represent the same variety over time. Increased uniformity resulted from pedigree selection in plant breeding, which served as a proxy for stability. Variety selection became more formal after official variety lists were introduced. Descriptions initially included both morphological and agronomic characteristics. Early varietal descriptions were based on agronomic characteristics, but in DUS, characters are sought that are least influenced by the plant's growing conditions. DUS is thus measured mainly on traits that have no agronomic value such as some anthocyanin coloration at the base of a leaf. The DUS criteria also support varietal traceability (certification) during the seed production phases.

The standards for DUS are both complex and flexible. Distinctness standards depend on what can be observed based on a list of standard descriptors. If two varieties appear to fit the same description but are clearly different with respect to another trait, the registrar may accept an additional trait in the description. Uniformity standards take into account the species reproduction system of the variety measured relative to the average uniformity of the existing set of varieties with similar breeding method of that (sub)species (an inbred line is regarded differently from an open-pollinated variety). With the development of more varieties to be distinguished, uniformity standards become stricter over time.

Novelty is an essential criterion to protect a given variety and since in Europe the same novelty criteria are used for first registration of field crop varieties, it is included here. A variety proposed for registration is first reviewed for prior commercialization anywhere in the world and, if so, since when. This check is done based on seed company catalogs and using expertise of those who know the market.

A variety must be named so that it does not create confusion in the market. It should not be too similar to existing names, should not consist only of numbers unless it is established practice, should not refer to specific qualities, and should not disrespect the morals of the community in any language. When new selections are proposed of an "umbrella variety," the variety name commonly consists of the name of the umbrella group plus a name to identify the specific variety. For example, the name 'Chantenay Red Cored' carrot was chosen for a variety of the 'Chantenay' type.

Variety Release: Value for Cultivation and Use

When a variety is accepted, its seed may not necessarily be marketed in Europe. Varieties of field crops need to prove their performance in terms of agronomic and product quality characteristics – their "Value for Cultivation and Use" (VCU) has to be assessed to prove that the variety will bring "progress."

The VCU testing system is meant to provide information to farmers to assist in their choice of seed. The compulsory system of VCU testing in Europe provides for an independent comparison of the agronomic and quality parameters of the variety such as baking quality of wheat. Because tests are commonly performed in different regions and trial management that represent the current practice under "good farmer" conditions, the trials demonstrate the yield potential under "good" management. Demonstrations thus create an incentive for farmers to use prescribed practices. Such tests also enable breeders to evaluate the competitors' varieties. Even though the applicants commonly have to pay the full or part of the trial costs, it is normally cheaper than for each breeder to conduct his or her own trials. This is thus beneficial, especially for smaller companies.

VCU experiments are organized through multi-local networks, which recognize the diversity of soil and climatic conditions in order to measure the productivity, agronomic, and technological characteristics. Such tests create benefits for farmers and the seed sector only if trials are performed effectively and efficiently and when decisions are made wisely (Tripp and Louwaars, 1998). When this is not the case, VCU regulations may create the opposite effect, resulting in fewer varieties available and less agricultural diversity and development. Official VCU trials have to represent actual farming conditions such as fertilizer levels, and the observed characteristics need to be relevant to the farmer. Countries differ widely in terms of farmer involvement, both with respect to representation and the influence of representatives. The system's effectiveness depends also on sufficient funds being available to do the work properly. Chronic lack of funds in certain areas may be why UK farmers tend to select varieties based on voluntary rather than regulated trials (Burgaud, personal communication).

Harmonization was achieved in Europe in 1970 through the "common catalog" of the EU for field crops and vegetables EEC Directives 70/457 and 70/458, respectively. Details were provided in 1972 regarding examination methods and standards for varieties to be included in the catalog. These did not have provisions for testing under organic conditions. The "common catalog" includes all varieties released in any of the member countries. National registration becomes EU-wide, which permits seed to be traded throughout the union unless a country explicitly refuses to permit such a variety. This gives European farmers potential access to 50,000 varieties of various crops. Not all varieties are suitable for all farmers, which means that farmers mainly rely on their own national lists.

In countries where VCU testing is not legally compulsory for variety registration – such as in the United States – such variety trials may be demanded by farmers' organizations. The outcome of these trials is, however, not binding on the decision of whether to release or market varieties. More varieties are assumed to be available to farmers. In practice, most farmers choose only varieties that perform well in these trials.

Seed Quality Control and Certification

Seed quality is a major concern for all farmers. Key elements are germination capacity and vigor, physical purity, contamination with diseases and weed seeds, and genetic identity and purity.

The first law in Europe to control the quality of seed is believed to be an ordinance passed in 1815 by the Republic of Berne, Switzerland, and it intended to establish the quality of sweet clover (*Melilotus alba*) and prevent adulteration by weeds or other species of clover (Rollin and Johnston, 1961).

The Rules for Truth-in-Labeling in the United States

The current structure of the Federal Seed Act [7 USC 1551, et seq.] has changed little from the 1939 reorganization and amendment. The seed industry called upon the U.S. Congress to codify, clarify, and harmonize the patchwork of federal and state laws that developed over the previous 34 years [P.L. 76-354](U.S. Code, 2010). The Federal Seed Act was established foremost as a truth-in-labeling law that now governs interstate commerce. Seed sales within a given state are exempt. States are permitted to have more strict seed laws, but all seed shipped between states must meet the federal minimum.

Seed sold in interstate commerce in the United States must bear the following for agricultural and vegetable seeds specified in the regulation (U.S. Code of Federal Regulations, 2010):

- The kind of crop at the species or subspecies level for all crops covered by the regulations, and variety of certain kinds. If a variety is not claimed, then the label must bear "Variety Not Stated." [7 CFR 201.10]
- Rate of germination and hard seed [7 CFR 201.6].
- Purity, including percentage by weight of seeds that meet the identified kind and variety, weed seeds, noxious weed seeds, and inert matter [7 CFR 201.7].
- Declaration of any substance applied to treated seeds [7 CFR 201.7a].

Apart from the market transparency and farmer-protection, these systems were to support agricultural development and food security through the extended use of "improved and tested varieties." These compulsory systems created a level playing field for the seed companies. An independent and publicly available review of the value of the varieties and the quality of the seed have reduced the dependence on the marketing powers of the larger seed companies. Grow-out tests to validate the variety identity and field inspections preceded the concept of certification through a generation system. A reliable seed certification scheme requires a detailed morphological description for each variety. Identification of off-types can be done well only in uniform crops. Seed certification thus provides an additional incentive toward varietal uniformity.

Operation of Seed Quality Control and Certification

Seed in the market must conform to the standards (Europe) or the label (the United States) with respect to varietal identity and uniformity and physiological, physical, and sanitary quality. Varietal issues are taken care of by certification through a generation system where at the base, the highest quality breeder's seed is continuously selected (maintained) and is then multiplied to sufficient quantities of certified seed. Some loss of varietal uniformity is permitted with each generation of multiplication. Field inspections that are necessary to perform such certification also provide a basis for some of the other seed quality aspects, such as the presence of diseases and the general status of the crop.

Other seed quality parameters are checked on the seed itself through standardized procedures laid down by the International Seed Testing Association (ISTA), prescribing in detail how to take samples, how to do germination tests, etc. Such quality control can be done by an official (government) certification agency, an accredited agency, or by the seed producer.

Conflicting seed regulations and time-consuming procedures in different countries could potentially harm the international seed trade and present a barrier to varieties or seeds. The Organization for Economic Co-operation and Development (OECD) developed "seed schemes" to harmonize regulations, but many countries in different regions of the world have simultaneously developed their own harmonized systems.

However, borders can create significant problems. The U.S. National Organic Program (NOP) requires untreated conventional seed (or those with approved treatments only), but phytosanitary controls in Mexico require all imported seed to be chemically treated. Raw organic seed can be imported when accompanied by organic certification, country of origin documentation, and phytosanitary certificate. A company can obtain a raw organic seed import permit if they provide the Mexican authorities with a detailed outline of their disease prevention and post-control procedures and host an auditor to verify their documentation. Mexican organic growers importing seed from the United States appear out of compliance with one law or another.

Special Needs for Organic Agriculture

Introduction

In Europe, specific rules apply to organic seeds (Council Regulations Nos. 834/2007 and 889/2008), and, in practice, varieties must have a double certification: The first to be marketable and the second to be labeled as organic. The EU-member states maintain an online database to facilitate information about the availability of organic seeds so as to match offer and demand. Suppliers can enter organically produced seeds and seed potatoes that are available for purchase in these national lists (http://ec.europa.eu/agriculture/organic/eu-policy/seed-databases_en).

If a farmer wants to plant a variety not available in the database or not certified as organic, he or she must seek derogation from the national certification body before purchasing non-organic, untreated, seed (Gaile, 2005). Organic farmers have difficulties in sourcing organic seed (GRAIN, 2008), mainly because seed companies are not very interested in this small market segment. This is supported by the need for 30,782 derogations in France in 2009 and 30,379 in Italy, mainly due to a limited range of organic variety seeds (Anonymous, 2010). In other countries however, the number of derogations is decreasing (e.g., in the Netherlands in 2009, 1,456 derogations were granted; Lammerts van Bueren and Ter Berg, 2009).

In the United States, the NOP standards for organic seed that were implemented by the USDA in 2002 were seen within the organic movement as a great first step in encouraging the growth of organic seed. This set of rules requires that certified organically grown seed be used by organic farmers when commercially available. All seed used in organic production must be untreated, or treated only with substances that are either non-synthetic or on the National List of synthetic substances allowed for organic production. Commercial availability is based on ability to obtain the seed in an appropriate form, quality, or quantity, as reviewed by the certification agent (Navazio, 2010).

VCU Testing in Organic Context

In theory, the EU allows Member States to choose the agronomic conditions for VCU testing, and nothing prevents members from assessing traits and characteristics important to organic agriculture, such as plant architecture and resistance to diseases. However, official variety tests reflect the practices of "advanced" – conventional – farmers. Apart from yields under such conditions, priority traits are mainly those of interest to the food industry, e.g., wheat quality for pasta and bread. Moreover, the regulatory costs to the breeder at current rates for the national tests may be prohibitive where the target is a very local market.

For organic breeding to succeed, VCU testing protocols may thus need to be modified in order to identify the varieties that perform best under organic farming conditions (Rey et al., 2008). Government supports coordinated variety trials under organic conditions next to the regular VCU tests in, e.g., Canada, Switzerland, Austria (Löschenberger et al., 2008), and a special protocol for evaluating spring wheat varieties under organic growing conditions has been developed in The Netherlands (Osman and Pauw, 2005).

DUS Testing in Organic Context

A second issue in variety registration is the DUS requirements for official release of arable crops and for certification of all crops in Europe. Some of the resilience needed in organic farming may be obtained by introducing genetic diversity in the field. This is the basis of alternative breeding methods leading to, for example, multiline-varieties, and top-cross hybrids. Such varieties probably fail the uniformity standards of the DUS test when compared with conventional ones. A famous example is the wheat variety aptly named 'Tumult' presented for release in the late 1970s (Louwaars, 2001). The multiline consisted of several lines that were the same except for resistance genes against yellow rust. The registration was initially turned down for lack of uniformity and finally the registration of the individual components was accepted which allowed the variety to be marketed as a "varietal mixture." However, by that time, the variety had been "overtaken" by newer releases and it never became a commercial success. Similar challenges are faced by registration authorities when variety mixtures and composite cross populations are presented for release (Wolfe, 2008; see Chapter 5, this book).

Organic Seed Market and the Current Rules

While many of the regulations are consistent with organic farmers' needs, some are not. Organic farmers need – as do their conventional colleagues – quality seed that is properly identified and free from diseases and weed seeds. However, seed regulatory systems have been modified to accommodate changes in agricultural technologies. Seed qualities of high yielding varieties needed to match requirements of mechanization and precision agriculture. Vegetable growing in computerized greenhouses requires not only genetic uniformity of varieties, but also uniform germination. Such varieties may not be optimally adapted to organic production and different seed quality parameters are emphasized – in organically produced seed it is not primarily germination but seed vitality and seedling vigor that are important.

Seeds need to germinate similarly for organic farmers as they do for conventional farmers. However, because seed producers have to follow the rules of organic farming, and they cannot rely on chemical control, measures need to be taken to avoid the build-up of seed transmitted diseases. The central concept of plant health aims at increasing the self-regulating capacity and the amplification of tolerance to pests and diseases instead of regulation with chemical protectants. Good growing conditions with minimal stress will enhance the natural tolerance of plants to competitors and parasites. Also the presence of diseases in a crop – even if they are not transmitted through seed – can result in smaller and shrivelled seeds that may have reduced germination and/or seedling vigor. This aspect reinforces the necessity of specific varieties that are able to produce healthy seeds. The F1 hybrid parents are often inbred lines, altered in their vigor by consanguinity, and unfavored for organic seed production.

Field inspection and seed testing may be more critical and complex for organic seed compared to conventional lots. Seed standards may need to be adjusted to achieve balance between what is preferred and what is practical.

The Evolution of Seed Standards in Europe

The EU "Organic Revision" project, initiated in 2007 (http://www.organic-revision.org/), has recommended improvements to organic seed certification and production:

- To consider specific variety traits requested by organic farming systems in the test for variety inclusion into the registers of varieties;
- To include specific local demands from organic farmers in the variety trials;
- To promote public breeding for organic farming and to support spin-off of seed companies dedicated to organic seed production; and
- To offer a low-cost option, which may be Participatory Plant Breeding (PPB), that combines the advantages of introducing requested varietal traits and facilitates dissemination.

All these concerns are included in the ongoing evaluation of EU seed legislation, which will establish new objectives and priorities for European seed legislation in the coming years (FCEC, 2008). As stated in this report in Brussels:

"At the moment the commercial breeding systems strongly influence the interpretation of both DUS and VCU characteristics and the testing systems. With environmental issues becoming more important there is a stronger case now for a wider interpretation of both DUS and VCU for varieties that are going to be used for organic and low input agriculture."

A Recent Development in Europe: Conservation Varieties

As indicated above, the basic rules on variety registration in the EU are very strict. They make exchange and marketing of seed from varieties that are not registered in the common catalog illegal. However, countries differ widely in their implementation of the term "marketing." For example, France holds closely to the rules; seed regulation authorities in The Netherlands openly allow some level of deviation from the rules while in England, the interpretation allows "clubs" through which members can exchange and even sell seed without violating the law. For example, the Henry Doubleday Research Association, maintains the Heritage Seed Library (http://www.hdra.org.uk/hsl.htm), which contains over 700 varieties of interesting and traditional vegetables that cannot legally be traded in Europe. "These seeds cannot be purchased but Members of the Library can receive free seed packets each year" (http://homepage.tinet.ie/~merlyn/seedsaving.html).

Some European countries have special parallel lists for the registration of varieties that would otherwise not satisfy the European common catalog and that would therefore risk falling out of use and disappearing. Moreover, in 1998 Europe opened the catalog to the so-called conservation varieties, with the aim of conserving agricultural biodiversity. We briefly discuss the lists held by France (national/crop approach) and Italy (regional approach), and then describe the new requirements for a conservation variety to be listed in the catalog.

France

France opened a register for old vegetable varieties for home gardeners. To be included in this list, the variety has to have been published, e.g., the variety has to have appeared in a publication (a seed catalog or something else). The register included 300 varieties of about 30 species in 2009, including squash, pumpkin, lettuce, tomato, and melon. Such varieties are registered at a minimal cost (about €100), based on a description provided by the applicant (Zaharia and Kastler, 2003). The varieties are allowed to be commercialized only among amateur gardeners in France and marketed only in small packets, labeled "standard seed" and marked with the statement "old variety exclusively reserved for home gardeners." This assumes that the produce is consumed at home. The seed quality parameters – germination, purity, and seed health – are the same as in the regular market. France also has a national list of old fruit tree varieties for home gardeners established in 1952 with more than 1,000 apple, pear, plum, hazel, and walnut varieties. Since 1993, there is a list that includes five old potato varieties for specific use in France. These varieties had been on the market for a long time and did not need additional VCU tests. The objective was to control the crop health risks in producing seed of these varieties and to regulate variety maintenance.

Italy

In Italy, regions are responsible for agricultural policies, so many of them promote local varieties (and animal breeds) to reduce the loss of agricultural biodiversity and promote the marketing of regional products. The main regional mechanisms are (1) the establishment of a regional repertoire for traditional varieties and breeds; (2) the creation of a "network of conservation and security" for both on-farm and ex situ conservation; and (3) the recognition that the heritage of these varieties and breeds belongs to the local communities (Bertacchini, 2009).

The framework for saving and exchanging farmers' seeds provided by regional regulations addressed a very specific situation in which only small numbers of farmers still grow and manage traditional varieties. They thus create legal space for protecting and promoting traditional products based on specific varieties. The region of Emilia Romagna is a good illustration of this framework. The Emilia Romagna legislation recognizes different classes of varieties, such as those originating from the region; those that have been introduced long ago and integrated in the traditional agriculture; and those that have been selected from Emilia Romagna and are no longer grown there and maintained elsewhere. This legislation specifies under which conditions seeds of such varieties can be exchanged.

A national inventory of all the regional landraces is in preparation under the framework of the National Plan on Agrobiodiversity (Porfiri et al., 2009). According to the needs of the growers of these varieties, they could be freely listed in the new conservation varieties catalog and therefore be sold on the market as seeds.

EU-Rules on "Conservation Varieties"

EU member states signed the 1992 Convention on Biological Diversity and the 2001 International Treaty on Plant Genetic Resources for Food and Agriculture, which required them to promote the conservation and sustainable use of plant genetic resources. Civil society organizations urged them to implement these agreements at the level of crop varieties. In 2008, the first directive on conservation varieties created legal openings in the European Community seed regulations (three different directives constitute the core of the legislation on conservation varieties: The directives

62/2008 on agricultural plant species, 145/2009 on vegetables, and 60/2010 on fodder plant seed mixtures as "preservation seed mixtures".) This new legislation aims at supporting *in situ* management of varieties threatened by genetic erosion in their regions of origin, thus mitigating genetic erosion. The production and sale of seed from such varieties is being legalized, thereby also enabling the reintroduction of old varieties into the farming systems.

These directives attempt to strike a new balance between the need for transparency and quality in the seed market and conservation objectives. Key issues in the debate included:

- How to assess whether a variety is "threatened by genetic erosion."
- How much flexibility could or should be introduced on DUS standards to accommodate the varieties that are the intended subject of the directive?
- How much flexibility can or should be introduced on quality control, production, and marketing of seeds of these varieties?

It is questionable whether this concept will bring about a sustainable conservation of crop genetic diversity on-farm, and whether it will increase the diversity of farming systems in Europe. A major risk is that administrative requirements become burdensome and costly for very small seed markets.

The Meaning of "Threatened by Genetic Erosion"

It is difficult to define and quantify levels of genetic erosion in plant genetic resources. First a census or a list of the local varieties still grown by farmers is needed to estimate the risk of inter-varietal erosion (Negri et al., 2009). Second, the variability within local varieties has to be measured – these are often fairly heterogeneous populations – to estimate the risk of intra-varietal erosion.

The absence of a preliminary field survey makes it difficult to indicate the risk of genetic erosion of a specific resource. Furthermore – once a variety is marketed it is no longer at risk (Bocci et al., 2010). The threat of genetic erosion is expected to be clarified during national implementation (Lòpez Noriega, 2009).

Region of Origin

Only varieties that can be linked to a particular region can be called conservation varieties, including genetically diverse landraces and old varieties that have disappeared from the common catalog but are adapted to a particular region. The rules exclude the possibility of listing new varieties that are not completely uniform even if adapted to certain niche conditions, such as varieties based on participatory breeding. The commercial seed sector was concerned that allowing new farmers' varieties (Salazar et al., 2007) as "conservation" varieties would create a "back door" opportunity for the registration of new varieties that do not meet the regular standards. Some conservationists and farmers' groups, on the other hand, lobbied hard to have a broader definition of the term "conservation varieties," particularly with regard to the possibility of using such varieties outside their region of origin and allowing such varieties to be improved through selection (Chable et al., 2010). Experiments demonstrated that some landraces gave very good – even superior – results for certain productivity traits outside their zone of origin or natural adaptation (Farm Seed Opportunities, 2007–2010[1]).

Member states address this issue differently with opposite approaches. For example, Norway and The Netherlands each consider their whole country as one region, while Italy leaves this choice to regions within its country. In this case, a particular conservation variety could be restricted to a specific region or municipality.

DUS Criteria of the "Conservation Varieties"
Countries may also interpret the uniformity rules for conservation varieties according to their own needs. Thus, landraces may be registered on the basis of a limited description. They must be maintained within such description and are thus not allowed to evolve too much. Instead of a detailed morphological description, some member countries simply ask the users what makes their local variety special (and thus distinct), which then forms the basis for formal registration. In order to allow for evolution to take place, there are also possibilities to re-describe and thus re-register conservation varieties (e.g., every five years).

Quality Control, Production, and Marketing
Some specific rules have been developed with regard to seed production. A conservation variety must be linked to its "region of origin" and maximum amount of seed that is allowed in the market has been defined. For field crops, each variety is limited to 0.5% of the total seed market or to the quantity of seed required for 100 hectares, whichever is larger; all conservation varieties of a single crop should not exceed 10% of the total (directive 62/2008). In the case of vegetables, quantities for each species are listed in directive 145/2009.

Intellectual Property Rights and Plant Breeding

The Historical Origin

The emergence of scientific breeding in the first decades of the twentieth century led to a call to protect the intellectual property of plant breeders, similar to the protection of inventions. Living organisms were excluded from patentability because of ethical, legal, and political reasons (LeBuanec, 2006). Variety registration procedures succeeded in fixing a name to a given cultivar, but did not give exclusive rights to the breeder. The need to protect breeders' commercial investments was discussed at the Horticultural Congress in Paris in 1919 (Bos, 1920). Also in this aspect, the views in Europe and the United States differed initially. The two approaches – plant breeder's rights and utility patents – now create overlapping rights, which leads to considerable debate.

Breeder's Rights: From Europe to the United States
Europe developed various *sui generis* protection systems for plant varieties in the early twentieth century to promote investment in plant breeding and seed production. Germany protected plant breeders with a "breeder's seal" (Leskien and Flitner, 1997); The Netherlands provided prizes to the best breeders during the national agricultural shows. The first intellectual property protection law for plant varieties was established in The Netherlands in 1941 (Rangnekar, 2000). Other European countries followed suit and harmonized their systems in the International Convention for the Protection of New Varieties of Plants (1961; www.upov.int). They became the first members of the International Union for the Protection of New Varieties of Plants (UPOV). The requirements for protection were formalized: Distinctness, uniformity, stability, novelty, and name (DUS-NN), and standardized methodologies for establishing these requirements were developed based on the existing registration systems. Membership in UPOV gradually expanded with the United States joining in 1970 and Canada joining 20 years later: 67 states are parties of the UPOV as of November 2010.

Apart from the requirements, Plant Breeder's Rights includes important exceptions that are rooted in agriculture. The rights include the use of the protected variety for further breeding and there are privileges for farmers to reuse their seed without the consent of the breeder. In Europe this is currently only allowed for cereals (except maize) and potato, and farmers have to pay a royalty to the breeder on such re-use of the protected variety. Smallholder farmers are, however, excluded from such payment.

The legal protection of varieties thus led to the creation of yet another national variety list in addition to the lists of registered and recommended varieties. Both protection and registration for market regulation required a clear description of the variety. Both also required that the variety had to be stable and did not change after repeated reproduction. The use of the same DUS standards both for registration and protection allowed one set of variety trials to serve both procedures.

Patents: From the United States to Europe
In 1906, the first piece of legislation in the United States that sought to establish patent rights for plant breeders was introduced, but legislators were reluctant then to extend such protections over higher organisms (Kloppenburg, 1988). It was not until 1930 that Congress passed the Plant Patent Act. Patents could only be applied to asexual reproduction, such as grafting, budding, cuttings, layering, and division. Edible root and tubers were explicitly excluded, and Congress confirmed that the rules could not apply to seed crops in the Patent Reform Act.

While Congress had long resisted awarding utility patents to plants, it saw merit for some form of protection for breeders of sexually reproducing plants. As such, Congress passed the Plant Variety Protection Act (PVPA) in 1970. However, the United States applied a very liberal privilege for the farmer to reuse seed. Until 1994 farmers were allowed to sell significant quantities of their farm-saved seed without the consent of the breeder. Since the 1994 amendments, such sales are not allowed, but farmers may save any amount for their own use without royalty payment, which accounts for a significant loss in potential royalty income for breeders. Breeders largely consider the U.S. breeder's rights system to be weak and giving insufficient returns.

In 1980, the landmark Supreme Court case of *Diamond v. Chakrabarty* allowed the first utility patent on a living organism when the U.S. Patent and Trademark Office (USPTO) denied a patent for a genetically engineered bacterium and was sued by the inventor. In 1985, it became clear through *Ex parte Hibberd* that the decision also extended to new plant varieties irrespective of the existence of the Plant Patent Act and the PVPA (Roberts, 2003). In 2001, the Supreme Court provided judicial clarification in *J.E.M. Ag Supply, Inc. vs. Pioneer Hi-Bred International* when it concluded that plants were not explicitly excluded in section 101 of the Patent Act.

The lack of protection in the biotechnology sector led to the "Biotechnology Directive" in Europe in 1998, which stated that plant varieties are not patentable but that biotechnological inventions can be patented. It is now clear that if an invention is not limited to one variety, it may be patented. This means that plant traits and biotechnological methods may be patented and that protection may extend to all varieties where such traits are expressed.

Developments at the Global Level
The developments in the United States and Europe, with respect to IPRs, have consequences at the global level. The World Trade Organization's (WTO's) Agreement on Trade-Related Aspects of Intellectual Property Rights (TRIPS) – concluded in 1994 – introduced intellectual property rules into the multilateral trading system for the first time (Drahos and Braithwaite, 2002). Global seed markets expanded following the WTO trade and IP rules, creating opposition from developing countries that harbored the origins of the majority of the world's agricultural biodiversity.

Industrialized countries considered the TRIPS requirements too low and demanded – in bilateral and regional trade negotiations – even higher standards of IP protection. There is pressure on developing countries to become members of UPOV and to recognize biotechnology patents even though TRIPS allows countries to exempt plant and animals and methods to produce them and allows for any "effective *sui generis* system" to protect plant varieties.

Recent Evolution

Opponents of patents on plants continue to argue that U.S. Congress should amend the PVPA to be the exclusive means of protection for sexually reproducing plants, pointing to the extensive control utility patent holders exercise over access and use of protected varieties. The strategic use of patent protection, by claiming both biotechnological and breeding methods, genes, plant parts, plants, and even the products harvested by farmers, triggered broad patents and reach-through patents related to important commercial crops providing strong bargaining positions with competitors (Boyd, 2003). Plant breeders find it increasingly difficult to obtain germplasm from companies, consequently ultimately removing valuable resources from genetic commons that breeders rely on for improving agricultural crops (Price, 1999).

Utility patents have also made it difficult for public institutions to ask critical questions of protected products. In 2009, 26 university scientists submitted a joint complaint to the U.S. Environmental Protection Agency regarding the veto power patent-holding biotechnology firms exercise over independent researchers, stating: "No truly independent research can be legally conducted on many critical questions" (Anonymous, 2009). These scientists pointed to situations in which biotechnology companies were keeping universities from fully researching the effectiveness and environmental impact of the industry's genetically engineered crops. Courts found U.S. regulations un-enforced and insufficient to protect the environment (*Center for Food Safety v. Vilsack,* 2010; Platt, 2010).

Private research has also been affected. Royalties tied to licensing agreements are costly and serve as a barrier to new companies entering the plant breeding industry. This trend has facilitated consolidation in the seed industry (Louwaars et al., 2009). Increased concentration of financial and genetic resources has led to buyouts where larger firms increase market share and lock up greater amounts of genetic resources through acquisitions. Four transnational firms now control more than half of the world's proprietary seed market (Hubbard, 2009). The number of independent seed companies in the United States and Europe, especially family-operated businesses, has dramatically declined in the last few decades.

The rights are specifically beneficial for companies that can afford the legal counsel to enter into lengthy court procedures. Seed companies have sued each other over conflicting patent claims and infringement (*Monsanto v. Syngenta,* 2007; *Monsanto v. Bayer,* 2008). Seed companies also sue farmers who have saved patented varieties for replanting (*Monsanto v. McFarling*, 2004; *Monsanto v. Scruggs,* 2006). Seed companies increasingly use contract law to limit the freedom of farmers to use their seed (Monsanto, 2009). Among other restrictions, the licensing agreement terminates farmers' legal rights to save seed from their harvest. The pursuit of farmers for patent infringement has led to thousands of investigations and more than 100 lawsuits (Center for Food Safety, 2005). These investigations and legal actions are costly and have bred distrust, suspicion, and less information sharing among farmers. Louwaars et al. (2009) assert that companies with extensive patent portfolios spend more money on legal counsel than on research.

The debate is also heated in the seed industry itself. In Germany, government is formally aimed at stopping the patenting of life forms, and in The Netherlands the seed association Plantum.nl has stated that it wants a full breeder's exemption to be included in patent law. This would free the genetic building blocks for use in breeding and, in turn, minimize the commercial value of some classes of patents (www.plantum.nl).

Opposition to extensive use of IPR may be shifting the pendulum back in the other direction. Although since 1980 more and more biotechnology and breeding patents were approved, there are now signs that courts and patent agencies are starting to more critically assess patenting requirements. Patents on gene sequences are not approved any longer because of lack of industrial application (failure to meet the usefulness criterion of utility patents) and the publication requirements (*In Re Fisher*[2]). A ruling on a patent on a (human) gene is even more restrictive, based on a perceived lack of inventiveness (see *In re Kubin*[3]), and the patenting of genes that exist in nature may be restricted because of lack of novelty following a ruling on a human breast cancer gene. In Europe, a recent ruling against Monsanto argued that a patent on a plant gene is exhausted when it does not perform a function anymore (in this case, herbicide resistance gene in soybean flour). The major patent offices are increasingly critical of patent applications.

Plant Breeders Rights and Organic Agriculture
Organic farmers need good varieties like any crop grower, and incentives may be needed to create a sustainable interest in investing in organic breeding (Almekinders and Jongerden, 2002). However, strong IPRs appear to stimulate breeding for large "recommendation domains." Breeding for specific adaptation may be better organised through Participatory Plant Breeding (PPB, see Chapter 6 of this book), which could be considered a non-monetary benefit sharing mechanism according to the International Treaty on Plant Genetic Resources for Food and Agriculture (Halewood et al., 2007). PPB is the process by which farmers are substantially involved in a plant breeding program with opportunities to make decisions throughout (Ceccarelli et al., 2009).

PPB is seen in the Netherlands as the most efficient strategy to address the needs of organic potato producers because commercial potato breeders select varieties mainly for conventional farming (Lammerts van Bueren et al., 2008; see Chapter 14, this book). In Portugal, PPB was initiated with two objectives not specifically linked to organic agriculture: The conservation of the white maize populations for traditional bread and the maintenance of evolutionary processes in the farmers' fields, which could be valuable for continuing adaptation to future environmental conditions (Patto et al., 2008). PPB simultaneously began in 2001 in three areas in France within groups of farmers of cauliflower, maize, and durum wheat (Chable et al., 2008; Desclaux, 2005). PPB of several other species is under way and farmers and others have created networks to organize the collective actions in the context of the current regulation (Chable et al., 2009).

Farmers involved in PPB commonly value the sharing of the fruits of their work among farmers (Salazar et al., 2007) even though there could also be arguments for the protection through breeders' rights so that farmers share some of their benefits with the farmer-researchers.

Discussion

European Union and United States Comparison

The private versus public origins of breeding programs may explain the different approaches to regulation taken on different sides of the Atlantic. In Europe, the governments were called to protect farmers in the nineteenth century from commercial seed producers misrepresenting the value of their varieties, and governments did this through obligatory variety lists and seed certification. In the United States, where public breeding programs produced most of the varieties, these quality control systems remained voluntary in most states. There are exceptions for specific crops and/or states for marketing or quality control purposes. One example is the San Joaquin Valley Cotton

Board (SJVCB) in California, which only allows extra-long staple acala and pima cotton varieties to be planted. The SJVCB has been challenged in court several times over jurisdictional issues. Organic farmer and colored-cotton breeder Sally Fox unsuccessfully challenged the predecessor organization, the Acala Cotton Board, for the right to plant colored varieties in the San Joaquin Valley. However, this one program is an exception.

Another reason for the absence of compulsory variety lists may be that such regulation does not fit within a free market paradigm that more strongly developed in the United States, compared to Europe, in the first half of the twentieth century. Furthermore, the history of seed regulation in Europe is a history of a joint concern by governments, farmers' organizations, and the emerging seed producers for food security and sustainable rural development. Food security was a vital element in U.S. policies in the nineteenth century when the agricultural experiment stations were first established, but much less so in most of the twentieth century. In Europe, on the other hand, food security remained important due to the World Wars that ravaged the continent. More recently, even though sustainability is prioritized over yield maximization, seed regulations have proved hard to change.

Compulsory seed certification in Europe strictly limits the freedom of farmers to find seed of desired varieties in the market and to exchange seed with neighbors. Absence of the legal requirement to certify seed in the United States allows for all kinds of parallel seed systems, including the Seed Savers Exchange, Native Seed Search, and other networks that are important sources of genetically diverse heirloom varieties. Such networks would be illegal by European regulations.

IPRs in plant breeding have had very different historic trajectories on either side of the Atlantic. In Europe, where plant breeders' rights have been predominant, patents on traits and biotechnologies increasingly impact breeding and seed supply. IPRs have had a greater impact in the United States, not only because varieties can be patented unlike in Europe, but also because seed can be sold under contracts that give the seed company extensive powers. In both territories the roles of the patent system seem to be changing as a result of subsequent court cases and changes in the laws themselves. The Netherlands parliament has decided that a breeder's exemption needs to be included in patent law.

Organic Farming, Seed Policy, and Agricultural Biodiversity

The new EU rules for conservation varieties solve only part of the need for genetically diverse varieties for the organic sector (Bocci, 2009). Countries are allowed to register old landraces and amateur varieties of vegetables, but farmers who improve such varieties or breed new ones need to follow the conventional registration rules.

The registration of new heterogeneous varieties could potentially be approved within the current rules that include uniformity as a relative measure. A new variety has to be uniform when compared to the existing varieties of the crop. This means that heterogeneity could be approved if a special class of varieties is recognized for organic production. New organically bred varieties could be compared within that class if registration authorities are flexible.

A new approach is needed to meet organic seed demand. An informal seed system where farmers play a role to conserve agricultural biodiversity and improve varieties through PPB programs is seen as pivotal. In fact, due to importance in organic agriculture of genotype × environment interaction, varieties should be adapted to the different growing conditions and the best way to do it is through a decentralized and participative research approach (see Chapter 6, this book). The result of PPB experiences is not only useful varieties, but also mainly autonomy of the farmers who become progressively independent of researchers.

Agriculture and seed policies should recognize the role of farmers in seed production, bearing in mind "[that]it is impossible to replace farmers' seed systems completely and it would be unwise to try. Farmers' seed systems provide an important component of food security, a vital haven for diversity and space for further evolution of PGR [plant genetic resources]" (FAO, 2009).

Conclusions

Seed certification schemes and plant breeders' rights systems all involve some form of variety registration, which has developed over time alongside plant breeding and seed production. In some countries, such as the United States, registration is voluntary. In most countries however, including Europe and most developing countries, registration is compulsory for taking seed to market, and as a result, the impact of the rules on farmers is much larger.

Registration requirements and procedures have been harmonized in several regions partly in an attempt to stimulate international trade and partly as a result of international pressure to create globally harmonized trade and intellectual property rights systems. Distinctiveness and stability are basic of any registration system, and uniformity is considered important as a proxy for stability. Uniformity of important agronomic characteristics furthermore supports mechanized farming and processing. The protection of IPRs leads to further increased uniformity standards also for characteristics that have no agronomic advantage and creates dependence on purchased inputs. Diversity, by contrast, can support crop resilience. Agronomic performance testing is often an additional component of the registration process. Ineffective implementation of this system with regard to organic farming can create a bottleneck for the number of varieties available to organic farmers and may thus decrease diversity in the field.

Organic rules for the availability of organically produced seed appear difficult to implement. The concentration of the global seed sector as the result of technological development, globalization, and, particularly, the expansion of patent protection in the life sciences leaves few professional suppliers of organic seed in the market. Incentives are needed to produce quality seed for the organic sector.

In countries with compulsory variety registration systems, special derogations may need to be specified in the law, as with conservation varieties in Europe, or the scope of the seed regulatory law framework may need to be limited to only the formal seed sector. Different countries have different farming and seed systems and will arrive at different solutions.

Countries need to balance different policy objectives under the overall goal of promoting agricultural growth. For example, they need to (1) create a transparent market for seeds with a level playing field for competing companies; (2) provide farmers with suitable protection with respect to the qualities of seed in the market; (3) promote sustainable agriculture; and (4) support the conservation and use of genetic resources by continuing to allow farmers to use and create genetically diverse varieties. Regulatory systems evolved at a time when food security was the leading objective, and the systems responded to these needs. Now that diverse policy objectives are to be pursued, regulatory change is imperative.

Notes

1. Farm Seed Opportunities: STREP project, contract no. 044345, under the 6th Framework Programme, priority 8.1, "Specific Support to Policies". http://www.farmseed.net.

2. United States Court of Appeals for the Federal Circuit (CAFC) 7 September 2005, *Re Fisher*, 421 F.3d 1365 (Fed Circ 2005). http://ftp.resource.org/courts.gov/c/F3/421/421.F3d.1365.04-1465.html.
3. United States Court of Appeals for the Federal Circuit (CAFC), April 3, 2009, *In*: re Marek Z. Kubin and Raymond G. Goodwin. http://www.cafc.uscourts.gov/opinions/08-1184.pdf.

References

AASCO. 2010. Association of American Seed Control Officials. Accessed September 26, 2010. http://www.seedcontrol.org/.

Almekinders, C., and J. Jongerden. 2002. On visions and new approaches. Case studies of organisational forms in organic plant breeding and seed production, Working Paper Technology and Agrarian Development, Wageningen University, The Netherlands.

Anonymous. 2009. Response to US EPA Docket EPA-HQ-OPP-2008-0836, February 9. Accessed April 14, 2011. http://www.regulations.gov/#!documentDetail;D=EPA-HQ-OPP-2008-0836-0043/.

Anonymous. 2010. Fonctionnement de la base de données sur les variétés disponibles en semences issues du mode de production biologique – Synthèse annuelle des dérogations accordées, en application du Règlement (CE) no. 889/2008. RAPPORT DE SYNTHESE – ANNEE 2009 – France – Commission nationale semences du CNAB-INAO. Accessed December 2010. http://www.semences-biologiques.org/pages/Rapport-semences-bio-2009-final.pdf.

Bertacchini, E. 2009. Regional legislation in Italy for the protection of local autochthonous varieties. *Journal of Agriculture and Environment for International Development* 103:51–63.

Bocci, R. 2009. Seed legislation and agro-biodiversity: Conservation varieties. *Journal of Agriculture and Environment for International Development* 103:31–49.

Bocci, R., V. Chable, G. Kastler, and N. Louwaars. 2010. Policy recommendations – Farm Seed Opportunities. Accessed November 2010. www.farmseed.net.

Bonneuil, C. 2006. Mendelism, plant breeding and experimental cultures: Agriculture and the development of genetics in France. *Journal of the History of Biology* 39:281–308.

Bonneuil, C., and F. Thomas. 2009. *In*: Gènes, pouvoirs et profits. Recherche publique et régimes de production des savoirs de Mendel aux OGM. Editions fph Quæ.

Bos, H. 1920. Patenteering van nieuwigheden op tuinbouwgebied [Patenting of novelties in horticulture], *Floralia* 20:14–18.

Boyd, W. 2003. " Wonderful Potencies?" *In*: Schurman, R.D., D. Doyle, and T. Kelso (eds.). 2003. Engineering trouble, biotechnology and its Discontents. p. 38-39. Berkeley: University of California Press.

Ceccarelli S., E.P. Guimarães and E. Weltzien (eds). 2009. Plant breeding and farmer participation. Rome: FAO.

Center for Food Safety. 2005. Monsanto v. U.S. farmers. A Report by the Center for Food Safety. Accessed April 2011. http://www.centerforfoodsafety.org/pubs/CFSMOnsantovsFarmerReport1.13.05.pdf.

Center for Food Safety v. Vilsack. 2010. WL 3222482 (N.D. Cal. Aug. 13, 2010). Accessed January 4, 2011. http://graphics8.nytimes.com/packages/pdf/business/20100813Order.pdf.

Chable, V., F. Le Lagadec, N. Supiot, and N. Léa. 2009. Participatory research for plant breeding in Brittany, the Western region of France: Its actors and organisation. *In*: Proceedings of the 1st IFOAM conference on Organic Animal and Plant Breeding, Breeding Diversity. pp. 363-370, Santa Fe, New Mexico, August 25–28, 2009.

Chable, V., M. Conseil, E. Serpolay, and F. Le Lagadec. 2008. Organic varieties for cauliflowers and cabbages in Brittany: From genetic resources to participatory plant breeding. *Euphytica* 164:521–529.

Chable, V., A. Thommens, I. Goldringer, T. Valero Infante, T. Levillain, and E.T. Lammerts van Bueren. 2010. Report on the definitions of varieties in Europe, of local adaptation, and of varieties threatened by genetic erosion. Accessed December 2010. http://www.farmseed.net/home/the_project/download_the_project_documents/D1.2.pdf.

Culver, J.C. and J. Hyde. 2000. American dreamer: The life of Henry A. Wallace. New York: Norton.

Desclaux, D. 2005. Participatory plant breeding methods for organic cereals, *In*: Proceeding of the COST-SUSVAR/ECO-PB workshop on Organic Plant Breeding Strategies and the Uses of Molecular Markers. pp. 17–19. Driebergen, The Netherlands, January 2005.

Drahos, P., and J. Braithwaite. 2002. Information Feudalism. London: Earthscan.

Duvick, D.N. 2001. Biotechnology in the 1930s: The development of hybrid maize. *Nature Reviews Genetics* 2:69–74.

Einarsson, P. 1992. The urge to merge. Ref: seedling|seed-92-10-4. Accessed December 2010. http://www.grain.org/seedling/?id=375.

FAO. 2009. Second report on the state of the world's plant genetic resources for food and agriculture. Rome: FAO. Accessed April 14, 2011. http://www.fao.org/docrep/013/i1500e/i1500e00.htm.

FCEC. 2008. Evaluation of the Community acquis on the marketing of seed and plant propagating material (S&PM). Food Chain Evaluation Consortium Final Report, European Commission Directorate General for Health and Consumers. Accessed December 2010. http://ec.europa.eu/food/plant/propagation/evaluation/s_pm_evaluation_finalreport_en.pdf.

Fowler, C. 1994. Unnatural selection: Technology, politics, and plant evolution. New York: Gordon & Breach Science Publishers.

Gaile, Z. 2005. Organic seed propagation: Current status and problems in Europe. Report under the 6th Framework Program FP-2003-SSA-1-007003 ENVIRFOOD.

Gilstrap, M. 1961. The greatest service to any country. Seeds: The USDA yearbook of agriculture. pp. 18–27. Washington D.C.: U.S. Government Printing Office.

GRAIN. 2008. Whose harvest? The politics of organic seed certification, GRAIN Briefing. Accessed April 2011. http://www.grain.org/briefings/?id=207.

Halewood, M., P. Deupmann, B. Sthapit, R. Vernooy, and S. Ceccarelli. 2007. Participatory plant breeding to promote farmers' rights. Rome, Italy: Bioversity International.

Hjalmar-Nilsson, N. 1898. Einige kurze Notizen über die Schwedische Pflanzen-Veredlung zu Svalöf. Malmö, Skånska Lithografiska Aktiebolaget. 14 p.

Hubbard, K. 2009. Out of hand: Farmers face the consequences of a consolidated seed industry. Accessed December 2010. www.farmertofarmercampaign.org.

Jaffee, D., and P.H. Howard 2010. Corporate cooptation of organic and fair trade standards. *Agriculture and Human Values* 27:387–399.

J.E.M. Ag Supply v. Pioneer Hi-Bred. 2001. 534 U.S. 124. Accessed October 2010. http://www.law.cornell.edu/supct/html/99-1996.ZS.html.

Kloppenburg, J.R. 1988. First the seed: The political economy of plant biotechnology. New York: Cambridge University Press.

Lammerts van Bueren, E.T., M. Tiemens-Hulscher, and P.C. Struik. 2008. Cisgenesis does not solve the late blight problem of organic potato production: Alternative breeding strategies. *Potato Research* 51:89–99.

Lammerts van Bueren, E.T., and C. ter Berg. 2009. Biologisch uitgangsmateriaal voor 2010: Eindrapportage met het advies van de expertgroepen voor het ministerie van LNV. Rapport LT33. Driebergen: Louis Bolk Institute. Accessed April 18, 2011. http://www.louisbolk.org/index.php?page=medewerker&ht_employeeID=1642.

LeBuanec, B. 2006. Protection of plant-related innovations: Evolution and current discussion. *World Patent Information* 28: 50–62.

Leskien, D., and M. Flitner. 1997. Intellectual property rights and plant genetic resources: Options for shaping a *sui generis* system. Issues in Genetic Resources 6, International Plant Genetic Resources Institute, Rome.

Lòpez Noriega, I. 2009. European legislation in support of home gardens conservation. *In*: Bailey A., P. Eyzaguirre, L. Maggioni (eds.) Crop genetic resources in European home gardens. Rome, Italy: Bioversity International.

Löschenberger, F., A. Fleck, G. Grausgruber, H. Hetzendorfer, G. Hof, J. Lafferty, M. Marn, A. Neumayer, G. Pfaffinger, and J. Birschitzky. 2008. Breeding for organic agriculture – the example of winter wheat in Austria. *Euphytica* 163:469–480.

Louwaars, N.P. 2001. Regulatory aspects of breeding for diversity. *In*: H.D. Cooper, C. Spillane, and T. Hodgekin (eds.) Broadening the genetic base of crop production. pp. 105–114, Wallingford: CABI Publishing; Rome: FAO and IPGRI.

Louwaars, N.P., H. Dons, G. van Overwalle, H. Raven, A. Arundel, D. Eaton, and A. Nelis. 2009. Breeding business. The future of plant breeding in the light of developments in patent rights and plant breeder's rights. CGN-Report 2009–14. Wageningen, The Netherlands: Centre for Genetic Resources.

Monsanto v. Bayer. 2008. 514 F.3d 1229. Accessed January 4, 2011. http://caselaw.findlaw.com/us-federal-circuit/1132723.html.

Monsanto Co. v. McFarling. 2004. 363 F.3d 1336. Accessed January 4, 2011. http://caselaw.findlaw.com/us-federal-circuit/1167421.html.

Monsanto Co. v. Scruggs. 2006. 459 F.3d 1328. Accessed January 4, 2011. http://caselaw.findlaw.com/us-federal-circuit/1066894.html.

Monsanto v. Syngenta. 2007. 431 F.Supp.2d 482. Accessed January 4, 2011. http://caselaw.findlaw.com/us-federal-circuit/1466371.html.

Monsanto Co. 2009. "Monsanto's Technology/Stewardship Agreement." Accessed October 13, 2010. www.westernfarmservice.com/pdf/Corn/2009MTSA.pdf.

Navazio, J. 2010. Why organic seed? Organic Seed Alliance and Washington State University. Accessed March 2011. www.extension.org/article/18339.

Negri, V., N. Maxted, and M. Vetelainen. 2009. European landraces conservation: An introduction. *In*: Vetelainen, M., V. Negri, and N. Maxted (eds.). 2009. European landraces: On-farm conservation, management and use. Rome, Italy: Bioversity International.

Organic Seed Alliance. 2011. State of organic seed. Accessed April 2011. www.seedalliance.org.

Osman, A., and J.G.M. Pauw. 2005. Zomertarwerassen voor de bioteelt (in Dutch). Accessed April 2011. http://library.wur.nl/WebQuery/wurpubs/348362.

Patto, M.C.V., P.M. Moreira, N. Almeida, Z. Satovic, and S. Pego. 2008. Genetic diversity evolution through participatory maize breeding in Portugal. *Euphytica* 161:283–291.

Platt, A. 2010. Center for Food Safety vs. Vilsack: Roundup Ready Regulations. *Ecology Law Quarterly* 37:773–780.

Porfiri, O., M.T. Costanza, and V. Negri. 2009. Landrace inventoring in Italy: The Lazio regional law case study. *In*: Negri V., Vetelainen, M. and N. Maxted (eds.) European landrace conservation. Rome, Italy: Bioversity International.

Price, S.C. 1999. Public and private plant breeding. *Nature Biotechnology* 17:938.

Rangnekar, D. 2000. Intellectual property rights and the agriculture: An analysis of the economic impact of plant breeders' rights. http://www.actionaid.org.uk/doc_lib/ipr_agriculture.pdf

Rey, F., L. Fontaine, A. Osman, and J. Van Waes (eds.). 2008. Value for cultivation and use testing of organic cereal varieties: What are the key issues? *In*: Proceedings of the COST SUSVAR-ECOPB workshop in Brussels/Belgium, February 28–29, 2008. Paris, France: ITAB. Accessed December 2010. http://www.ecopb.org.

Roberts, M.T. 2003. J.E.M Ag Supply, Inc. v. Pioneer Hi-Bred International, Inc.: Its meaning and significance for the agricultural community. *Southern Illinois Law Journal* 28:91–128.

Rollin, S.F., and F.A. Johnston 1961. Our laws that pertain to seeds. Seeds: The USDA yearbook of agriculture. pp. 482–492. Washington, D.C.: U.S. Government Printing Office.

Salazar, R., N. Louwaars, and B. Visser. 2007. Protecting farmer, new varieties: New approaches to rights on collective innovations. *Plant Genetic Resources World Development* 35:1515–1528.

Smith, B.D, and R.A. Yarnell. 2009. Initial formation of an indigenous crop complex in eastern North America at 3800 B.P. *Proceedings of the National Academy of Science* 106:6561–6566.

Tripp, R., and N. Louwaars. 1998. Seed regulation: Choices on the road to reform. *Food Policy* 22:433–446.

US Code of Federal Regulations. 2010. Federal Seed Act Regulations. 7 CFR 201 et seq. Accessed October 5, 2010. http://ecfr.gpoaccess.gov/cgi/t/text/text-idx?c=ecfr&sid=a68a82e5c99b80dc9f77c26594d35774&tpl=/ecfrbrowse/Title07/7cfr201_main_02.tpl.

US Code. 2010. Federal Seed Act. 7 USC 1511 et seq. Accessed October 5, 2010. http://www.law.cornell.edu/uscode/html/uscode07/usc_sup_01_7_10_37.html.

Wolfe, M. 2008. How can we fit adaptive populations into a wider interpretation of the regulations? *In*: Rey, F., L. Fontaine, A. Osman, and J. van Waes (eds.) Value for Cultivation and Use testing of organic cereal varieties: What are the key issues? *In*: Proceedings of the COST SUSVAR-ECOPB workshop in Brussels/Belgium, pp. 49–51, February 28–29, 2008. Paris, France: ITAB.

Zaharia, H., and G. Kastler. 2003. La directive européenne 98/95/CE: Une avancée législative européenne pour les semences paysannes? Accessed October 4, 2010. http://www.semencespaysannes.org/dossiers/bip/fiche-bip-12.html.

Webliography

www.plantum.nl: Plantum NL is the Dutch association for breeding, tissue culture, production and trade of seeds and young plants.

www.hdra.org.uk/hsl.htm: Garden Organic – the national charity for organic growing.

www.organic-revision.org: Research to support revision of the UE regulation on Organic agriculture.

http://homepage.tinet.ie/~merlyn/seedsaving.html: Seedsaving and Seedsavers' resources.

http://ec.europa.eu/agriculture/organic/eu-policy/seed-databases en: links to databases for organic seeds and propagating materials in the different Member States.

www.upov.int: UPOV was established by the International Convention for the Protection of New Varieties of Plants, in Paris in 1961.

Section 2
Organic Plant Breeding in Specific Crops

9 Wheat: Breeding for Organic Farming Systems

Matt Arterburn, Kevin Murphy, and Steve S. Jones

Introduction

Wheat is a self-pollinated crop, and in contrast with many crops, seed recovery at maturity takes little effort. Breeding wheat for organic systems utilizes the standard techniques of pure line breeding toward homozygous stands. The main traits of interest in breeding for conventional systems – yield, mass market end-use quality, and disease resistance – are also of importance in organic breeding, but there is an additional urgency for selecting genotypes based on weed competition, nutrient-use efficiency, nutritional value, and end-use for local artisan markets.

Here we concentrate on less common breeding techniques for traits that are of specific interest to organic growers and marketers. We also focus on empowering growers and providing direct roles for them in the breeding of improved lines for their particular environments and specific needs. Farmers have been selecting wheat lines, in a variety of locales, systems, and cultures, for about 10,000 years. Our approach is to build on that legacy for the betterment of organic wheat production.

Methods

When selecting for traits in organic systems, the fidelity of the selection system is critical, as micro-environmental effects (highly localized environmental fluxes throughout a field), such as hydrological conditions, disease reservoirs, and nitrogen "hotspots" in the soil can often exert greater influence than in systems where high inputs create a more uniform set of conditions throughout a field plot. Without such inputs, environmental variances become more pronounced and the need to select genotypes that can better resist environmental flux become more critical (Wolfe et al., 2008). Breeding strategies must be modified to achieve gains in this new context.

Mass Selection

Plant breeders seeking superior genotypes suitable for organic cultivation will frequently use mass selection, a basic and straightforward breeding method, in which plants with superior performance

---

Organic Crop Breeding, First Edition. Edited by Edith T. Lammerts van Bueren and James R. Myers.
© 2012 John Wiley & Sons, Inc. Published 2012 by John Wiley & Sons, Inc.

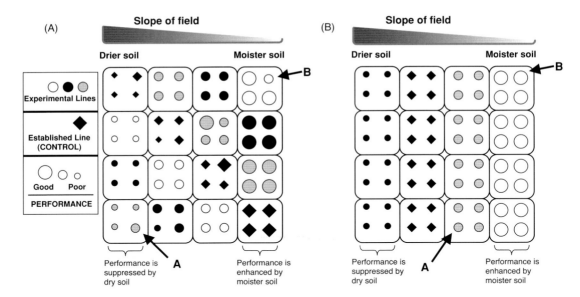

Figure 9.1 Randomized Complete Block Design (RCBD). Genotypes are randomized in grids across both axes of a field (Panel A). Plant A's genotype is superior, but micro-environmental effects in the field make B perform comparably. When RCBD design is not employed, environmental effects obscure genotypic effects (Panel B). Line B gives the false appearance of a superior genotype.

are selected under field conditions and their offspring are advanced, in bulk, to the next generation (Acquaah, 2007). After several generations of selection, the refined population consists of a heterogeneous mixture of plants that are optimized for cultivation in organic systems. In the case of self-fertilizing crops like wheat, plants are homozygous at essentially all loci, although heterogeneous at the population level (Allard, 1999). Individual plants can be isolated and advanced further to generate pure-line varieties. Mass selection is a very accessible selection method and practical for any investigator, from professional breeder to amateur gardener, and thus has potential for use in participatory plant breeding efforts involving local growers, co-ops, or indigenous groups (Witcombe et al., 1996). Because plants of superior phenotypes are selected at random via walking transects of a field, selection is highly influenced by micro-environmental impacts. In order to prevent selection of lines based largely on environmental variation, a randomized complete block design (see fig. 9.1) is implemented with a judicious system of controls. Even with these methods, mass selection is most effective when the traits being selected for have high heritability, and are influenced more by genetic variance than environmental factors (Bernardo, 2002).

Cross-pollination, by wind or animal pollinators, is a potential complication in selection efforts. Pure-line breeding of self-pollinating crops such as wheat requires homozygosity for consistent performance under the desired conditions. A common method to isolate homozygous lines is the ear-to-row method of progeny testing (see fig. 9.2) in which single ear of grain is sown in a row to evaluate if segregation for a trait is occurring.

Bulk Selection and Evolutionary Breeding

Bulk population strategies permit natural selection to exert its effects on segregating populations in the field. Inferior germplasms are culled passively from the population, as poor performance

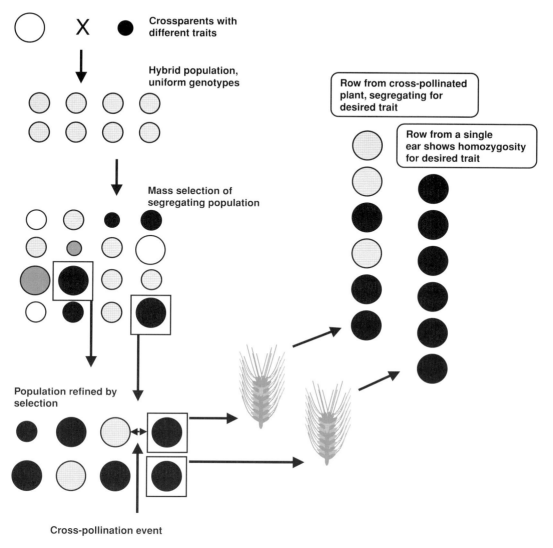

Figure 9.2 Progeny testing using ear to row method. In this simplified example, circle diameter and depth of shading indicate desired traits for selection.

results in a proportional decrease in representation in the next generation. This represents a de facto selection process with little breeder input, although initial rounds of natural selection can be followed by active selection in later generations. Seed is bulked and a random portion is planted the next generation to advance the lines. In later generations, superior lines can be isolated and grown as progeny rows. Large population sizes must be generated to avoid bottleneck effects, and seed must be randomized after each generation, to minimize the loss of desirable genotypes through random sampling, a concept known as genetic drift (see fig. 9.3).

The passive nature of bulk breeding methods is a logical complement to organic cultivation systems, which likewise employ less external interference (Dawson et al., 2008b). Genotypes with

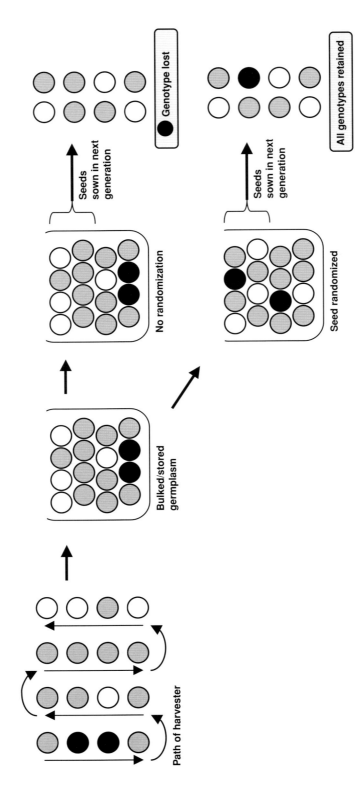

Figure 9.3 Bulk breeding can cause unintended loss of genotypes (genetic drift) unless seed is randomized between generations.

superior response to biotic and abiotic factors naturally rise to prominence in the population, a concept known as evolutionary breeding (EB; Suneson, 1956). Such selection methods are most effective in populations that exhibit broad genetic heterogeneity; a notable example is the improvement of barley lines resulting from composite crosses (involving multiple, genetically disparate parents) in an EB effort at the University of California Davis (Allard, 1990). Bulk breeding based on natural selection can be used in adapting lines for specific environments, such as marginal lands or in selecting for lines with strong performance in a variety of environments (Lammerts van Bueren et al., 2008; Wolfe et al., 2008).

Combined and Dedicated Approaches

Many breeding programs use a combination of methods, designed to either expedite the breeding process or to permit the maintenance of traditional and organic breeding efforts within the same program. Organic breeding can be time consuming when compared with conventional breeding, particularly in breeding programs where winter nurseries are not employed to advance generations in the off-season. Such methods are often eschewed when all selection efforts must be carried out under field conditions. This is in part the rationale for combining traditional and organic breeding methods, where initial rounds of selection are under traditional conditions, followed by selection of advanced generations under organic conditions. Choosing the stage at which to transition from conventional to organic conditions is critical. Earlier transitions lengthen the duration of breeding efforts, while later transitions run the risk of losing genotypes that would thrive under organic conditions, but exhibit unexceptional performance in early stages of conventional selection. The first decision that must be made is whether initial crosses are based on expectations of performance under both conventional and organic regimens, or under solely organic conditions. Three commonly used approaches are (1) limiting organic trials to user-end efforts only, (2) using selection under conventional regimens in early generations and organic conditions in advanced populations, and (3) a wholly organic program in which crosses and selection efforts in every generation are carried out within the context of organic systems (Wolfe et al., 2008). It must be noted however, that variations in input levels in different systems can have an impact on selection efforts and data acquired for a number quantitative traits such as yield. Even under strictly organic conditions, the quantity of permissible inputs can vary based on grower preference and national standards for organic production. Löschenberger et al. (2008) found success breeding for both organic and low-input agriculture simultaneously, for select traits, including protein content and yield stability. Indirect selection in low-input systems should focus on highly heritable traits, such as tillering capacity and resistance to specific diseases, while some studies suggest that traits that are more specific to organic agriculture, including weed suppression ability and nutrient efficiency, should be selected for within such systems (Löschenberger et al., 2008). This idea is supported by data indicating that direct selection in organic systems can produce wheat yields from 5 to 31% higher than the yields resulting from indirect selection (Murphy et al., 2007). Few breeding programs, public or private, conduct efforts solely under organic conditions, although this could change if demand for organic products increases proportionally.

Evolutionary Participatory Breeding

A relatively novel approach to organic wheat breeding combines the social and logistical framework of participatory plant breeding (PPB) approaches (see Chapter 6) with the selection strategies of EB.

This method is termed evolutionary participatory breeding (EPB) and involves the use of natural selection in growers' fields, to select genotypes adapted to their intended cultivation site. In essence, this method promotes the production of "modern landraces," including a large diversity of genotypes adapted through natural selection to a specific local ecosystem (Murphy et al., 2005). Initial crosses between parent lines are generally conducted at research facilities and seed increased in the F2, after which F3 seed is distributed to the grower for several generations of selection at their farm. Various levels of farmer-based selection, such as roguing of poor performers or screening of seed, can be applied. Depending on end-use quality needs, breeders or farmers can isolate high-performing plants and conduct cultivar trials and generate pure lines, or may simply repeat bulk selection indefinitely to generate a heterogeneous, high-performing population (Murphy et al., 2005). Parent genotypes should be chosen with the needs and goals of the grower in mind, but the degree of heterogeneity in the hybrid population is also of great importance. Populations with relatively low variability, such as those derived from a single cross between disparate genotypes, speed the refinement process but contain fewer unique genotypes. Populations resulting from composite crosses invite greater complexity and length of breeding efforts, lower the proportional representation of optimal genotypes, and thus increase the potential for micro-environmental effects to obscure optimal genotypes. Population heterogeneity is also proportional to required population size, which must be tailored to the land- and time-commitment of the farmer.

Traits for Selection in Organic Breeding Programs

Using the aforementioned selection methods, investigators have made considerable gains in breeding and research efforts to isolate genotypes of optimal value for organic systems. Desirable traits in organic systems include those shared with conventional cropping systems, such as yield and end-use quality, and those that are of heightened importance in organic systems, including competition with weeds, seed-borne disease resistance, and nitrogen-use efficiency (Lammerts van Bueren et al., 2010; Wolfe et al., 2008). Various studies on these traits have helped shape the current view of the challenges and strategies for optimizing genotypes for organic systems.

Weed Suppression and Competition

Wheat varieties are genetically variable in their ability to compete with a variety of weeds, and grain yields have been found to be significantly ($p < 0.001$) affected by weedy conditions (Huel and Hucl, 1996). In a study evaluating grain yield and weed suppression ability (WSA) of 63 historical and modern spring wheat varieties, top-ranked varieties suppressed weed weight per plot by 573% over those with lowest WSA (Murphy et al., 2008a). This indicates wide genetic variance in WSA among wheat varieties, most likely due to lack of selection for WSA due to herbicide use. With future selection efforts oriented toward WSA (a logical goal for organic systems), this genetic variability implies great potential for improvement of this trait.

Because organic systems rely more heavily on mechanical weed control, selection of genotypes that exhibit tolerance for these methods may prove a promising strategy for direct selection of weed competitiveness. Physical damage to wheat plants from mechanical weed control may cause significant yield reduction (Faustini and Paolini, 2005). The ability to tolerate damage and rapidly recover following mechanical weed treatments is therefore an important trait for varieties used in organic systems (Donner and Osman, 2006; Murphy et al., 2008a). Such traits would be of particular benefit in reduced tillage systems, which require more frequent application of mechanical weed control (Berner et al., 2008; Krauss et al., 2010). In breeding studies under different systems, significant variety × tillage interactions for grain yield were evident, indicating that selection under no tillage conditions could optimize genotypes for such systems (Hall and Cholick, 1989). Additionally, in a study on the effect of mechanical harrowing on spring wheat, a genotype × treatment interaction was found with weed weight per plot as a response variable (Murphy et al., 2008a).

Resistance to Seed-Borne Disease

Organic farming standards prohibit most chemical seed treatments; hence, selection for resistance to seed-borne diseases in organic production systems is of great importance. Common bunt (*Tilletia tritici* syn. *T. caries*), the most important disease of wheat prior to fifty years ago (Line, 2002), is obviated by fungicide seed treatments in conventional farming but which are prohibited under organic certification standards (Matanguihan et al., in press). Employment of organic seed treatments has shown varying degrees of effectiveness but it generates additional inputs and increases production costs (El-Naimi et al., 2000; Smilanick et al., 1994). Breeding of varieties with resistance, or at least tolerance, to common bunt and dwarf bunt for would promote crop health and provide an economical advantage for organic wheat cultivation (Blazkova and Bartos, 2002). Wheat varieties with enhanced resistance to common bunt have been developed by introgression of major race-specific resistance genes (Bt1–Bt13) and three quantitative trait loci associated with common bunt resistance were recently detected (Ciuca and Saulescu, 2008; Fofana et al., 2008; Martynov et al., 2004).

Nitrogen-use Efficiency

Due to an abundance of available nitrogen (N) in high-input systems, wheat varieties selected under conventional conditions may not possess traits in which the nitrogen-use efficiency of the plant is optimized (Dawson et al., 2011). Nitrogen-use efficiency is a genetically complex trait, and in addition to a variety of gene loci within the plant, external biotic and abiotic factors such as soil quality and fertility, precipitation levels, temperature flux, weed presence, and disease and insect pressure play roles in the chemical reactions that condition this trait (Dawson et al., 2008a). However, this same complexity and influence of biotic factors provide several avenues for selection efforts. Suggested strategies for refining nitrogen-efficient varieties from populations include optimizing symbiosis with free-living soil microorganisms such as *Azospirillum* or arbuscular mycorrhizal fungi (Azcon et al., 2001; Hetrick et al., 1992; Hoagland et al., 2008), the ability to out-compete weeds for available N (Giambalvo et al., 2010), and the ability to maintain photosynthesis under N stress (Dawson et al., 2008a).

End-use Quality

Desirable wheat quality characteristics differ based on the intended market class and baking product. Comparatively, few studies have focused on quality aspects of organic wheat production (Gooding et al., 1999; Kihlberg et al., 2004; Nass et al., 2003; Starling and Richards, 1990). More emphasis is also needed on the development of tests to assess bread making quality for whole meal bread and artisan bread-making processes (including sour dough–based bread production) often used by smaller bakeries (Kihlberg et al., 2004; Gelinas et al., 2009). Grain protein content is a critical factor in determining overall bread-making quality. Some studies have found no difference in grain protein content between varieties selected from organic versus conventional systems (Mason et al., 2007; Shier et al., 1984), while other studies have shown higher levels of protein in conventional compared with organic systems (Baeckstrom et al., 2004; Poutala et al., 1993). Grain protein content is highly dependent on climatic conditions and available soil-N, especially late in the growing season during grain filling. Thus there is potential for much greater environmental variance for these traits in organic systems.

Sodium dodecyl sulfate (SDS) sedimentation rate, a measure of gluten strength, has been shown to be lower in organic compared to conventional crops (Mason et al., 2007) and to increase proportionally with N supply (Ames et al., 2003; Gooding et al., 1993; Loveras et al., 2001). However, while total protein content and gluten strength is lowered, other studies have indicated that organic fertilization regimes, while reducing protein content, can improve other bread-making quality related parameters, including protein composition, gliadin-to-glutenin ratio, starch quality and length of amylopectin chains, acetic acid soluble proteins, diameter of starch granules, pentosan content, amylase activity, and water absorption (Fredriksson et al., 1998; Guttieri et al., 2002). However, reports from bakers in several EU countries suggest that these gains do not compensate fully for the reduction in protein content (Lammerts van Bueren et al., 2010).

Nutritional Value

Nutritional quality in organic wheat is frequently dependent on specific management practices and particularly on soil environment (Hornick, 1992). Soil organic matter, pH, the bioavailability of soil minerals, and other environmental and soil conditions can also affect grain micronutrient content (Fageria and Zimmerman, 1998; Fageria et al., 2002; Sims, 1986). Low soil pH has been shown to reduce plant uptake of the essential macronutrients calcium and magnesium, while increasing uptake of the micronutrients zinc, manganese, and iron (Fageria et al., 2002). Selection under specific soil conditions may allow further optimization of nutritional quality.

Despite a large role for environmental impact on nutritional content, there is a significant genetic contribution to these phenotypes. Differences in concentration of iron, zinc, and other micronutrients have been reported among various wheat genotypes (Frossard et al., 2000; Murphy et al., 2008b; Shen et al., 2002). Wild species, in particular, appear to be valuable genetic resources for traits associated with enhanced nutritional content (Murphy et al., 2009). Wild emmer wheat, for instance, exhibits grain protein, zinc, and iron concentrations that are often twice as high as domesticated modern wheat varieties (Chatzav et al., 2010; Uauy et al., 2006), though this could be due to the dilution effect rather than to actual higher levels of nutrients in the bran. Wild species are likely to be critical genetic resources for introgression of these traits, just as they are for disease resistance phenotypes.

A Case Study for EPB: Lexi's Project

The power and versatility of the EPB technique is illustrated on a non-organic farm, but is nonetheless a relevant demonstration of what a single farm family can accomplish in breeding their own wheat line. Twelve year-old student Lexi Roach was looking for something to do on her grandfather's farm and decided that breeding her own wheat variety would be a fun project. She contacted breeders at Washington State University, selected parents with promising traits for her family's farm and made initial crosses. Within two years, Lexi was cultivating F3 populations on the family farm. Choosing emergence as her a key selection criterion while growing wheat in dry areas, Lexi planted the seed very deep, over 200 mm. She removed diseased lines each year, harvested in bulk and planted a random sample of the next generation. She maintained three populations from three separate crosses on approximately 1 hectare each year over a period of six years. In 2010, one of Lexi's bulk lines, WA0008094, was the highest yielding wheat in the official Washington State University wheat yield trials for Douglas County and topped 59 other lines from 11 public and private breeding programs in the United States. This project, executed by a grade-school student, illustrates the versatility, utility, and efficacy of EPB methods.

A Case Study for Breeding within a Supply Chain Approach: Peter Kunz and Sativa

In collaboration with biodynamic and organic farmers in Switzerland and Germany, grain breeder Peter Kunz developed multiple varieties of wheat and spelt over a 20-year period (see www.peter-kunz.ch). Currently, Peter Kunz has released 10 winter wheat and 5 spelt varieties. Wheat varieties include Pollux, Ataro, Scaro, and Tengri among others, and spelt varieties include Alkor, Titan, Sirino and Tauro (Hildermann et al., 2009). These varieties emphasize their flavor and taste and are marketed and sold advertising these qualities. To assist with the marketing of these varieties, Kunz works with an independent organization, Sativa Vermehrung organization (Sativa) to multiply and sell his wheat. The second largest and a "socially-fair" supermarket chain in Switzerland, COOP, sells bread that has been baked from Kunz's spelt and wheat varieties under the trademark "Sativa." A farmer cooperative of 20 to 30 members grows the grain and Kunz himself tests for quality control and varietal identity for COOP. It is this type of breeder-farmer-marketer collaboration that will prove invaluable to the future sales and distribution of wheat varieties bred through participatory methods. Kunz and Sativa have helped pave the way for long-term success in collaborative varietal development that can be used as a template to other breeders of organic wheat varieties across the world breeding for limited areas (Osman, 2007).

Conclusion

Breeding wheat under organic regimens cannot be approached in the same manner as conventional wheat breeding. Selection methodologies and logistics must be altered to isolate competitive genotypes for organic systems. Refinement of populations based on natural selection shows promise. Critical traits for successful organic varieties differ significantly from those selected for in conventional systems, including a heightened need for weed suppression ability and nitrogen-use efficiency in organic varieties. Data acquired in a variety of studies supports separation of organic and conventional breeding programs, with dedicated organic breeding efforts potentially capable of yielding great success. Efforts to date suggest that selection conditions should reflect the input levels and field conditions employed by the grower as closely as possible, due to higher environmental variation in organic systems. For this reason, participatory and evolutionary participatory breeding initiatives are gaining traction in the organic arena. With increasing market demand, and growing interest in low-input agriculture for ecological reasons, organic wheat production efforts are certain to increase, representing a bold challenge for plant breeders willing to contribute to this promising new system.

References

Acquaah, G. 2007. Principles of plant genetics and breeding. Malden, MA: Blackwell.

Allard, R.W. 1999. Principles of plant breeding. Hoboken, NJ: Wiley.

Ames, N.P., J.M. Clarke, J.E. Dexter, S.M. Woods, F. Selles, and B. Marchylo. 2003. Effects of nitrogen fertilizer on protein quantity and gluten strength parameters in durum wheat (*Triticum turgidum* L var. *durum*) cultivars of variable gluten strength. *Cereal Chemistry* 80:203–211.

Azcon, R.L., and R. Rodriguez. 2001. Differential contribution of arbuscular mycorrhizal fungi to plant nitrate uptake (^{15}N) under increasing N supply to soil. *Canadian Journal of Botany* 79:1175–1180.

Baeckstrom, G., U. Hanell, and G. Svensson. 2004. Baking quality of winter wheat grown in different cultivating systems 1992–2001 – a holistic approach. *Journal of Sustainable Agriculture* 3:63–83.

Bernardo, R. 2002. Breeding for quantitative traits in plants. Woodbury, MN: Stemma Press.

Berner, A., I. Hildermann, A. Fliessbach, L. Pfiffner, U. Niggli, and P. Mäder. 2008. Crop yield and soil fertility response to reduced tillage under organic management. *Soil and Tillage Research* 101:89–96.

Blazkova, V., P. Bartos. 2002. Virulence pattern of European bunt samples (*Tilletia tritici* and *T. laevis*) and sources of resistance. *Cereal Research Communications* 30:335–342.

Chatzav, M., Z. Peleg, L. Ozturk, A. Yazici, T. Fahima, I.Cakmak, Y. Saranga. 2010. Genetic diversity for grain nutrients in wild emmer wheat: Potential for wheat improvement. *Annals of Botany* 105:1073–1080.

Ciuca, M., N.N. Saulescu. 2008. Screening Romanian winter wheat germplasm for presence of Bt10 bunt resistance gene, using molecular markers. *Romanian Agricultural Research* 25:1–5.

Dawson, J.C., K.M. Murphy, D.R. Huggins, and S.S. Jones. 2011. Evaluation of winter wheat breeding lines for traits related to nitrogen use under organic management. *Organic Agriculture* 1:65–80.

Dawson, J.C., D.R. Huggins, and S.S. Jones. 2008a. Characterizing nitrogen use efficiency in natural and agricultural ecosystems to improve the performance of low input and organic agricultural systems. *Field Crops Research* 107:89–101.

Dawson, J., Murphy K., and Jones S. 2008b. Decentralized selection and participatory approaches in plant breeding for low-input systems. *Euphytica* 160:143–154.

Donner, D., and A. Osman. 2006. Cereal variety testing for organic and low input agriculture handbook, COST860-SUSVAR, Wageningen, Netherlands.Driebergen, The Netherlands: Louis Bolk Institute.

El-Naimi, M., H. Toubia-Rahme, and O.F. Mamluk. 2000. Organic seed-treatment as a substitute for chemical seed-treatment to control common bunt of wheat. *European Journal of Plant Pathology* 106:433–437.

Fageria, N.K., V.C. Baligar, and R.B. Clark. 2002. Micronutrients in crop production. *Advances in Agronomy* 77:185–268.

Fageria, N., and F. Zimmerman. 1998. Influence of pH on growth and nutrient uptake by crop species in an oxisol. *Communications in Soil Science and Plant Analysis* 29:2675–2682.

Faustini, F., and R. Paolini. 2005. Organically grown durum wheat (*Triticum durum* Desf.) varieties under different intensity and time of mechanical weed control. *In*: Proceedings of the 13th EWRS Symposium. Bari, Italy, June 20–23, 2005.

Fofana, B., D.G. Humphries, S. Cloutier, C.A. McCartney, and D.J. Somers. 2008. Mapping quantitative trait loci controlling common bunt resistance in a doubled haploid population derived from the spring wheat cross RL4452 x AC Domain. *Molecular Breeding* 21:317–325.

Fredriksson, H., L. Salomonsson, R. Andersson, and A.C. Salomonsson. 1998. Effects of protein and starch characteristics on the baking properties of wheat cultivated by different strategies with organic fertilizers and urea. *Acta Agriculturae Scandinavica Section B-Soil and Plant Science* 48:49–57.

Frossard, E., M. Bucher, F. Machler, A. Mozafar, and R. Hurrell. 2000. Potential for increasing the content and bioavailability of Fe, Zn, and Ca in plants for human nutrition. *Journal of the Science of Food and Agriculture* 80:861–870.

Gelinas, P., C. Morin, J.F. Reid, and P. Lachance. 2009. Wheat cultivars grown under organic agriculture and the bread making performance of stone-ground whole wheat flour. *International Journal of Food Science and Technology* 44:525–530.

Giambalvo, D., P. Ruisi, G. DiMiceli, A.S. Frenda, and G. Amato. 2010. Nitrogen use efficiency and nitrogen fertilizer recovery of durum wheat genotypes as affected by interspecific competition. *Agronomy Journal* 102:707–715.

Gooding, M., W. Davies, A. Thompson, and S. Smith. 1993. The challenge of achieving bread making quality in organic and low input wheat in the UK – A review. *Aspects of Applied Biology* 36:1183–1190.

Gooding, M.J., N.D. Cannon, A.J. Thompson, and W.P. Davies. 1999. Quality and value of organic grain from contrasting bread making wheat varieties and near isogenic lines differing in dwarfing genes. *Biological Agriculture and Horticulture* 16:335–350.

Guttieri, M.J., R. McLean, S.P. Lanning, L.E. Talbert, and E.J. Souza. 2002. Assessing environmental influences on solvent retention capacities of two soft white spring wheat cultivars. *Cereal Chemistry* 79:880–884.

Hall, E.F., and F.A. Cholick. 1989. Cultivar × tillage interaction of hard red spring wheat cultivars. *Agronomy Journal* 81:789–792.

Hetrick, B.A.D., G.W.T. Wilson, and T.S. Cox. 1992. Mycorrhizal dependence of modern wheat varieties, landraces and ancestors. *Canadian Journal of Botany* 70:2032–2040.

Hildermann, I., M. Messmer, P. Kunz, A. Pregitzer, T. Boller, A.Wiemken, and P. Mäder. 2009. Cultivar × site interaction of winter wheat under diverse organic farming conditions. Tagungsband der 60. Tagung der Vereinigung der Pflanzenzüchter und Saatgutkaufleute Österreichs, November 24-26, 2009. pp. 163–166. Austria: Raumberg-Gumpenstein.

Hoagland, L., K. Murphy, L. Carpenter-Boggs, and S. Jones. 2008. Improving nutrient uptake in wheat through cultivar specific interaction with *azospirillum*. *In*: Neuhoff, D. et al. (eds.) Cultivating the future based on science. Volume 1: Organic crop production. Proceedings of the Second Scientific Conference of the International Society of Organic Agriculture Research (ISOFAR), held at the 16th IFOAM Organic World Conference in cooperation with the International Federation of Organic Agriculture Movements (IFOAM) and the Consorzio Modena Bio in Modena, Italy, June 18–20, 2008. pp. 562–565. Bonn, Germany: ISOFAR.

Hornick, S.B. 1992. Factors affecting the nutritional quality of crops. *American Journal of Alternative Agriculture* 7:63–68.

Huel, D.G., and P. Hucl. 1996 Genotypic variation for competitive ability in spring wheat. *Plant Breeding* 115:325–329.

Kihlberg, I., L. Johansson, A. Kohler, and E. Risvik. 2004. Sensory qualities of whole wheat pan bread – influence of farming system, milling and baking technique. *Journal of Cereal Science* 39:67–84.

Krauss, M., A. Berner, D. Burger, A. Wiemken, U. Niggli, and P. Mäder. 2010. Reduced tillage in temperate organic farming: Implications for crop management and forage production. *Soil Use and Management* 26:12–20.

Lammerts van Bueren, E.T., H. Østergård, I. Goldringer, and O. Scholten. 2008. Plant breeding for organic and sustainable, low-input agriculture: Dealing with genotype–environment interactions. *Euphytica* 163:321–322.

Lammerts van Bueren, E.T., S.S. Jones, L. Tamm, K.M. Murphy, J.R. Myers, C. Leifert, and M.M. Messmer. 2010. The need to breed crop varieties suitable for organic farming, using wheat, tomato and broccoli as examples: A review. *NJAS Wageningen Journal of Life Sciences* 58:193–205.

Löschenberger, F., A. Fleck, H. Grausgruber, H. Hetzendorfer, G. Hof, J. Lafferty, M. Marn, A. Neumayer, G. Pfaflinger, and J. Birschitzky. 2008. Breeding for organic agriculture: The example of winter wheat in Austria. *Euphytica* 163:469–480.

Line, R.F. 2002. Stripe rust of wheat and barley in North America: A retrospective historical review. *Annual Review of Phytopathology* 40:75–118.

Loveras, J., A. Lopez, J. Ferran, S. Espachs, and J. Solsona. 2001. Bread-making wheat and soil nitrate as affected by nitrogen fertilization in irrigated Mediterranean conditions. *Agronomy Journal* 93:1183–1190.

Martynov, S.P., T.V. Dobrotvorskaya, and O.D. Sorokin. 2004. Comparative genealogical analysis of the resistance of winter wheat to common bunt. *Genetika* 40:516–530.

Mason, H., A. Navabi, B. Frick. J. O'Donovan, D. Niziol, and D. Spaner. 2007. Does growing Canadian Western Hard Red Spring wheat under organic management alter its bread making quality? *Renewable Agriculture and Food Systems* 22:157–167.

Matanguihan, J.B., K.M. Murphy, and S.S. Jones. 2011. Control of common bunt in organic wheat. *Plant Disease* 95:92–103.

Murphy K., D. Lammer, S. Lyon, B. Carter, B., and S. Jones. 2005. Breeding for organic and low-input farming systems: An evolutionary-participatory breeding method for inbred cereal grains. *Renewable Agriculture and Food Systems.* 201: 48–55.

Murphy K., K. Campbell, S. Lyon, and S. Jones. 2007. Evidence of varietal adaptation to organic farming systems. *Field Crops Research* 3:172–177.

Murphy, K., J. Dawson, and S.S. Jones. 2008a. Relationship among phenotypic growth traits, yield and weed suppression in spring wheat landraces and modern cultivars. *Field Crops Research* 105:107–115.

Murphy, K.M., P.R. Reeves, and S.S. Jones. 2008b. Relationship between yield and mineral nutrient concentration in historical and modern spring wheat cultivars. *Euphytica* 163:381–390.

Murphy, K.M., L.A. Hoagland, P.G. Reeves, B. Baik, and S.S. Jones. 2009. Nutritional and quality characteristics expressed in 31 perennial wheat breeding lines. *Renewable Agriculture and Food Systems* 24:285–292.

Nass, H.G., J.A. Ivany, and J.A. MacLeio. 2003. Agronomic performance and quality of spring wheat and soybean cultivars under organic culture. *American Journal of Alternative Agriculture* 18:67–84.

Osman, A. 2007. Sativa bread in Switzerland: Collaboration between an organic cereal breeder, farmers and a retailer. *In*: Osman, A.M., K.-J. Müller, K.-P. Wilbois (eds.) Different models to finance plant breeding. Proceedings of the ECO-PB International workshop on Different Models to Finance Plant Breeding, February 27, 2007, Frankfurt, Germany. Accessed January 26, 2011. www.eco-pb.org.

Poutala, R., J. Korva, and E. Varis. 1993. Quality of commercial samples of organically grown wheat. *Journal of Applied Biology* 36:205–209.

Shen, J.B., F.S. Zhang, Q. Chen, Z. Rengel, C.X. Tang, and C.X. Song. 2002. Genotypic difference in seed iron content and early responses to iron deficiency in wheat. *Journal of Plant Nutrition* 25:1631–1643.

Shier, N., J. Kelman, J. Dunson. 1984. A comparison of crude protein, moisture, ash and crop yield between organic and conventionally grown wheat. *Nutrition Reports International* 30:71–76.

Sims, J. 1986. Soil pH effects on the distribution and plant availability of manganese, copper and zinc. *Soil Science Society of America Journal* 50:367–373.

Smilanick, J.L., B.J. Goates, R. Denisarrue, G.F. Simmons, G.L. Peterson, D.J. Henson, and R.E. Riji. 1994. Germinability of *Tilletia* spp. teliospores after hydrogen-peroxide treatment. *Plant Disease* 78:861–865.

Starling, W., and M.C. Richards. 1990. Quality of organically grown wheat and barley. *Aspects of Applied Biology* 25:193–198.

Suneson, C.A. 1956. An evolutionary plant breeding method. *Agronomy Journal* 48:188–191.

Uauy, C., A. Distelfeld, T. Fahima, A. Blechl, and J. Dubcovsky. 2006. A NAC gene regulating senescence improves grain protein, zinc, and iron content in wheat. *Science* 314:1298–1301.

Witcombe, J.R., A. Joshi, K.D. Joshi, and B.R. Sthapit. 1996. Farmer participatory crop improvement. I. Varietal selection and breeding methods and their impact on biodiversity. *Experimental Agriculture* 35:445–260.

Wolfe, M.S., J.P. Baresel, D. Desclaux, I. Golringer, S. Hoad, G. Kovacs, F. Löschenberger, T. Miedaner, H. Østergård, and E.T. Lammerts van Bueren. 2008. Developments in Breeding Cereals for Organic Agriculture. *Euphytica* 163:323–346.

10 Maize: Breeding and Field Testing for Organic Farmers

Walter A. Goldstein, Walter Schmidt, Henriette Burger, Monika Messmer, Linda M. Pollak, Margaret E. Smith, Major M. Goodman, Frank J. Kutka, and Richard C. Pratt

Introduction

A large German seed company working with a university and a group of U.S. public breeders acting in partnership with each other and with farmers' organizations have been testing or breeding maize for organic farming. This report presents an overview on their efforts and strategies and presents preliminary results on the value of selection under organic conditions.

In the first part of this chapter we will review results that address many of the issues associated with breeding corn for organic systems. These include what kind of maize organic farmers want, alternatives to single cross hybrids, obtaining enough testing data to make results interesting to seed companies and farmers, benefits to breeding under organic conditions, and traits that are necessary to test for under organic conditions.

The second part of the chapter will be devoted to describing ongoing breeding programs, methods, and future directions.

What Kind of Maize do Organic Farmers Want?

The issue of breeding maize hybrids specifically adapted to the needs of organic farmers seemed to surface as a question for larger numbers of organic farmers in the United States in 2002 and 2003. At that time, transgenic contamination of maize cultivars and the effects of seed industry consolidation had become major threats to the availability of suitable seed. Michael Fields Agricultural Institute (MFAI) had multiple listening sessions with organic farmers throughout the upper Midwest at that

---

Organic Crop Breeding, First Edition. Edited by Edith T. Lammerts van Bueren and James R. Myers.
© 2012 John Wiley & Sons, Inc. Published 2012 by John Wiley & Sons, Inc.

time. Farmers clearly wanted more than access to non-transgenic seed. They also wanted cultivars that would do well under conditions in which competition with weeds is an issue and nitrogen is obtained mainly from soil organic matter or organic manures. Some of the farmers did not have easy access to animal manure and depended on soils and rotations for guaranteeing sufficient N for their corn crops. Farmers who fed the maize to their own livestock were interested in nutritional value and palatability. They were aware that modern breeding had reduced grain quality (see also Duvick et al., 2004). Considerable interest was expressed in obtaining new varieties with increased protein content and protein quality, more minerals, vitamins, and better flavor. However, most farmers generally wanted to continue using hybrids and few seemed willing to take more than a 10% reduction in yield for a high quality hybrid.[1]

Are There Viable Alternatives to Single Cross Hybrids?

Most farmers are presently growing single cross hybrids. Such hybrids are produced by crossing two inbred lines belonging to different but complimentary heterotic groups. However, for organic farmers in the United States there may be room for open pollinated varieties or for alternative types of hybrids that may be cheaper and easier to produce due to weed problems. Some farmers are interested in growing open pollinated populations because they want to save their own seed. Alternative hybrids could include sister line crosses or other three-way crosses, double crosses, topcrosses between diverse open pollinated populations and inbreds, and the use of modern synthetic open pollinated populations.

Production of hybrids made by crossing weed competitive, open pollinated varieties with each other or with inbreds rather than by crossing inbreds was tested and proposed by the midwestern team as an alternative, because it resulted in higher seed yields and less weeds and some of the alternative hybrids were competitive with commercial hybrids for yield (see fig. 10.1). However, the response from seed companies was that hybrids made with populations would not satisfy their customers' needs for uniformity and might thereby damage the companies' reputations.

Synthetic varieties are populations that are generally created by intermating a set of proven inbred lines. Synthetics can achieve higher yield levels than older open pollinated varieties that have received little systematic breeding for yield. They can also be propagated for many generations with little loss of yield.

However, numerous studies have shown that synthetics currently produce lower yields that are probably not acceptable to most farmers in the United States (Kutka and Smith, 2007). To test the viability of synthetics under organic conditions, Burger (2008) developed a set of synthetics using results from yield trials on organic farms in Germany. Information on general combining ability between heterotic groups was used to create three different flint × dent synthetics (involving 44 × 44, 22 × 22, and 11 × 11 crosses). A fourth synthetic was produced by recombining the broad-based synthetic that had previously been made with 88 lines. Figure 10.2 shows the performance of the four synthetics compared to the single hybrids, each synthetic having been tested four times. The yield of the synthetics suggests half of the supplementary heterosis between the dent and the flint pool was lost in the synthetics. Furthermore, the results show enormous differences in yield between the individual hybrids. The best hybrids surpass the best synthetics by more than 3 Mg ha^{-1}. For a farmer who prefers hybrids, this means around €900 more income per hectare, calculating with the prices achieved in Germany in January of 2009. Even if the entire cost for organically produced hybrid seed is included in the calculation (around €205.00 per ha), the organic farmer still obtains a surplus profit of around €700 per ha.

Figure 10.1 An experiment comparing conventional inbreds and populations for hybrid seed production. Detasseled "female" plants of the conventional inbreds (left two rows) showed greater weed infestations and produced approximately half as much seed as the open pollinated females (right two rows) (photo courtesy of Walter Goldstein).

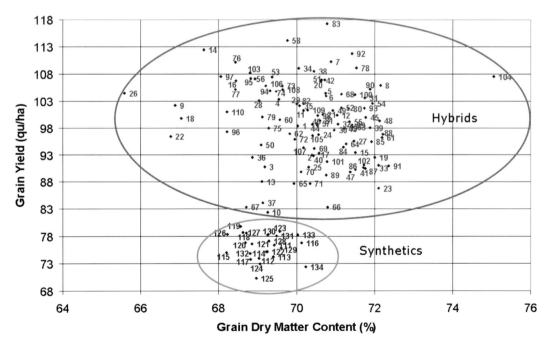

Figure 10.2 Performance comparison of four synthetics (each synthetic tested four times) and a set of inter-pool hybrids developed from the same starting samples of germplasm, including some commercial check-hybrids (according to data from Burger (2008)).

Testing and Using Alternative Hybrids

In the United States, interest by small companies that produce non-transgenic hybrids for farmers has increased as consolidation of the seed industry has continued. Progressively fewer elite inbreds are available to these farmers from the major breeding companies that do not use transgenic methods or treat with fungicide. On the other hand, insufficient testing of public inbreds has slowed the use of these cultivars because companies are not sure that they are reliable. At an annual meeting of the midwestern project in February 2009, a set of companies agreed to engage in a common project together with the public breeders to test both private and public hybrids. An organization called the U.S. Testing network was started. This organization is coordinated by Practical Farmers of Iowa (PFI) and has statistical help from United States Department of Agriculture Agricultural Research Service (USDA-ARS) and overview from an advisory board consisting of members from seed companies and public breeders. In 2009, trials took place at numerous organic and conventional sites ranging from Texas to New York to North Dakota. In 2010, this project grew to include contributions from 20 companies and 7 public breeders. The trials took place in 11 states on 37 sites. These sites spread over three different zones: Early (FAO 200–390), medium (FAO 400–480), and late (FAO 490-550).[2] The trials allow for comparison of elite and experimental hybrids from both private and public breeders under conventional and organic conditions.[3]

Are There Benefits for Breeding under Organic Conditions?

Weeds are presently a major problem in producing organic hybrid seed because inbreds lack vigor and do not compete well with weeds. Weeds can also be a major problem in hybrid maize when weather does not allow for timely mechanical control. One potential solution is to breed maize that competes better with weeds. Research with weeds tested numerous hybrids and populations of maize that were bred under organic conditions or under conventional conditions. Maize cultivars were grown in replicated trials on an organic farm. Weed foliage density scores in maize stands were almost two times higher for the conventional hybrids than for hybrids based on varieties selected under organic conditions, and late planted sunflowers into maize stands produced twice as much biomass in the mixture with the conventional hybrids as in mixture with the organically bred cultivars (Goldstein et al., 2006; see fig. 10.3).

Weeds are not the only issue. The poor vigor of conventional inbreds under organic conditions is a problem that is driving some organic seed producers in Wisconsin to seek permission to use fungicide treated seed. For a number of seasons the MFAI and USDA-ARS breeders have noticed an *"Inbred Culture-Shock Syndrome."* Some inbred cultivars perform poorly (e.g., they display yellow-green leaves and are relatively stunted) the first year they are grown under organic conditions, but when they are multiplied a second year under organic conditions their color and vigor improves. Explanations for this phenomenon, if it truly exists, could include undetected outcrosses, epigenetic changes, shifts in the relationship of plants to microbes including the balance of endophytes in seed, or subtle effects of selection practices or complex interactions between these factors. Though maize inbreds are commonly thought to be genetically homozygous, genetic variation has been shown to exist within the same inbreds grown in different generations (Bogenschutz and Russell, 1986; Russell and Vega, 1973) or locations (Gethi et al., 2002). Growing corn on organic sites may affect genetic and epigenetic variation and/or the ability to distinguish such variation.

In 2010, eight inbred or partly inbred cultivars that had been grown and selected under organic or conventional conditions were compared by MFAI (unpublished results, Goldstein, 2010). Four were

Figure 10.3 Walter Goldstein in front of KWS experiments in Germany. The experiments evaluate the effects of competition on different maize hybrids. Sunflowers are used as a proxy for weeds. In similar trials at MFAI, the sunflowers were planted in the row with the maize (photo courtesy of Walter Schmidt).

conventional commercial inbreds which either had been produced at the organic MFAI farm or were the original conventionally produced inbreds. The other four were derived from conventionally grown S3 lines and further bred either under organic conditions at MFAI or under conventional conditions by USDA-ARS. The resulting lines, most of which had been under numerous years of selection, were grown side by side in 2010 with replications on an organic field without fertilizer. Plants from seed that had been grown or bred under organic conditions had consistently higher chlorophyll content of leaves in early September during grain fill with an average positive increase of 16% (range 9 to 28%) relative to the conventionally grown/bred lines (difference significant at $P < 0.001$). These results suggest that the organically grown/bred maize is more efficient at obtaining N because chlorophyll content is linearly equated to the N content of plants (Schepers et al., 1992).

For Which Traits Is It Necessary to Test under Organic Conditions?

As pre-selection occurs primarily under conventional conditions, breeders at Klein Wanzleben Saat company (KWS) have utilized the following formula[4,5] to examine which information on traits may be considered helpful for their organic breeding program.

$$\frac{CR_{org}}{R_{org}} = \frac{i_{con} \cdot h_{con} \cdot r_{con/org} \cdot \sigma_{org}}{i_{org} \cdot h_{org} \cdot \sigma_{org}} = \frac{i_{con} \cdot h_{con} \cdot r_{con/org}}{i_{org} \cdot h_{org}} \tag{10.1}$$

The indirect selection gain (CR_{org}) is the selection gain obtained by selecting indirectly under conventional cropping conditions for organic farming. The direct selection gain (R_{org}) is achieved if selection is done directly in the organic environment (see also Chapter 2, section 1 of this book).

The direct selection gain, R_{org}, is composed multiplicatively by the factors selection intensity, i_{org}, the square root of heritability, h_{org}, and the genetic standard deviation, σ_{org} – all factors derived from or inherent in organic trials. If, however, varieties for organic farming are indirectly selected in conventional trials, the analogous factors i_{con} and h_{con} in the numerator, become effective, as well as another additional factor $r_{con/org}$. This latter factor is the genotypic correlation between the values of varieties under organic and conventional cropping conditions. As σ_{org} is present in the numerator as well as in the denominator, it is of no relevance for the comparison of direct and indirect selection. However, this only true if the same breeding material is used for a conventional as well as an organic breeding program. If breeders have access to unlimited resources coming from a large breeding company, σ_{org} may be greater, compared to a locally acting breeder of organic varieties with a more limited access to new elite cultivars.

In general, it is assumed that the 10% best candidates are selected under both conventional and organic testing, which is a selection intensity of $\alpha = 10\%$ or an i-value of 1.755. In practice, i-values are much higher in conventional breeding programs because more candidates are tested and there is a higher selection pressure.

To estimate h_{con} and h_{org} heritability values, the varieties were tested under conventional and organic conditions in the same places and with the same number of replications. Table 10.1 presents the values derived from four series of test crosses of 90 dihaploid flint lines and 90 dihaploid dent lines in 2008. The lines were crossed each with a tester of the opposite pool. All of the four series were carried out in three different sites with two replications under conventional as well as organic farming conditions.

The results were surprising. As maize grown under organic farming conditions usually suffers from weed competition which is irregularly spread all over the field, one usually reckons with a significantly higher experimental error under such conditions and subsequently with significantly lower heritability estimates. Numerous series indeed showed the expected higher experimental error, but surprisingly, their heritability estimates were only slightly lower (as shown in table 10.1: $h_{con} = 0.83$ versus $h_{org} = 0.78$). A closer look at the variance components (not shown here) that enter into the calculation of heritability provided an explanation. The variance between the entries in organic trials tended to be higher, which to a great extent compensates for higher experimental errors. Under organic conditions many hybrids fail, which increases genetic variance. It may be concluded that the heritability estimates in organic trials are similar to those in conventional trials. This applies even more to highly heritable characteristics like grain maturity (see table 10.1), quality criteria, or resistance to diseases. However, the usually higher budgets of conventional breeders allow them to test on more sites and/or with more replications. This is, of course, reflected in a higher h^2_{con}-value.

Table 10.1 Estimates for the heritabilities h^2_{con} and h^2_{org} (defined in text) of four trial series (DT = dent tester; FT = flint tester) for the characteristics grain yield and dry matter content (DMC) of the grain in % (data from Messmer et al., 2009)

Genotypes	Series	h^2 Grain Yield		h^2 Grain DMC %	
		h^2_{con}	h^2_{org}	h^2_{con}	h^2_{org}
90 Flint × DT	151	0.54	0.49	0.95	0.95
90 Flint × DT	152	0.68	0.54	0.92	0.90
90 Dent × FT	153	0.75	0.72	0.91	0.92
90 Dent × FT	154	0.78	0.67	0.93	0.92
	Mean h^2	0.69	0.61	0.93	0.92

Table 10.2 Estimates for phenotypic r_p and genotypic correlations r_g between the performance under organic and conventional conditions, calculated based on the data of four trial series on grain yield and dry matter content (DMC) of the grain % (data from Messmer et al., 2009)

Genotypes	Series	$r_{\text{Grain Yield}}$		$r_{\text{Grain DMC \%}}$	
		r_p	r_g	r_p	r_g
90 Flint × DT	151	0.36	0.56	0.92	0.96
90 Flint × DT	152	0.39	0.68	0.89	0.94
90 Dent × FT	153	0.43	0.54	0.93	0.97
90 Dent × FT	154	0.31	0.38	0.91	0.94
	Mean r	0.37	0.54	0.91	0.94

Because more emphasis is put on selection under conventional conditions there are higher selection intensities, higher heritabilities, and potentially higher variances, confirming the fact that for large breeding companies, conventional breeding programs are an ideal source for the development of organic varieties. However, higher selection intensities, higher heritabilities, and higher variances in numerous conventional trials do not help at all if the rank of varieties turns the other way around or strongly differs in organic trials. In the end, the genotypic correlations $r_{\text{con/org}}$ are decisive for whether selection gains achieved in conventional tests apply to organic conditions.

Genotypic correlation coefficients r_g are unbiased but nonetheless have relatively high errors of estimation. Therefore, the phenotypic correlation coefficients r_p are often taken into account to assess correlative relations, as their errors of estimation are lower.

Table 10.2 lists such estimates for the correlation coefficients of the varieties under organic and conventional conditions. They have been ascertained in the same trial as the heritability estimates see in table 10.1. Table 10.2 shows that the correlation coefficients are extremely high ($r_p = 0.91$ and $r_g = 0.95$) for grain dry matter content (DMC)%. There is an almost functional relationship between the maturity under organic and conventional cropping conditions. Experiences acquired in many trials show that this similarity of characteristic values in organic and conventional conditions also applies to all resistance characteristics. However, in grain yield, which is the most important agronomic characteristic, the estimates for the correlations were only low to medium. This was true for the four trial series illustrated in table 10.2 as well as in other numerous series of tests carried out with a total of more than 4,000 varieties. A genotypic correlation of $r_g = 0.54$ means that in the end, about only 50% of the selection gain achievable under conventional conditions was also found under organic conditions. Similar results with lower phenotypic correlations between organic and conventional grain yields than grain dry matter were reported by Burger et al. (2008).

Therefore, good organic varieties need organic testing to be identified unerringly. These results corroborate multiple years of testing in New York, Iowa, Wisconsin, and Germany, which have consistently shown that the ranking of hybrids often differs if they are grown on organic or conventional sites.

Choice of Parents for Breeding Programs

Though yields and grain moisture are critical for most farmers, competitiveness with weeds, performance under N-limited conditions, and grain quality are also important for organic farmers and should be taken into consideration in future heritability and selection studies. In this regard, the choice of initial parents may strongly influence results of studies on breeding for organic conditions. For example, Lorenzana and Bernardo (2008) crossed two common U.S. inbreds (B73 × Mo17) and examined grain yields of test crosses of the recombinant inbreds with a commercial tester under organic and conventional farming. They found a much higher correlation between the yield performance of the test crosses under organic and conventional conditions than is reported here and concluded that it was not necessary to have separate breeding programs for organic and conventional farming. There are concerns with this study on purely methodological grounds.[6] But aside from those concerns, B73 and Mo17, though established and important "classical" inbreds, may not be the best choices as parents for breeding for organic farming because they may not contribute the necessary traits. For example, MFAI trials compared competitiveness of different maize cultivars with giant ragweed (*Ambrosia trifida* L.), a highly competitive weed for both organic and conventional farmers (Goldstein, unpublished data, 2002). Studies took place under replicated mini-plot conditions with equal numbers of maize and weeds in population stands. Cultivars included numerous F_2 populations derived from crosses between inbreds B73 or Mo17 with different landraces, as well as F_1 crosses between the organically bred Nokomis Gold population with the same (F_1) inbred × landrace crosses. Dry matter yield of the weed in the pure Nokomis Gold plots was 3.1 t ha^{-1}. Weed yields for the B73-or Mo17-based populations or the Nokomis Gold × B73– or Mo17-based populations averaged 4.6 t ha^{-1} and 3.3 t ha^{-1}, respectively (difference significant at $p < 0.0001$). Crossing with Nokomis Gold appeared to reduce dry matter production of the weed by approximately one-third. Where the dosage of Mo17 and B73 in the F_2 populations went from 50 to 75%, the dry matter production by the weed increased from 4 to 4.6 t ha^{-1} for B73 populations and from 3.6 to 4.7 t ha^{-1} for Mo17 populations (differences significant only for the Mo17 populations at $p < .1$). Furthermore, multiple years of observation of B73 or commercial B73 derived inbreds under MFAI nursery conditions suggests that though they may be useful for contributing to yield, they have especially poor competitive ability with weeds per se.

Breeding Programs

Germany–KWS

The breeding method now applied at KWS SAAT AG and directed by Walter Schmidt is the result of research carried out in close cooperation with the University of Hohenheim (H.H. Geiger) from 2004 through 2008. The program integrates testing under conventional and organic conditions to produce cultivars that benefit both farming systems.

The program works in the following way for a cycle of selection: In both dent and flint breeding pools, 10,000 new, homozygous, dihaploid lines (DH-lines) are developed in the conventional breeding program by means of in vivo haploid induction. These lines are reduced down to 2,000 lines by visual selection with a selection intensity of $\alpha = 20\%$ based on performance for easily identified characteristics (field emergence, cold tolerance, tendency to tillering, early maturity, productivity, resistance to lodging, and disease resistance). This is done on conventional farms, as the correlations between response under organic and conventional farming conditions are high for these characteristics. Then a general combining ability test (GCA) is carried out by crossing each line with a single tester in the opposite heterotic group and evaluating performance of the resulting hybrids. This test is done on conventionally farmed fields, even though the correlation between the performance under organic and conventional conditions is only low to medium for yield. This is not optimal for maximum selection gain for organic farming but necessary in light of the market situation as 99.3% of the area planted to maize in Germany is conventionally farmed. The selection intensity is $\alpha = 10\%$, leaving 200 lines in each pool.

The next step for those lines is the Specific Combining Ability (SCA) test in which each line is crossed with several testers of the respective opposite pool. The best specific hybrid combinations are identified on four conventional sites and four organic sites. Only 50 lines, which had good results under both organic and conventional conditions, are then selected for further work. These remaining lines are recombined within pools to initiate a new cycle of reciprocal recurrent selection from which new DH lines will be produced.

At the same time, the remaining individual lines are used to produce additional hybrid combinations with a higher number of opposite testers. In the following year, these hybrid combinations are tested on at least twice as many sites before the best hybrids are submitted to the official tests for approving and registering varieties. Hybrids resulting from such lines benefit both farming systems. These varieties have high and stable yields both under organic conditions without pesticides and herbicides and higher input conventional conditions.

Northeastern United States

The maize breeding program at Cornell University, directed by Margaret Smith, is breeding maize for marginal glacial soils and cool, stressful growing conditions. Part of the program is directed to serve organic farmers. These farmers need maize varieties suitable for planting at later dates, which can readily obtain soil N from organic matter and thrive without seed treatment and herbicides. The program is developing varieties with multiple disease and insect resistance. Yield trials carried out over five years on several organic sites have identified two Cornell hybrids that are well adapted to organic conditions and competitive with commercial checks. One of these is a modified single cross and the second is a double cross. These hybrids have been sold commercially as organic seed in 2009 and 2010.

Pollen drift contamination from transgenic maize is a threat to organic farmers as it raises concerns with product de-certification and potential loss of income. Contamination problems increased as transgenic seed stocks become the norm. Farmer participants in the Cornell Organic Production and Marketing Program Work Team specifically asked for help with this in 2005. To avoid pollen drift, organic farmers have resorted to planting their maize later than conventional maize, but late plantings are associated with riskier, sub-optimal yields, lower financial returns, and greater expenses associated with drying maize. Contamination in maize can be controlled by the use of special maize genes called *Ga1-S* (Nelson, 1994) and *Tcb-1* (Evans and Kermicle, 2001), which cause "gametophytic cross incompatibility." When a maize plant has one of these alleles only pollen that carries the same allele will grow normally on its silks.[7] Cultivars are being developed by Cornell with the *Ga1-S* allele (Nelson, 1994), and they are exploring white capped seed and brown midrib leaf traits as markers for detecting contamination.[8]

Southeastern United States

The breeding program at North Carolina State University, led by Major Goodman, is in the process of developing seed to meet the needs of organic farmers in the Southeast together with the North Carolina Organic Grain Project, farmers, and Rural Advancement Foundation International (funded by USDA-NIFA-OREI). Their strategy is to find cultivars that will (1) reduce problems with weeds associated with growing inbreds under organic conditions and (2) reduce contamination from adventitious pollen of transgenic maize. To alleviate problems with weeds during seed production the group is presently developing and testing four-way crosses. The most promising is DKHBA1/NC476//NC320/NC368, which appears to produce yields competitive with commercial hybrids. Dr. Goodman has been a pioneer in the United States in developing yellow kernelled inbred lines possessing *Ga1-S* and an effort is being made to develop double crosses that will be homozygous for the *Ga1-S* gene. However, his lines are closely related and lack hybrid vigor when crossed with each other, so an effort is being made to find complimentary lines to raise the yield level of such hybrids.

Midwestern United States: Wisconsin and Iowa

Michael Fields Agricultural Institute's (MFAI) effort to develop high quality maize for organic farmers began in 1988 in response to a request from a group of organic farmers who were concerned about a loss in nutritional value and flavor in commercial varieties. They wanted open-pollinated populations with good agronomic characteristics and nutritional value. In the following decade, MFAI bred several open pollinated populations, one of which (Nokomis Gold) was released in 1998 and grown by farmers in the organic movement.

In 2004, funding was obtained, which allowed for a joint project that involved a USDA-ARS corn-breeding program in Ames in Iowa, MFAI, and Practical Farmers of Iowa (PFI) in breeding high nutritional value maize for sustainable farming. The USDA-ARS and MFAI programs breed and select maize under organic conditions and actively exchange breeding lines. Both programs utilize standard inbreeding and selection programs and early yield testing with test crosses. PFI manages multiple field days each year in Iowa, and MFAI has an annual field day in Wisconsin. Farmers participate in strip trials, practice their breeding skills to adapt populations, participate in making hybrids and growing increase seed, and donate land for yield trials. The USDA-ARS unit does yield trials for its program and for the MFAI program on the farms of cooperators and at different research stations on a mix of organic and conventional sites.

The USDA-ARS sustainable breeding program adapts maize to Central Maize Belt climatic conditions (FAO 500–600). The program has focused on: (1) testing and developing elite synthetic varieties and inbreds under organic conditions; (2) developing elite high protein inbreds with excellent agronomic traits; and (3) developing gametophytic cross-incompatible cultivars containing the *Ga1-S* gene. The program led by Linda Pollak builds on the Latin American Maize Project which was concerned with identifying promising landraces from Latin America and the Germplasm Enhancement of Maize (GEM) program, which utilized these landraces by crossing them with adapted breeding lines provided by commercial companies. The recent outcomes of the USDA-ARS program include a set of robust inbreds with proven yields in hybrid combination equivalent to commercial hybrids. Several of these inbreds are derived from crosses between landrace Cuba 117 and "Stiff Stalk" inbreds. A new set of high yielding, high protein inbreds is becoming available.

The MFAI breeding program, led by Walter Goldstein with advice from Kevin Montgomery, focuses on developing maize for the Northern Maize Belt (FAO 350–500). Prime interests have been (1) developing high yielding maize with higher protein quality; (2) enhanced ability of maize inbreds to compete with weeds; and (3) enhanced nitrogen efficiency under conditions where soil-N is not readily available. Selection is for vigorous inbreds grown only under organic conditions, complete with weeds.

Both the USDA-ARS and the MFAI program have focused on improving protein content in maize. Conventional breeding has emphasized increasing grain yield but this causes a progressive increase in starch and decrease in the protein content of the grain (Duvick et al., 2004). The protein content and quality of maize becomes especially important to produce balanced rations for organic livestock production in light of the high price of organic protein.

Methionine and lysine are generally regarded as being primary limiting amino acids for humans, hogs, poultry, and dairy cattle. For poultry, the sulfur-containing amino acid methionine is commonly regarded as being the first limiting amino acid for overall health and egg production (Bertram and Schutte, 1992), and lysine the second (Anonymous, 1994). Maize is the major ingredient of poultry food but it is naturally low in in lysine and the sulfur-containing amino acids methionine, cysteine, and cystine. This deficiency is commonly made up by combining maize with soybean meal and supplementing with synthetic DL methionine. Due to national restrictions on its use, organic poultry producers will start to reduce the use of synthetic methionine in poultry feed and completely replace it after 2015 (Federal Register, 2010).

Breeding sources with high amounts of lysine and methionine in grain have been identified and are being used to develop cultivars by the MFAI program. This work involves known genes such as *dzr-1, fl-2,* and others. Some high methionine cultivars have seed that have normal translucent hard endosperms, but others have seed with soft endosperms. Some soft endosperm cultivars have approximately 50% more lysine and methionine than conventional cultivars (Goldstein et al., 2008). Soft-kernelled high methionine and lysine maize bred by the MFAI program successfully replaced the need for synthetic methionine in poultry feeding trials carried out with broilers by Organic Valley (Levendoski, 2006) and with layers by the University of Minnesota (Jacob et al., 2008).

One of the issues being addressed is the tradeoff between high grain yield and high protein yield. A recent set of high protein, high-methionine, hard endosperm hybrids (HM) developed by MFAI under organic conditions were tested in 2009 within U.S. Testing Network and USDA-ARS sites. HM hybrids in the USTN trials averaged 87% of the average yield of normal elite hybrids on nine organic sites (87%) but only 81% of the average yield on conventional sites. However the HM hybrids produced up to one-third more protein per acre where quality was measured on one organic site (Goldstein, 2009 unpublished).

Breeding for high methionine varieties has been greatly assisted by the development of new near infrared spectroscopic (NIRS) calibrations for two models of transmission-type NIRS units (Hardy et al., 2009). MFAI's calibration is the result of work with the Iowa State University Grain Quality Lab (Charles Hurburgh, Connie Hardy, and Glen Rippke). This calibration is based on a unique set of MFAI samples that enabled breaking the inherent correlation between the content of protein and the content of its constituent amino acids lysine and methionine. The calibration is presently licensed through the Bruins Instrument company[9] for Bruins and Foss Infratec NIRS machines.

Midwestern United States

The Ohio State University (OSU) corn-breeding program initiated soil-building practices in the late 1980s at a site that had received massive fertilizer inputs and had been planted to maize after maize, for at least 20 years. The intent was to select for tolerance to pathogens that would prevail in this environment. One spring a newly emerging stand died. That caused a fundamental shift in attitude about what kind of systems a public breeding program should breed for. The first change was to reduce conventional inputs by 15%, introduce red clover and alfalfa into the rotation, and reduce tillage practices. Later, large quantities of composted manure were applied, and the program entered into organic transition and then certification. Selection is for maize breeding lines that can perform well in the organic environment. Emphasis is on multiple disease resistance (field resistance to northern corn leaf blight, gray leaf spot, and more recently rust) due to the humid environment of the Eastern Corn Belt. In addition, the program emphasizes selecting to ensure that protein and oil levels do not drop while advancing other traits. More recently research has focused on kernel color pigments (anthocyanin pigments associated primarily with purple and blue color; and carotenoid pigments associated with yellow and orange color). Other research continues to point toward health benefits associated with these pigments. The program will release breeding germplasm and inbred parents that will produce competitive, non-transgenic hybrids.

Future Directions

In addition to progressive improvement of the agronomic value of organic hybrids, the KWS effort is developing large seeded, vigorous cultivars that emerge better from greater depths allowing for less damage from crows and greater competition with weeds. KWS is also breeding "energy-maize" for organic farms to enable them to become independent of fossil fuels through biogas production.

Recently in the United States, USDA-NIFA-OREI funded a cooperative project involving USDA-ARS (Ames, Iowa), PFI, MFAI, Cornell, Ohio State University, and New Mexico State University to breed maize for organic farmers. The common breeding goals are to develop cultivars with reliably high agronomic performance under low-N conditions that also have high nutritional value, excellent disease and insect resistance, seed integrity, resistance to cold and wet soils during germination, and superior ability to compete with weeds. The partners have begun working together using an organic winter nursery in Puerto Rico. Screening for disease and insect resistance and seed integrity under cold and heat stress will take place at a centralized location. Care will be taken to breed cultivars that fit into heterotic groups used by the seed industry in order to optimize hybrid vigor for grain yields. The Cornell, New Mexico, and Ohio State University efforts will focus on disease resistance while the USDA-ARS project and MFAI will focus on protein quality. MFAI will also work on N efficiency in conjunction with microbial endophytes. The project should produce a constant flow of hybrids to farmers and seed companies for testing and increase.

Notes

1. A more recent survey of organic farmers (Organic Seed Alliance "State of Organic Seed – 2010" report) showed that 1) field crop farmers produced more maize than any other crop; 2) most (82%) farmers strongly agreed that varieties bred for organic system management are important to the overall success of organic agriculture; 3) yield, quality and emergence were the most important traits for maize; and 4) farmers want to be part of the breeding process.
2. FAO ratings indicate maturity. In the U.S. system of relative maturity, FAO 200, 350, 390, 400, 480, 490, 500, 550, and 600, corresponds approximately to 80, 95, 99, 100, 108, 109, 110, 115, and 120 day corn, respectively.

3. Information on performance of hybrids in the USTN is held confidential by the participants. Sarah Carlson, Practical Farmers of Iowa, Ames, Iowa, 515-232-5661 ext. 305, is the contact person for this organization.
4. Formula modified after Falconer (1981).
5. Heritability was estimated using plot values according to the following standard formula:

$$h^2 = \sigma_g^2/(\sigma_g^2 + \sigma_{gs}^2/S + \sigma_e^2/SR),$$

where σ_g^2 is the genotypic variance; σ_{gs}^2 is the genotype × site interaction variance; σ_e^2 is the error variance for plot values; S is the number of sites; and R is the number of replications.
6. Though the question remains open for more research, the conclusions of Lorenzana and Bernardo (2008) seem founded on a weak statistical basis. A one-year experiment with only two locations is not much, as one of the organic sites was still in the process of certification. Large estimation errors were evidenced by large differences between the phenotypic and the genetic correlations. Furthermore, in their table 3, there was a negative variance component (Variance of the Test Cross × Location) for grain yield under organic conditions. This negative component is grounds enough for questioning the statistical results. Other reservations on the strength of their data include whether they used the most appropriate formula and statistical procedures for comparing selection in different environments (see Holland, 2006).
7. The use of *Ga1-S* in yellow dent and flint hybrids and inbreds was patented by Thomas Hoegemeyer in 2005 (U.S. Patent 6,875,905). However, the patent's validity is controversial as the idea was not new (Thomas, 1955) and yellow dent inbred lines with *Ga1-S* existed before the patent. Yellow kernelled *Ga1-S* maize genetic stock MGS 25776 (also called 401D) was donated to USDA in 1993 by Jerry Kermicle (MGSC, 2010). Major Goodman from North Carolina State University released several yellow dent inbreds with *Ga1-S* before the patent including NC302 (1994), NC338 (1997), NC354 (1998), and NC440 (2002). Furthermore the double pollination technique of using colored aleurone *ga-1* stock as testers, an integral part of the patent, is not new, having been described and illustrated by Ashman (1975). According to the Cerrado Natural Systems Group (CNSG), which holds the patent and has marketed under the name Puramaize, the intent of the patent was to protect the gene from contamination with transgenic material (Personal communication Thomas Hoegemeyer and Stephan Becerra, 2011). CNSG is willing to allow others to develop and use inbred lines and hybrids containing Ga1-S at no charge, provided that: (1) The licensee institutes a stewardship program that insures that in fields neighboring seed production fields volunteer plants of Ga1-S/GMO adventitious crosses are not allowed to live to flower and contaminate a subsequent Ga1-S containing seed crop the following season; (2) that such a stewardship program is transferred with any inbred and hybrid seed; (3) that buyers of such seed also institute the stewardship program all the way through to commercial grain production; (4) that licensees agree to meet or exceed specific quality characteristics to insure the integrity of the trait; and (5) no license is granted for the use of the name PuraMaize or similar nomenclature under U.S. trademark laws.
8. Recessive traits with marked effects on plant appearance in homozygous state can easily show whether out-crossing has occurred during seed multiplication because they alter the appearance for subsequent generations. Seed trait examples include white endosperm (recessive), white cap (partially dominant), and waxy endosperm (recessive). There are several recessive brown midrib traits that show clearly in leaves. Outcrosses to normal maize would produce offspring with green midribs that could easily be identified and rogued out. Other genes that

might work as markers include *rd1* and *br1*, which are reduced plant-height and brachytic traits. Outcrosses not carrying these recessive genes would be taller than the rest of the line and could be rogued out before pollen shed. Plants homozygous for *rd1* and *br1* can be of a nearly normal size if bred into the right backgrounds, but outcrosses would still be taller generally.
9. Bruins Instruments, contact Karen Waystack, P.O. Box 1023, Salem, NH, 03079 (www.bruinsinstruments.com).

References

Anonymous, 1994. Nutrient Requirements of Poultry: Ninth Revised Edition. National Academies Press.

Ashman, R.B., 1975. Modification of cross sterility in maize. *Journal of Heredity* 66:5–9.

Bertram, H.L., and J.B. Schutte. 1992. Evaluation of the sulfur containing amino acids in laying hens, *In*: Proceedings of the 19th World's Poultry Congress. pp. 606–609. Sept. 19–24, 1992, Amsterdam, The Netherlands. Wageningen, The Netherlands: Ponsen and Looijen.

Bogenschutz, T.G., and W.A. Russell 1986. An evaluation for genetic variation within maize inbred lines maintained by sib-mating and self-pollination. *Euphytica* 35:403–412.

Burger, H. 2008. Comparison of methods for developing optimal cultivars of maize for ecological farming (German) Dissertation, University of Hohenheim, Germany.

Burger, H., M. Schloen, W. Schmidt, and H.H. Geiger. 2008. Quantitative genetic studies on breeding maize for adaptation to organic farming. *Euphytica* 163:501–510.

Duvick, D.N., J.S.C. Smith, and M. Cooper. 2004. Long-term selection in a commercial hybrid maize breeding program. *Plant Breeding Reviews* 24:109–151.

Evans, M.M.S., and J. Kermicle. 2001. Teosinte crossing barrier 1, a locus governing hybridization of teosinte with maize. *Theoretical and Applied Genetics* 103:259–265.

Falconer, D.S. 1981. Introduction to quantitative genetics. New York: Longman Group.

Federal Register, 2010. Federal Register/Vol. 75, No. 163. August 24, 2010, Rules and regulations. pp. 51919–51920. Retrieved from http://www.ams.usda.gov/AMSv1.0/getfile?dDocName=STELPRDC5086261&acct=noprulemaking.

Gethi, J.G., J.A. Labate, K.R. Lamkey, M.E.Smith, and S. Kresovich. 2002. SSR variation in important U.S. maize inbred lines. *Crop Science* 42:951–957.

Goldstein, W., A. Wood, and B. Barber. 2006. Methods to breed field maize that competes better with weeds on organic farms. Report to the Organic Farming Research Foundation. Accessed February 27, 2011. https://ofrf.org/funded/reports/goldstein_06s07.pdf.

Goldstein, W., L.M. Pollak, C. Hurburgh, N. Levendoski, J. Jacob, C. Hardy, M. Haar, K. Montgomery, S. Carlson, and C. Sheaffer. 2008. Breeding maize with increased methionine content for organic farming, *In*: Sohn, S.M., and U. Köpke (eds.) Organic agriculture in Asia proceedings of the regional ISOFAR conference in the Republic of Korea.

Hardy, C.L., G.R. Rippke, C.R. Hurburgh, and W.A. Goldstein. 2009. Calibration of near-infrared whole grain analyzers for amino acid measurement in corn. *Cereal Foods World* (supplement)54:A45(abstract). Accessed April 4, 2011. http://meeting.aaccnet.org/2009/default.cfm verified 4 Apr. 2011.

Holland, J. 2006. Estimating genotypic correlations and their standard errors using multivariate restricted maximum likelihood estimation with SAS Proc MIXED. *Crop Science* 46:642–654.

Jacob, J., N. Levendoski, and W.A. Goldstein. 2008. Inclusion of high methionine maize in pullet diets. *Journal of Applied Poultry Research* 17:440–445.

Kutka, F.J., and M.E. Smith. 2007. How many parents give the highest yield in predicted synthetic and composite populations of maize? *Crop Science* 47:1905–1913.

Levendoski, N. 2006. Alternatives to synthetic methionine feed trial. *In*: Gopalakrishnan, O., L. Melotti, J. Ostertag, and N. Sorensen (eds.) Proceedings of the 1st IFOAM International Conference on Animals in Organic Production. August 23–26, 2006, St. Paul, Minnesota. Accessed April 13, 2011. http://www.ifoam.org/press/Publications_Teasers_20110331_ILB_for-webshop_small.pdf.

Lorenzana, R.E., and R. Bernardo. 2008. Genetic correlation between corn performance in organic and conventional production systems. *Crop Science* 48:903–910.

Maize Genetic Stock Center listings (MGSC). 2010. Accessed February 27, 2011. http://www.maizegdb.org/cgi-bin/displaystockrecord.cgi?id=25776.

Nelson, O.E. 1994. The gametophytic incompatibility factors in maize. *In*: Freeling, M., and V. Walbot (eds.) The maize handbook. pp. 496–502. New York: Springer-Verlag.

Organic Seed Alliance. 2011. State of organic seed 2010 report. Accessed February 23, 2011. http://www.seedalliance.org/Publications/.

Russell, W.A., and U.A. Vega. 1973. Genetic stability of quantitative characters in successive generations in maize inbred lines. *Euphytica* 22:172–180.

Schepers, J.S., D.D. Francis, M. Vigil, and F.E. Below. 1992. Comparison of maize leaf nitrogen concentration and chlorophyll meter readings. *Communications in Soil Science and Plant Analysis* 23:2173–2187.

Thomas, W.I. 1955. Transferring the GaS factor for dent-incompatibility to dent-compatible lines of pop maize. *Agronomy Journal* 47:440–441.

11 Rice: Crop Breeding Using Farmer-Led Participatory Plant Breeding

Charito P. Medina

Introduction

Rice is the staple food of more than half of the world's human population. It is a major determinant of food security in many countries; thus, it is a political commodity. In Asia, rice is life because 80% of the rural population depends directly on rice crops for livelihood, especially the region's small farmers. The crop has deep social and cultural significance because it is the foundation of nutrition, health, and food security. Often, rice landholdings define a farmers' social status, and "to eat" is almost always synonymous with "to eat rice."

In addition to being seed keepers, farmers were also seed developers. The central role of rice in the lives of farmers has contributed to the evolution and diversification of rice wherein more than a hundred thousand varieties are estimated to exist. These diverse varieties are the result of farmers collecting, propagating and selecting germplasm in a myriad of agro-ecological conditions since the dawn of agriculture some 10,000 years ago. Thus, many rice varieties are adapted to a wide range of agro-ecological conditions such as dryland or paddy, upland or lowland, or deepwater or saline environments. Rice also occurs in different shapes, colors, and flavors because of farmers' broad ranges of preference.

Conventional Rice Breeding and Its Impacts

In the 1960s, formal rice breeding by scientists at the International Rice Research Institute (IRRI) produced dwarf varieties that are resistant to lodging and have higher harvest index. These dwarf varieties are high yielding varieties (HYVs) mainly because they were bred as highly responsive to chemical fertilizers. But the context of this high chemical input farming was only practical during the 1960s when energy prices were still low. Also, the lesson learned from highly intensified chemical farming was that sustainability associated with depletion of soil organic matter, acidification, associated nutrient imbalances, mineral deficiencies, and toxicities was lacking.

Widespread planting of a few varieties developed in the formal sector, coupled with a narrow genetic make-up of such modern varieties displaced the diverse traditional rice varieties. For example, farmers historically cultivated some 30,000 rice varieties in India, but now 75% of their rice

---

Organic Crop Breeding, First Edition. Edited by Edith T. Lammerts van Bueren and James R. Myers.
© 2012 John Wiley & Sons, Inc. Published 2012 by John Wiley & Sons, Inc.

comes from just 10 varieties. In Indonesia, some 1,500 rice varieties disappeared between 1975 and 1990 (Ryan, 1992). In Burma, Cambodia, Indonesia, Philippines, and Thailand, about 85% of rice fields had been planted to HYVs by 2002 (WRI, UNEP, and IUCN, 2002). The narrowing of genetic diversity in rice agro-ecosystems also caused outbreaks of pests and diseases.

Moreover, Green Revolution (GR) technology brought uneven benefits across social classes, with small scale and poor farmers adversely affected, particularly due to the need for expensive inputs required by HYVs. Because the seeds and technology were developed in research and breeding stations that were socially isolated, rice farmers even complained that they themselves "forgot how to grow rice." New rice varieties, inputs, and management practices were developed by research institutions and disseminated by extension workers with farmers as passive recipients of technology (Chambers, 1993; Pimbert, 1994).

MASIPAG and Participatory Rice Breeding

MASIPAG literally means *industrious* in Pilipino and is used here as acronym for a phrase meaning "Farmer Scientist Partnership for Agricultural Development." This organization was established as a partnership among some scientists, NGO workers, and smallholder farmers in the Philippines dissatisfied with the GR technology. The desire of farmers to breed rice in a discernible scientific fashion, described by Frossard (1998) as peasant science, was motivated by the threat to their survival as farmers and was supported by a few volunteer scientists from the academe. Regional assessments of the impacts of GR in the early 1980s culminated in a national forum on rice issues in 1985 (Bigas Conference, 1985) with a resulting objective of recovering traditional rice varieties and improving them for adaptation to small scale and resource poor farmers' conditions.

Local rice varieties were difficult to find in the 1980s having been largely displaced by modern varieties. The initial rice collection was donated by farmers from remote farming areas untouched by the GR, collected by academic friends of MASIPAG whenever they travelled, and NGOs in their respective areas. A total of 47 rice varieties were initially collected in 1986 and this was the starting germplasm base for breeding at MASIPAG.

Since the landmark *Bigas* conference on rice issues, a program on collection, identification, multiplication, maintenance, and evaluation (CIMME) of traditional rice varieties has been in place (MASIPAG, 2000; Medina 2002, 2004). This active collection of traditional rice varieties (TRVs) and the documentation of characteristics of different varieties was the foundation of the rice breeding activities.

The MASIPAG rice breeding project began in 1987, as an integrated component of its bottom-up approach and farmer empowerment goals. It was meant for transformation and emancipation of farmers and it sought to demystify science for the farmers. The focus was to address small-scale farmers' desires for varieties that suited their economic capacity, were adapted to diverse agro-ecological conditions of mostly marginal areas, and fit into their indigenous farming systems. Farmers decided that rice was to be bred without the inputs of chemical fertilizers and pesticides as to avoid expensive production costs, health hazards, and environmental pollution. It was clear that MASIPAG rice seeds to be developed were organic seeds because they were not subjected to chemical inputs including the intended ultimate production at the farmers' fields.

In conventional breeding, the scientist breeder identifies the objectives of breeding, which are almost invariably high yield with broad geographic adaptability (dependent on chemical fertilizers and pesticides to provide near ideal conditions in most instances; Desclaux, 2005). In participatory plant breeding (PPB), farmers can be involved in different stages of the breeding process. There have been different modes of participation by farmers, from farmer-centered to scientist-centered (Halewood et al., 2007; see also Chapter 6 of this book; Morris and Bellon, 2004). Thro and Spilane (2000) and Desclaux (2005) differentiated two types of approaches: Functional PPB focused on getting adapted materials, improving local adaptation of varieties, and promoting genetic diversity, and process PPB aimed at empowering farmers in developing their plant breeding skills.

MASIPAG PPB is both function- and process-oriented. Aside from their own selections of parental materials, farmers helped in crossing other varieties chosen by farmers during field days. Their work also included the selection of plants from plots of segregating populations. All the breeding work components were implicitly building the self-confidence of farmers that led to their feelings of empowerment. Indeed it was PPB in practice even before the term itself was coined and before the mainstream scientists believed that farmers could do breeding.

Beyond PPB: Farmer-Led Rice Breeding

The beginning of PPB in MASIPAG started with hands-on training with a combination of lectures, demonstration, and practicum. Actual rice specimens in the flowering stage were used during the training. Emasculation and pollination were easy for the farmers and they were excited about their new knowledge (see fig. 11.1). After the training, the trainees had to wait for one month if new seeds were formed to indicate their success.

Farmers who successfully produced the F_1 seed from their training sessions were encouraged to join the breeding work of staff at the MASIPAG back-up farm. Aside from their own selections of parent materials, they helped in crossing other varieties chosen by farmers during field days. Their work also included the selection of plants from plots populated with segregating plants. It was in this selection process that the farmers put into actual practice what was learned in the training.

Figure 11.1 Jojo Paglumotan, a farmer-breeder, doing rice emasculation in preparation for breeding (MASIPAG).

Rice breeding in the MASIPAG network has evolved beyond participatory plant breeding because farmers are now breeding rice by themselves on their own farm: They maintain germplasm on their own farms, or at least on their organizations' trial farms, they set their own objectives, select the parental materials themselves, do cross pollination, make selections of segregating lines, evaluate their selections, and share developed varieties with other farmers. Currently, there are 67 farmer rice breeders. Many farmers got interested in breeding because of their feeling of reinforcement by other farmers, the scientists, and NGO workers. The tangible results of their breeding work was not only a solution to their farming problems, but also a source of pride when other farmers planted the rice that they bred.

When asked why they do their own breeding, farmer breeders give a variety of answers: They want to develop their own rice variety that is appropriate in their farm (edaphic and agro-ecological settings); and they have a desire for varieties that do not need chemical fertilizers, that are resistant to pests and diseases, and are early maturing and have high yield. Some farmer breeders also wanted to develop varieties based on household priorities: For example, types that are aromatic and possess good eating quality. Other farmers do breeding because they want to develop seeds adapted to different climatic variations (climate change), so they could have their own seeds every season and not have to buy from the marketplace. In addition, there was an altruistic motive in that they wanted to share the varieties they had developed with other farmers.

Role of Back-Up and Trial Farms

One unique aspect of MASIPAG is its network of back-up farms and farmer-managed trial farms. These facilities serve as the farmers' seed bank (Medina, 2004), making breeding materials available to farmers. The varietal characterization conducted by MASIPAG is distributed to farmers who are interested in breeding, in order to facilitate selection of parental varieties. Field days prior to harvest in trial farms are also used as a method of selecting preferred varieties for breeding.

The Breeding Process

Setting the Objectives

In participatory rice breeding, the objectives are set by the farmers. In 1986, and prior to the start of rice breeding in MASIPAG, a workshop was conducted to allow farmers to identify what characters a good variety must possess (Medina 2004), and the following were their criteria:

- High yield without chemical fertilizers
- Early maturing
- Resistant to pests and diseases
- Resistant to drought
- Medium height
- Strong stalk; non-lodging
- High tillering capacity
- Long panicles
- More grains per panicle
- Long grains

- Small seed embryo (high milling recovery)
- No unfilled grains
- Good eating quality

These criteria are not ranked according to priority but according to a combination of traits will be chosen by each farmer depending on the objectives of each farmer breeder and the context of their individual farms.

Identification of Parent Materials

Prior to breeding, potential parent materials need characterization, and the data should be available to identify the best materials. For example, a candidate parent material must possess the trait(s) that can satisfy the objective(s) of breeding program. If the goal is to create a variety with improved productivity, the candidate parent materials must have traits related to productivity, such as having more tillers, a high percentage of productive tillers, long and dense panicles, a high percentage of filled grains, and so on. If a variety has high potential yield but is susceptible to pests, then a variety resistant to pests is selected to be cross bred to transfer pest resistance traits.

The variety that possesses the traits that fit the needs of the main objective serves as the "mother" because it contributes 60% of the genes to the progeny in single crosses. The male parent should be a variety that bears a character desired to be incorporated into a variety for improvement. Farmer breeders source their immediate parent materials for breeding from their own farm but they can source their collection of potential parents from the trial and back-up farm.

Planting to Synchronize Flowering, Emasculation, and Pollination

Different rice varieties have different flowering dates as well as maturity dates. For cross pollination to be effected, the flowering of intended parental varieties with different flowering dates must be synchronized by adjusting the date of sowing. When crossing is done with many varieties and where the information on flowering date is not available, synchronization is accomplished by estimating flowering dates of the earliest variety and the latest variety in the group, based on maturity data.

Rice has perfect flowers and is mainly self-pollinated, although, as a rule of thumb, about 10% outcrossing is possible. Self-pollination can occur even before the anthers emerge because anthers mature as they reach the apex of the lemma and palea. Thus, the six anthers per flower must be removed before the pollen matures. A mother plant is selected from the breeding nursery where the parent materials are grown, then potted and brought to the shade where emasculation and cross pollination is performed. For details on emasculation and crossing, see Coffman and Herrera (1980). If pollination is successful, seeds are formed, semi-crescent in shape, tapering to the tip, with a partially burned appearance. The seeds mature in 28 to 30 days from pollination.

Care and Management of F_1 Seeds and Seedlings

The whole panicle is harvested after 30 days from pollination and gradually sun-dried for three days with the original glassine bag intact. The seeds remain dormant for about 15 to 20 days, after which they can be sown again.

For the succeeding generations, several seeds are sown in moistened paper-lined Petri dishes, which are then covered and allowed to germinate. In about a week the young (germinated) seedlings may attain a height of 3 to 5 cm. They are transferred to a small pot with puddle soil (structureless saturated soil) and grown to 15 to 18 cm before transplanting them finally in single rows about 3 m long in the nursery. Ten to 15 healthy seedlings are selected and planted singly, with 30 cm between plants and between rows. Seeds from the mother plant are also sown and transplanted simultaneously with the F_1 plants, occupying one row on the left side of the F_1 row. This serves as a check if the F_1 plants are really new or are results of self-pollination.

Indications of successful hybridization include the following observations: Seeds are formed, seeds formed are firm or hard, seeds will germinate, germinated seeds will grow to maturity and bear seeds, and, finally, the new plants are distinctly different from the mother plant in all observable features or traits.

Selection and Handling of Segregating Populations

Selection is most complicated stage of breeding and proper recording is required. It constitutes about 90% of the total effort in breeding work. Starting from the F_2, plant types with a combination of traits would appear in the progeny as a result of the recombination of genes. However, if the progenies show no phenotypic variation and are exactly the same as the mother plant, then cross pollination was not successful and the resulting population is discarded.

The formal sector practices the pedigree method as the preferred selection method in rice. MASIPAG initially used the pedigree method, but it was too cumbersome to farmers in terms of plant selection and record keeping. After a few years, bulk method was used with some modifications. Farmer breeders in the MASIPAG rice breeding program practice bulk selection because this method is simpler, more convenient and practical (Vicente et al., 2010). The advantages of the Bulk Method are: Simple and convenient to monitor and practical for farmers, selections are more varied and have many sources of genes, the breeding process is shorter, and selections are stable, durable, and more broadly tolerance to pests and diseases.

Pedigree Method

Seeds from F_2 to F_6 are separately planted in each succeeding crop season, usually in plots of 100 to 200 plants or hills. Each plant is observed to monitor the flow of desirable heritable traits. Through the pedigree method, the purity and uniformity of the selection increases through the filial generations since variability is removed every season. At about F_6, a highly uniform pure line is readily discernible.

Bulk Method

The two most important plant traits indicative of variability are first considered in the early stages of bulking: Maturity and plant height. In the case of a single cross, when the plants are mature, the seeds of 10 plants are harvested and bulked to serve as a single source of seeds for the next planting season. Enough seed is prepared to grow about 200 plants in a plot.

During harvest time, the best plants are harvested and again bulked to serve as the first bulk sample. This sample may be considered as the early maturing group and may contain plants of varying heights but matured at the same time. After about 10 days, another bulk sample is obtained

Figure 11.2 Panicle selection in segregating F_3 bred rice (MASIPAG).

and designated as the second bulk which may constitute a medium maturing group, again containing short, medium, or tall plants. After another 10 days a third bulk sample is obtained if there are still productive plants. This will represent the late maturing batch (see fig. 11.2).

In the next cropping season, the three bulked samples are planted in separate plots. When bulk no. 1 is mature, it may be possible to separate the short, medium, and tall varieties and produce three sub-bulk samples of short-early, medium-early, and tall-early. In the same manner, bulk no. 2 and bulk no. 3 are sub-bulked. At the end of the harvest, there may be nine sub-bulked samples distinguished from one another by maturity and height combinations. In the ensuing generations, further sub-bulking may be done after assessing other distinctive variables including plant base color, tillering capacity, panicle length, hull color, and grain density, shape, color, and size, etc. These sub-bulks eventually create many possible selections to choose from in response to the original breeding objectives. Members of the organization involved in the breeding can freely choose their farmer selections based on their needs and preference.

Bulking tends to retain many variable traits within the population, such that the selection may appear non-uniform and aesthetically less attractive to most farmers who are accustomed to the uniformity exhibited by certified seeds of HYVs. To a great extent, bulk samples may be considered phenotypically uniform but genetically diverse.

Genetic diversity imparts stability and durability and increases the useful life of varieties. For example, resistance of new varieties to pests and diseases typically lasts an average of five years (Rubenstein et al., 2005). On the other hand, even the early MASIPAG selections mass produced 20 years ago are still being grown by farmers in the MASIPAG network across the Philippine rice landscape and they are still viable and productive.

Farmers' Selection in Segregating F_3 to F_5 Generation

Segregating F_3 to F_5 lines are distributed to trial farms in different locations, so that other farmers who are interested can pursue their own objectives and select for local adaptability as influenced by the interaction of genotype and environment. The ability to effect improvement and visible phenotypic change keeps farmers interested and motivated. The more time that a variety is under observation in the farmers' field, the more farmers can become acquainted with the characters of that variety. Familiarizing farmers with the important characters of new varieties is important to spread and acceptance. This way, the farmers are the ones making the selections based on their goals, preferences, and resources and in response to their physical, biological, and socio-economic environments. In addition, each rice cross can have much greater potential for impact than if a centralized breeding program was employed because the progeny from the cross are subjected to many more different environmental conditions than is possible with one or a few locations in a centralized program.

Outcomes of the MASIPAG Program

In rice breeding, one single cross can result in many lines with different characteristics. As an example, the cross between 'Elon-elon' and 'Abrigo' produced 12 different selections with different characteristics (see table 11.1). The height of the progenies ranged from 75 to 90 cm, with 11 to 15 productive tillers, differences in leaf structure, tolerance to pests and grain quality. These differential qualities were the result of selection to fit different conditions, which was achieved by distributing the segregating lines to different farmer groups in different agro climatic conditions. In this example, after having been distributed to several trial farms, four selections showing specific local adaptation were chosen by farmers in specific environments.

After 24 years of MASIPAG retrieving TRVs and conducting PPB, 1,100 TRVs had been collected and maintained in back-up farms and farmers' organization-managed trial farms. Continuous characterization of each TRV has been conducted at the national back-up farm in order to generate information on their unique characters for breeding. This collection is an enabling factor for farmer-led breeding because the varieties available as breeding materials are immediately accessible to them. From 398 crosses, 1,085 MASIPAG rice varieties have been developed. This wide array of rice varieties is readily available to farmers, with many of them already evaluated under diverse agro-ecological conditions. Many of the selected varieties in specific locations are different from those selected in other locations in 49 provinces of the Philippines where MASIPAG has members. This provides evidence that genotype × environment (G×E) interactions are of particular importance in breeding for adapted rice varieties.

In addition to the 1,085 varieties bred using PPB collaboration between researcher and farmers, another set of varieties has been bred independently by farmers, without active support from researchers. Sixty-seven active farmer–rice breeders in the MASIPAG network have produced 328 crosses, from which 508 farmer-bred varieties have been distributed and many were selected to be locally adapted in different locations. Many of the farmer-bred lines (FBL) are still in the early stages of selection; thus, it is expected that many more FBLs will be tested and selected in different locations for adaptability.

The characterization of TRVs and the breeding activities in the MASIPAG network has resulted in many rice varieties with desirable characteristics (see table 11.2). MASIPAG rice with high tillering capacity, long panicles, and resistance to pests and diseases are characters important in breeding

Table 11.1 Adaptability of selections from segregating lines of a single cross (M5: Elon-elon × Abrigo) in different locations

Parent Material of M5		
Female Characteristics	Male Characteristics	Breeding Objective(s)
ELON-ELON: tall, many tillers, long and dense panicles, small elongate grains, drought tolerant, prone to lodging but panicle can stand	ABRIGO: short, many tillers, long panicles, semi-erect leaves, slender grain	Reduce height of Elon-Elon, retain long and dense panicles, make it resistant to lodging
MASIPAG Selection	Feedback on Performance	Provinces where Selection became Locally Adapted (LAS)
M5-A-1	Not selected	
M5-A-2	Not selected	
M5-A-3	Not selected	
M5-A-5	Not selected	
M5-AS	80 cm tall, 11 productive tillers, resistant to lodging, resistant to tungro; tough grains	1) Negros Occidental (Cauayan, La Castellana); 2) Laguna (Pila); 3) Aklan; 4) Bohol (Mantacida); 5) Agusan del Norte (San Martin); 6) Cotabato (M'lang)
M5-B-1	Not selected	
M5-B-2	75 cm tall, 11 productive tillers, erect flag leaf	1) Camarines Sur (Pili); 2) Laguna (Pila)
M5-B-3	Not selected	
M5-BD-1	90 cm tall, 15 productive tillers, elongate grain; resistant to black bug	1) Negros Occ. (Cauayan); 2) Agusan del Norte (Valencia); 3) Lanao del Norte (Tenazs); 4) Cotabato (M'lang)
M5-C-1	Not selected	
M5-C-3	Not selected	
M5-C-S	85 cm tall, 12 productive tillers, elongate grain	Antique (Patnongon)

Table 11.2 Number of traditional rice varieties and MASIPAG-developed rice with farmer-preferred or environmentally adaptive characteristics

Characteristic/Adaptation	Traditional Rice Varieties	MASIPAG-Developed Rice
High tillering capacity	–	42
Good ratooning ability	–	24
Long panicles	–	11
Low fertility soils	12	36
Drought tolerance	8	9
Saltwater tolerance	7	12
Flood tolerance	1	7
Pest/disease resistance	6	17
Red/Black/Violet	152	79
Semi-glutinous	–	15

Table 11.3 Average rice yield (kg/ha) of MASIPAG organic rice and conventional rice farming in 2007 (n = 840 farmers), after Bachmann et al. (2009)

	Full Organic	MASIPAG in Conversion	Conventional
National average	3,424^{ns}	3,287^{ns}	3,478^{ns}
Luzon average	3,743^{ns}	3,436^{ns}	3,828^{ns}
Visayas average	2,683^{ns}	2,470^{ns}	2,626^{ns}
Mindanao average	3,767^{ns}	3,864^{ns}	4,131^{ns}
Maximum yield	8,710	10,400	8,070

ns – not significantly different.

as well as for production. Colored rice and semi-glutinous rice are sought-after characteristics in high-end markets. Moreover, many TRV and MASIPAG bred varieties were identified as drought, saltwater, and flooding tolerant, and these varieties have been classified as adaptation varieties with traits that can cope with environmental perturbations caused by climate change.

Comparing MASIPAG organic rice to that of conventional farming, the yields are not significantly different (see table 11.3). The comparable yield of the MASIPAG organic rice may be attributed to the breeding under organic condition as well as the selection for location specificity. This is already a positive benefit because it is commonly assumed that organic farming generally yields 10% lower than conventional farming.

The net income per hectare of fully organic farms was significantly higher compared to conventional farms even when there is no price premium for organic (Bachmann et al., 2009). Fewer expenses are incurred by organic farms when the need for chemical inputs and seeds has been eliminated.

Outlook

Rice-seed banking combined with farmer-managed trial farms and trainings in breeding has contributed to the success of rice breeding in the MASIPAG network in the Philippines. The participatory and farmer-led breeding programs have created a diversity of rice varieties suited for many different purposes and for diverse agro-ecological conditions. Scaling-up of this activity is also feasible because the network and infrastructure to facilitate seeds, technology, and training have been established, including support for organic certification and marketing (see box 11.1).

Participatory rice breeding is a very important approach in addressing the diversity of objectives of farmers that are often neglected by breeders in the formal sector. It is also important to maximize G×E interaction for location specificity of varieties that are developed – in particular, those adapted to soil fertility conditions, prevailing pests and diseases, water availability, and other unique agro-ecological conditions.

This approach returns the farmers to the original role of steward of genetic diversity – not in a static condition as in formal gene banks, but as an actively evolving collection. It also adds sustainability and availability to the seed system for small farmers, with their role not only to produce food but also to reproduce seeds that are a central biological resource in farming and food systems.

Participatory and farmer-led rice breeding is best conducted under circumstances of small scale farming in diverse and heterogeneous agro-ecosystems. With farmers breeding rice for themselves, they are not only the users of genetic resources, but they are also stewards and developers. This

> **Box 11.1. MASIPAG's broader role in fostering organic production in the Philippines**
>
> In addition to employing organic practices in its rice breeding efforts, MASIPAG has been more broadly involved in establishment and oversight of the organic industry in the Philippines, with the goal to ensure that small farmers do not get marginalized in this arena. MASIPAG has had efforts in three areas. First is the establishment of its own Participatory Guarantee System (PGS) which is called the MASIPAG Farmers Guarantee Systems (MFGS). This is an alternative, locally focused guarantee system that aims to: Ensure the organic integrity of farmers' products based on MASIPAG-developed guarantee systems; and strengthen farmers' control over the selling of their produce, while improving productivity and sustainability as well as achieving food self-sufficiency at local level. Second is the development of the MASIPAG Organic Standards, which is based on the IFOAM Basic Standards and adapted to farmers' practices and technologies. Third, MASIPAG has formed its Internal Quality Control Systems (IQCS) committees within respective farmers' organizations, which handles the inspection and certification of members. The MFGS is the main PGS in the Philippines, where farmers conduct inspection, assessment, and certification. It is the first to have organic standards developed by farmers in line with international standards. It is a strategy of farmers within MASIPAG to cope with the high cost of third party certification, which individual farmers cannot afford. Currently, there are 20 farmers' organizations within MASIPAG under MFGS production and marketing.

approach is also compatible with resource-poor farmers' diverse objectives, and with it, contributes to greater genetic diversity of rice varieties being developed. Moreover, participatory breeding develops and makes available organic seed, which is one of the constraints in organic farming.

References

Bachmann, L., E. Cruzada, and S. Wright. 2009. Food Security and Farmer Empowerment: A study of the impacts of farmer-led sustainable agriculture in the Philippines. pp. 152. Los Banos, Laguna, Philippines: MASIPAG.
Bigas Conference. 1985. *In*: Proceedings of the Bahanggunian sa Isyu ng Bigas. Unpublished report.
Chambers, R. 1993. Challenging the professions: Frontiers for rural development. London: Intermediate Technology Publications.
Coffman, W.R., and R.M. Herrera 1980. Rice. *In*: Fehr, W.R., and H.H. Hadley (eds.) Hybridization of crop plants. Madison, WI: American Society of Agronomy and Crop Science Society of America.
Desclaux, D. 2005. Participatory plant breeding methods for organic cereals: Review and perspectives. *In*: Lammerts van Bueren, E.T., I. Goldringer, and H. Østergård (eds.) Proceedings of the COST SUSVAR/ECO-PB workshop on organic plant breeding strategies and the use of molecular markers. pp. 17–23. January 17–19, 2005, Driebergen, The Netherlands. Driebergen, The Netherlands: Louis Bolk Institute.
Frossard, D. 1998. Asia's green revolution and peasant distinctions between science and authority. *In*: Fischer, M. (ed.) Representing Natural Resource Development in Asia: "Modern" versus "Postmodern" scholarly authority. pp. 31–42. Centre for Social Anthropology and Computing Human Ecology Series 1. Canterbury, UK: CSAC Monographs.
Halewood, M., P. Deupmann, B.R. Sthapit, R. Vernooy, and S. Ceccarelli. 2007. Participatory plant breeding to promote Farmers' Rights. Rome, Italy: Biodiversity International.
MASIPAG. 2000. Sowing the seeds of liberation. Unpubished Manuscript, pp. 144. Philippines: MASIPAG.
Medina, C.P. 2002. Empowering farmers for rural development: The MASIPAG experience. *Biotechnology and Development Monitor* 49:15–18.
Medina, C.P. 2004. The periphery as the center for biodiversity conservation: A case study from the Philippines. Currents 35/36:67–71. Stockholm, Sweden: Swedish University of Agricultural Sciences.
Morris, M.I., and M.R. Bellon. 2004. Participatory plant breeding research opportunities and challenges for the international crop improvement system. *Euphytica* 136:21–35.

Pimbert, M.P. 1994. The need for another research paradigm. *Seedling* 11:20–26.
Rubenstein, K.D., P. Heisey, R. Shoemaker, J. Sullivan, and G. Frisold. 2005. Crop genetic resources: An economic appraisal. *USDA, Economic Information Bulletin* 2.
Ryan, J.C. 1992. Conserving biological diversity. *In*: Brown, L., R. Lester et al. (eds.) State of the world 1992. New York: W.W. Norton.
Thro, A.M., and C. Spillane. 2000. Biotechnology-assisted participatory plant breeding: Complement or contradiction? Working document No. 4. April 2000, CGIAR.
Vicente, P.R., C.P. Medina, and R.A. Buena. 2010. Modified Bulk Selection Method of Breeding Rice for Broader Genetic Diversity. *Ecology & Farming* 47:25–29. Accessed April 20, 2011. http://www.bioforum.be/v2/cms/documenten/EF47_small.pdf.
WRI, UNEP, and IUCN. 1992. Global biodiversity strategy: Guidelines for action to save, study and use Earth's biotic wealth sustainably and equitably. Washington, D.C.: World Resources Institute.

12 Soybean: Breeding for Organic Farming Systems

Johann Vollmann and Michelle Menken

Introduction

Soybean is the most important oilseed crop in the world grown on 98.8 million ha and producing a harvest of over 222 million metric tonnes in 2009 (FAOSTAT, 2010). Major soybean producing countries include the United States, Brazil, Argentina, and China, whereas only 2% of the world soybean acreage is grown in Europe. Most of the recent increase in soybean growing area has been due to the spread of genetically modified (GM) soybeans exhibiting transgenic herbicide resistance traits, which were grown on 77% of the total acreage in 2009 (James, 2010). While certified organic soybeans are produced on less than 1% of the area in most countries at present, there is an increasing demand for organic soybeans, particularly for soy food manufacturing, and organic production can make up 10 to 20% of the area in some European countries. As a consequence, growers of organic soybeans are interested in varieties suitable for production under their specific farming conditions, in which competitiveness against weeds, enhanced nitrogen fixation, and food-grade quality are considered most important.

Soybean (*Glycine max* [L.] Merr.) is a legume crop originating in the northeast of China. As soybean is largely autogamous, all cultivars are homozygous pure-line cultivars from biparental crosses. Individual soybean genotypes are adapted to a narrow north-south latitude belt due to their photoperiod requirement, and soybean cultivars are classified into 13 maturity groups from 000 to X for high to low latitudes, respectively. The huge demand for soybean is due to its seed composition with 40% protein and 20% oil, which makes it a rich source of protein and a favorable raw material in many food and feed applications. Soybean breeding has repeatedly been dealt with in scholarly reviews emphasizing conventional breeding (Fehr, 1987), resistance to fungal, bacterial, or virus diseases and nematodes (Wilcox, 1983), the integration of molecular-based strategies (Orf et al., 2004), and advanced breeding objectives (Cober et al., 2009).

This review focuses on soybean breeding without genetic modification and the development of cultivars with favorable agronomic and seed quality features for organic farming systems. Improved crop competitiveness against weeds would support weed management, and good biological nitrogen fixation is important in yield formation and for contributing additional nitrogen to the crop rotation. Organically grown soybeans are utilized primarily for soy food production rather than for oil extraction or animal feed production; thus, specific characteristics such as high protein content or large seed weight are important features for marketing of food-grade soybeans.

Agronomic Characters

Competitiveness against Weeds

Weed control is one of the most challenging issues in organic crop production. Weeds reduce soybean yields primarily through competition for light, moisture, and nutrients. In addition, they may also reduce soybean seed quality, reduce harvesting efficiency, and increase costs by requiring additional grain cleaning prior to storage or sale. To manage weeds, section 205.206(a)(3) of the USDA NOP regulations states that producers must use management practices such as crop rotation and cultural practices that include *selecting varieties and species that are resistant to weeds* (NOP, 2002). Article 12(1)(g) of the EU organic regulations similarly says that *the choice of species and varieties shall be used to prevent damage from weeds* (EC, 2007). The development of varieties that grew without loss of yield in the presence of weeds, competed earlier and for a longer duration and therefore required less tillage, or even suppressed weeds and reduced weed seed production, would be of great benefit to organic production.

Studies have shown that there is a critical weed control period that is necessary to maximize yield. Once past this period, modern soybean varieties are effectively able to compete with weeds that emerge later without loss of yield. This critical control period varies with weed species, weed density, and moisture availability (Buhler and Hartzler, 2004). Time of emergence of weeds in relation to the crop and production practices also influence the critical control period (Iowa State University, 2000; 2007). Though study results vary, some of the soybean traits associated with competitiveness are early vigor, early canopy closure, branching, leaf-area expansion rate, and plant height (Callaway and Forcella, 1999; Pester et al., 1999).

Of interest to breeders is the information in several weed competition studies that there is variation in the ability of different soybean cultivars to maintain yield in the face of weed competition and even in some cases to suppress weed growth. In a test of six varieties grown with Johnsongrass (*Sorghum halepense* [L.] Pers.) and common cocklebur (*Xanthium pensylvanicum* Wallr.), one yielded significantly better in both weed-free and weedy conditions and was also a cultivar that allowed less Johnsongrass regrowth (McWhorter and Hartwig, 1972). Of five soybean cultivars grown with common cocklebur one variety yielded significantly better in 33-cm row spacing and another yielded better in 100-cm row spacing (McWhorter and Barrentine, 1975). Bussan et al. (1997) tested 16 soybean genotypes against 12 weed species commonly found in the U.S. soybean-growing region. The soybean varieties varied in their ability to maintain yield in the presence of weeds. They found varieties that yielded well in weed-free conditions and poorly in weedy conditions (indicating they were poor competitors) and varieties that yielded well in weedy conditions but performed poorly in weed-free conditions. Three of the 16 yielded well in weed-free and weedy conditions and were effective at limiting growth of weeds. They did not find any parameters (soybean canopy area, height, or volume) that consistently related to weed competitiveness. Since varieties varied in performance depending on the weed, they indicated that selection for weed competitiveness would need to be done under multiple weed conditions and environments.

Weed competitiveness can be expressed as tolerance to weeds, where crops maintain yield under weed pressure. Such a trait would be of use in organic production systems where early season weed control is subject to weather conditions that do not guarantee early tillage. The other mode of action is for a plant to actually suppress weeds through: Greater vigor, faster growth and therefore greater ability to shade out weeds, more efficient use of soil nutrients and moisture than weeds, or through allelopathy, either from root exudates of growing plants, or from compounds released from plant residues as they decay. The latter trait has been less commonly found in soybeans, but there are a few indicators that this might offer a breeding possibility. Massantini et al. (1977) in a greenhouse study tested 141 soybean lines against two weeds, bristly ox tongue (*Helminthia echioide*) and slender meadow foxtail (*Alopecurus myosuroides*); they found two lines that effectively suppressed growth in *Helminthia* (89.6% and 76.6%), none effective against *Alopecurus*, and one which promoted growth of *Helminthia* (67.3%) and *Alopecurus* (90.7%). Rose et al. (1983) screened 280 soybean lines in the field for ability to compete with weeds. Twenty cultivars of varying suppressive ability were selected to be further tested against foxtail millet (*Setaria italica* [L.] Beauv.) and velvetleaf (*Abutilon theophrasti* Medic) in a greenhouse study. They found exudates from the soybean roots reduced growth in velvetleaf but not foxtail millet. Ground soybean dry matter placed in the soil inhibited velvetleaf dry weight an average of 46% and foxtail millet germination and dry weight 82 and 65%, respectively. One cultivar inhibited velvetleaf germination and dry weight by 65 and 73%, and foxtail millet germination and dry weight by 95 and 87%, showing an outstanding ability to suppress growth of these two weeds. The authors speculate that the soybean cultivars possessed a chemical compound or plant metabolite that inhibited the weed growth. Only a small percentage of the germplasm collection has been screened for the ability to compete or suppress weeds, but given the positive results in the previously mentioned tests, breeders and agronomists should consider looking for other genotypes with these abilities.

Market quality traits must be the primary concern for soybean breeders developing new lines for organic production, but weed competitiveness would be a useful secondary trait. Conventional soybean breeders can generally work with a more predictable environment: Short set rotations (e.g., corn-soybean), standardized rates of fertilization using synthetic fertilizers, and the assumption of weed and insect control from pesticides. Attempting to breed for weed competitiveness in organic systems will be more of a challenge; in organic production little is standardized: Rotations, crop planting dates and densities, fertility practices and sources, tillage and planting methods and equipment, and environments can all vary as do annual weed densities, weed species, and moisture levels. Soybean aphids (*Aphis glycines*) have become an enormous problem in the United States. Ragsdale et al. (2011) have found that the common organic practice of delaying planting to allow for early season weed control (Delate, 2009) may allow for more damage later in the season from aphid attack. Since some studies have shown that earlier maturing lines may be more weed competitive (Vollmann et al., 2010), perhaps breeders should re-evaluate this trait both for weed and insect considerations.

The best practice may be to develop potential high-quality lines and use participatory farmer testing on as many fields as possible to find the wide adaptation that will ensure stability of yield and quality in weed-free and weedy environments.

Symbiotic Nitrogen Fixation

Biological nitrogen fixation by soybeans in symbiosis with root-nodulating rhizobia (*Bradyrhizobium japonicum* [Kirchner] Jordan) may generate up to 337 kg N/ha (average: 111 kg N/ha) in a soybean crop; however, nitrogen balance was negative for over 50% of cases in an N-fixation study due to the high N removal at harvest (Salvagiotti et al., 2008). Thus, breeding for enhanced nitrogen fixation has been proposed, as genetic variation in this characteristic has been repeatedly reported in soybean (Herridge and Rose, 2000). The level of nitrogen fixation could be determined through measuring N accumulation of plants, acetylene reduction activity of nitrogenase, or other N metabolism parameters; for screening large numbers of genotypes in plant breeding, more easily phenotyped traits such as number of nodules or nodule dry weight could be utilized as selection parameters, as they exhibit medium to high heritability (Nicolas et al., 2002).

Biological nitrogen fixation of soybean is highly complex and rapidly disrupted by low temperatures, drought stress, or high soil-nitrate concentrations (see fig. 12.1).

Figure 12.1 Difference in nodule formation on the roots of soybean cv. Apache at late flowering stage from two locations in Austria: Strong nodule formation on a plant grown in an organic field with an over 25 years soybean growing history (top); weak nodule formation in a conventional field with no soybeans previously grown and high soil nitrate concentration (bottom; photos courtesy of J. Vollmann, Vienna, Austria).

In addition, a soybean genotype by rhizobial strain interaction has been reported (Luna and Planchon, 1995). Moreover, indigenous strains of *Bradyrhizobium japonicum* with low nitrogen fixation capacity are present in many soils; those inefficient strains compete for root infection and might cause the development of soybean root nodules without nitrogen fixation (Champion et al., 1992), which means a loss of potential photosynthate energy to the host plant and a reduced yield potential. Some soybean genotypes are able to restrict nodulation to efficient bacterial strains only; this host plant characteristic is under simple genetic control (Qian et al., 1996) and could contribute to improved nitrogen fixation in many soils. For soils with mixed populations of efficient and inefficient rhizobial strains, Kiers et al. (2007) showed that older soybean cultivars have an enhanced capability of maximizing yield as compared to newer cultivars. They conclude that older cultivars could better discriminate between efficient and inefficient strains, either through strain-selectivity at the infection stage or through restriction of the resource supply for ineffective nodules at the post-infection stage; newer cultivars, which were selected in high-nitrogen environments, may have lost their discriminatory power with respect to rhizobial efficiency because of the reduced selection pressure on symbiotic nitrogen fixation.

Yield and Yield Stability

Average soybean yields ranged between 2000 and 3000 kg/ha for most soybean producing countries during the previous decade (FAOSTAT, 2010). Plant breeding has significantly contributed to yield increases. In the United States, soybean yields increased by 22 kg/ha per year from 1924 to 1997 with a faster increase (31 kg/ha per year) from 1972 to 1997 (Specht et al., 1999). Voldeng et al. (1997) reported an 11 kg/ha per year average yield increase for short-season soybeans in Ontario/Canada from 1932 to 1992, with a more rapid increase (30 kg/ha per year) for cultivars released after 1976. Continuous selection for grain yield brought about changes in various associated characters: Newer cultivar releases with higher yields had reduced lodging scores, reduced seed protein, and enhanced oil content (Voldeng et al., 1997). In addition, newer releases had significantly improved leaflet photosynthetic rate, higher stomatal conductance, lower leaf area index and leaf area ratio as well as an improved harvest index (Morrison et al., 1999). In China, yield progress in early maturity cultivars was associated with a higher number of seeds per pod, increased photosynthetic rate, reduced leaf area index, and improved harvest index, and included reductions in plant height, lodging, seed disease, and pest scores, whereas seed protein and oil content remained unaffected by selection (Jin et al., 2010). Long-term trends in characteristics associated with breeding for grain yield such as leaf-area index or seed protein content need particular attention in organic farming, as they may represent typical trade-offs, with respect to desirable features such as weed suppression through larger leaf area or food-grade quality through high protein content, respectively.

Selection for yield performance involves testing of advanced breeding lines in replicated plot experiments across different environments. Marker assisted selection through the utilization of yield quantitative trait loci (QTL) could support the identification of superior lines, but yield QTL have been identified in lower numbers than QTL affecting other characters due to the low heritability of seed yield and the extensive need for phenotyping to detect QTL with relatively small effects (Orf et al., 2004). Nevertheless, some of the many yield QTL identified in different mega-environments may still be transferred into elite germplasm through introgression (Palomeque et al., 2009).

Differences in yield stability of soybean lines as well as in stability of quality features such as protein or oil have frequently been reported with respect to adaptation to specific growing regions, e.g. Minnesota (Pazdernik et al., 1997), Wisconsin (Zhe et al., 2010), or Mato Grosso (Ferraz de Toledo et al., 2006). Yield stability was not improved in new and higher yielding Canadian germplasm (Voldeng et al., 1997), whereas Jin et al. (2010) reported increased yield stability over 56 years of Chinese soybean breeding, which they mainly attributed to improved stability in the number of pods per plant.

Yield stability is of special interest to organic farming: Organic environments are more diverse both within fields and crop rotations than conventional environments, which are more standardized and therefore reveal less genotype by environment interaction. For this reason, Desclaux et al. (2008) proposed a rethinking of the stability concept by predicting and valorizing genotype by environment interaction through participatory breeding instead of minimizing it. In other approaches, desirable cultivars are required to be robust and highly stable under both conventional and organic farming systems (wide adaptation), particularly in cases where organic breeding programs are not available. In soybean, many conventional breeding programs are presently converted to GM programs, which will limit the future availability of conventional germplasm for organic breeding programs, thus justifying the initiation of dedicated organic programs. In legumes such as lentils, stable ranking in grain yield under both organic and conventional systems as well as cross-over interaction were found for different genotypes revealing both wide and narrow adaptation (Vlachostergios and Roupakias, 2008). Preliminary analysis of grain yield for 550 soybean breeding lines grown under conventional and organic conditions is showing similar results (Menken, unpublished data).

In contrast to wide adaptation, Murphy et al. (2007) reported significant genotype by system interactions in wheat yield suggesting specific adaptation to organic conditions; they emphasize the need for direct selection in organic systems in order to be able to identify high-yielding cultivars through focusing on improved nitrogen efficiency, competitiveness against weeds, or adaptation to soil microbes. In soybean, selection in organic systems might similarly reveal advantageous effects such as specific adaptation to mycorrhiza (Douds et al., 1993) or enhanced nitrogen fixation through better accuracy in discriminating between efficient and inefficient rhizobial strains at the nodulation stage (Kiers et al., 2007); such traits could hardly be selected for by testing conventionally bred cultivars under organic conditions, but would instead require a whole breeding program to be carried out in organic target environments.

The soybean yield level is clearly lower than that of cereals such as corn or wheat. In contrast to cereals and other crops, however, soybean yield and economic return are not declining under organic management (Pimentel et al., 2005) because soybean is not as heavily dependent on fertilizer inputs as are non-legumes. Breeding of soybeans in organic systems might thus contribute to further yield increases, improved economic competitiveness as compared to other crops, and, subsequently, a higher rate of organically produced soybeans.

Seed Quality Features

In the worldwide soybean trade, soybean is an industrially processed commodity oilseed crop delivering two major products, vegetable oil and protein-rich meal. The main focus of conventional soybean breeding for harvest quality has traditionally been put on the improvement of soybean oil for human consumption and processing in the food industry: E.g., selection for low concentrations of saturated fatty acids for minimizing the risk of LDL-cholesterol increase in humans or reduction of polyunsaturated fatty acids such as linolenic acid (C18:3) to avoid the formation of trans-fatty acids (Cober et al., 2009).

In contrast, organic soybeans are mainly marketed as whole beans for the preparation of traditional or modern soy-foods rather than for oil production. As protein and other constituents present in whole soybeans play a major role in both nutrition and food technology, breeding for value-added traits such as seed protein content and food-grade quality will be covered in this section instead of oil. In addition, most of the fatty acid variants utilized at present are tracing back to induced mutations, which may not comply with future IFOAM standards or other regulations for organic variety breeding (Grausgruber, 2009).

Protein Content

Seed protein content of commodity soybean samples is about 40%, on average, based on dry matter. Generally, processors consider Chinese and Japanese soybeans higher in protein content and therefore better suited for production of traditional foods, whereas North American soybeans appear more suitable for oil extraction (Motoki and Seguro, 1994). This view is supported by studies of long-term trends in North American soybean breeding progress for higher yield, which was frequently associated with reduced protein content (Voldeng et al., 1997), whereas protein content of Chinese cultivars remained constant (Jin et al., 2010). Moreover, breeding for protein content is not a major objective in most selection programs, as soybeans are not priced based on seed composition (Fehr, 1987).

In breeding for protein content, a wide variation is available from various genetic sources. In the Chinese germplasm collection of cultivated soybeans, seed protein content in over 19,000 accessions ranged from 30 to 53% with the highest number of genotypes between 43 and 46% protein (Dong et al., 2004). Wehrmann et al. (1987) and Wilcox and Cavins (1995) reported the introgression of high protein content from plant introductions into elite germplasm through backcrossing, whereas Leffel (1992) selected soybean lines with protein content of 49 to 53% from biparental high-protein populations. Despite the widely reported negative correlation between grain yield and protein content, both Wehrmann et al. (1987) and Wilcox and Cavins (1995) succeeded in selecting high protein lines with fully recovered yield levels of the recurrent parent, which was explained by improved nitrogen metabolism: In high-protein lines, increased nitrogen fixation at normal nodule occupancy levels, a longer period of nitrogen fixation toward maturity and higher rates of nitrogen remobilization as compared to normal-protein genotypes were found (Leffel et al., 1992). Therefore, as a high level of nitrogen fixation is a desirable characteristic in organic soybean cultivars, seed protein content can be utilized as an indirect selection parameter for nitrogen fixation. Apart from being a breeding goal by itself, seed protein content can be selected for with less effort and higher through-put than symbiotic nitrogen fixation traits, e.g., from single-row plots (Pazdernik et al., 1996a) and using near-infrared reflectance spectroscopy (NIRS) for screening.

Genetically, seed protein content is a quantitative character with medium to high heritability, depending on the genetic background of the population investigated. Soybean protein and oil content are negatively correlated (Wilcox and Cavins, 1995), which makes it difficult to select for a simultaneous increase of both constituents. Seed protein content is considerably affected by environmental conditions such as soil parameters, nitrogen availability or precipitation rates, and temperature during the seed filling period (Vollmann et al., 2000). Significant genotype by environment interaction has also been described for protein content Sudarić et al., 2006).

The major fraction (70–90%) of total soybean seed protein consists of glycinin (11S, 320-380 kDa molecular mass) and β-conglycinin (7S, 180 kDa) storage proteins, which determine nutritional and functional properties of soy-foods; other protein fractions such as lipoxygenase, trypsin inhibitors, lectin, or urease have catalytic or anti-nutritional properties (Nielsen, 1996).

Food-Grade Soybean Quality

Soybean has manifold uses in human nutrition: Beside traditional Asian soy-foods, which are based on whole fermented or unfermented soybeans, isolated soybean components such as soybean oil, lecithin, fat-free meals, flakes, or protein isolates are utilized in thousands of different products in the modern food industry. As there are considerable differences between individual requirements, this review focuses on breeding aspects in particular foods frequently made from whole organic soybeans.

Edamame or vegetable soybean is harvested for immediate consumption or deep-freezing when pods and seeds are still green and tender. Edamame cultivars have very large seeds with a dry 1000-seed weight of over 250 g and up to 500 g; seed coat colors range from yellow to green, to brown or even black. Although most edamame cultivars are of Japanese origin, genetic diversity is higher in Chinese accessions (Mimura et al., 2007); recent edamame genotypes have also been developed by hybridizing Japanese cultivars with either North American or other large-seeded germplasm (Mebrahtu et al., 2005; Mimura et al., 2007). At the green harvesting stage, edamame genotypes are lower in oil content, have less off-flavor and a sweeter and nuttier taste as compared to conventional cultivars.

In soymilk and tofu production, large seed weight, yellow hilum, yellow and thinner seed coat, and high protein content are important quality parameters for processing and product color, whereas lipoxygenase-nulls and enhanced sucrose content can improve taste and consumer acceptance, particularly in non-Asian countries. Soybeans with large seed weight or yellow hilum and seed coat can easily be selected for in appropriate populations. Apart from high protein content, protein composition is essential for gel formation, high tofu yield, and desirable tofu properties. Higher ratios of 11S-to-7S protein fractions have been reported to improve consistency of soymilk and texture of tofu (Kim and Wicker, 2005). Effects of protein fraction subunit composition on tofu firmness and hardness have been reported which permit a fine-tuning of product properties through selection (Poysa et al., 2006). While the results discussed above are based on electrophoresis (SDS-PAGE) and other chemical and physical methods, small-scale methods for direct soymilk or tofu production have also been developed (Evans et al., 1997), and Pazdernik et al. (1996b) presented a NIRS method to determine the concentration of the 11S protein fraction, which permits the rapid screening of large numbers of breeding lines.

Generally, soy-foods such as soymilk, tofu, miso, and edamame are made of large seeds, whereas natto, soy sauces, tempeh, and soy sprouts are made of small seeds. Zhang et al. (2010) described the phenotypic diversity for each of the two seed size groups. In both small- and large-seeded groups they found higher oil and lower protein content in North American genotypes as compared to Asian soybeans; in addition, small-seeded North American genotypes had better characteristics for natto production, whereas large-seeded Asian genotypes had favorable tofu characteristics. Thus, the different genepools for food-grade soybeans might be utilized to introgress desirable genetic diversity into each other.

Product quality specifications for food-grade soybean cultivars are not fully standardized but depend on processing technology, regional preferences, and other factors. Thus, breeding strategies may either focus on a dedicated genotype for one particular product such as tofu, or on a soybean variety with a wider range of applications, e.g., edamame, soymilk, and tofu production, which may be preferable in the case of niche markets such as organic soybeans in Europe.

Soy-Food Health Characteristics

Numerous epidemiological studies have shown that consumption of soy-foods may have a number of health benefits such as lowering the risk for coronary heart diseases, breast and prostrate cancers, better bone health, relief of menopausal symptoms, and reduction of LDL-cholesterol levels (Messina, 2010; Xiao, 2008). As a consequence, food labeling of health claims for soybean protein foods has been approved by food authorities in several countries. However, while some health effects are well established, others still reveal inconsistencies (Xiao, 2008).

Soybean seed components such as protein, isoflavones, oligosaccharides, linolenic acid, tocopherols, or soy lecithin play particular roles in the health effects of soy-foods. Therefore, soybean breeding has utilized the genetic variation present in these components to develop lines with altered seed composition, e.g., increased isoflavone concentration. However, while higher isoflavone concentrations may be beneficial in some applications, they may be risky in others, such as infant soy food formulas. Having these controversial health discussions in mind with organic sectors' claims of producing healthy soy-foods from the whole soybean rather than isolated or highly concentrated ingredients, organic soybean breeding at this stage would probably do better in selecting for well-balanced seed composition than over-concentrating particular components. This view might also be in better agreement with basic concepts of organic plant breeding, such as integrity of plants and naturalness (Lammerts van Bueren et al., 2003).

Considerations on Breeding Methods

To date, most soybean cultivars are derived from single seed descent (SSD) populations which are utilized for a rapid generation advance toward homozygosity (Cober et al., 2009). This implies that segregating generations are passed in standardized artificial environments including greenhouse and off-season nurseries in the absence of selection pressure. For cultivars suitable for organic farming, however, there is a need for adaptation to weed pressure, varying soil conditions, and the rhizospheric microflora; the pedigree method or modifications thereof enable both direct human and natural selection in the target environment using a larger number of segregants than in SSD, which would likely improve weed suppression, nitrogen fixation, and stability of lines selected.

Due to the declining number of non-GM breeding programs and the low genetic diversity in Western germplasm, long-term yield progress might slow down considerably. Thus, strategies for sustaining future agronomic improvement of non-GM soybeans will need to be developed; they might include participatory plant breeding approaches, introgression of elite germplasm from divergent mega-environments, or focusing programs on particular soy-food products in close concurrence with regional processors and market needs.

At present, organic soybean breeding programs are being carried out at North Carolina State University (in the United States) and at the University of Hohenheim (in Germany). In both programs, competitiveness against weeds is a high-priority breeding target. While the number of conventional GM-free programs is declining both in North and South America, a few private breeders in the United States and Canada are continuing non-GM programs for developing food-grade soybeans for export markets, and some public breeding programs maintain GM and non-GM programs in parallel. European and Chinese soybean breeding programs are non-GM, and in some European countries, public performance trials for cultivar registration are partly carried out under organic farming conditions.

References

Buhler, D.D., and R.G. Hatzler. 2004. Weed biology and management. *In*: Boerma, H.R., and J.E. Specht (eds.) Soybeans: Improvement, production, and uses, 3rd ed. Agronomy, no. 16. pp. 883–918. Madison, WI: American Society of Agronomy.

Bussan, A.J., O.C. Burnside, J.H. Orf, and K.J. Peuttmann. 1997. Field evaluation of soybean (*Glycine max*) genotypes for weed competitiveness. *Weed Science* 45:31–37.

Callaway, M.B., and F. Forcella. 1993. Crop tolerance to weeds. *In*: Francis, C.A., and M.B. Callaway (eds.) Crop improvement for sustainable agriculture. pp. 100–131. Lincoln, NE: University of Nebraska Press.

Champion, R.A., J.N. Mathis, D.W. Israel, and P.G. Hunt. 1992. Response of soybean to inoculation with efficient and inefficient *Bradyrhizobium japonicum* variants. *Crop Science* 32:457–463.

Cober, E.R., S.R. Cianzio, V.R. Pantalone, and I. Rajcan. 2009. Soybean. *In*: Vollmann, J., and I. Rajcan (eds.) Oil crops. Series: Handbook of plant breeding, vol. 4. pp. 57–90. New York: Springer.

Delate, K. 2009. Organic grains, oilseeds, and other specialty crops. *In*: Francis, C. (ed.) Organic farming: The ecological system. Agronomy Monograph 54. pp. 113–136. Madison, WI: American Society of Agronomy.

Desclaux, D., J.M. Nolot, Y. Chiffoleau, E. Gozé, and C. Leclerc. 2008. Changes in the concept of genotype × environment interactions to fit agriculture diversification and decentralized participatory plant breeding: Pluridisciplinary point of view. *Euphytica* 163: 533–546.

Dong, Y.S., L.M. Zhao, B. Liu, Z.W. Wang, Z.Q. Jin, and H. Sun. 2004. The genetic diversity of cultivated soybean grown in China. *Theoretical and Applied Genetics* 108:931–936.

Douds, Jr., D.D., R.R. Janke, and S.E. Peters. 1993. VAM fungus spore populations and colonization of roots of maize and soybean under conventional and low-input sustainable agriculture. *Agriculture Ecosystems and Environment* 43:325–335.

European Commission (EC). 2007. EC Council Regulation No. 834/2007 of June 28, 2007, on organic production and labeling of organic products and repealing Regulation (EEC) No 2092/2091. Official Journal of the European Union L 189 (2007). pp. 1–23. Accessed October 1, 2010. http://eur-lex.europa.eu/LexUriServ/LexUriServ.do?uri=OJ:L:2007:189:0001:0023:EN:PDF.

Evans, D.E., C. Tsukamoto, and N.C. Nielsen. 1997. A small scale method for the production of soymilk and silken tofu. *Crop Science* 37:1463–1471.

FAOSTAT 2010. FAO Statistical Database, ProdSTAT module. Rome, FAO. Accessed September 6, 2010. http://faostat.fao.org.

Fehr, W.R. 1987. Breeding methods for cultivar improvement. *In*: Wilcox, J.R. (ed.) Soybeans: Improvement, production, and uses, 2nd ed. Agronomy, no. 16. pp. 249–293. Madison, WI: American Society of Agronomy.

Ferraz de Toledo, J.F., C.G. Portela de Carvalho, C.A. Arrabal Arias, L. Alves de Almeida, R.L. Brogin, M.F. de Oliveira, J.U. Vieira Moreira, A.S. Ribeiro, and D.M. Hiromoto. 2006. Genotype and environment interaction on soybean yield in Mato Grosso State, Brazil. *Pesquisa agropecuária Brasileira* 41:785–791.

Grausgruber, H. 2009. Organic plant breeding – A general overview. *In*: Østergård, H., E.T. Lammerts van Bueren, and L. Bouwman-Smits (eds.) Proceedings, workshop on the Role of Marker Assisted Selection in Breeding Varieties for Organic Agriculture. pp. 10–13. February 25–27, Wageningen, The Netherlands.

Herridge, D., and I. Rose. 2000. Breeding for enhanced nitrogen fixation in crop legumes. *Field Crops Research* 65:229–248.

Iowa State University. 2000. Early-season weed competition, integrated crop management. IC-484(8), May 8, 2000:66–67

Iowa State University. 2007. Protecting soybean yields from early-season competition, integrated crop management. IC-498(3), March 26, 2007:76.

James, C. 2010. 2009 ISAAA report on global status of biotech/GM crops. *International Service for the Acquisition of Agri-biotech Applications*. Accessed September 6, 2010. http://www.isaaa.org.

Jin, J., X. Liu, G. Wang, L. Mi, Z. Shen, X. Chen, and S.J. Herbert. 2010. Agronomic and physiological contributions to the yield improvement of soybean cultivars released from 1950 to 2006 in Northeast China. *Field Crops Research* 115:116–123.

Kiers, E.T., M.G. Hutton, and R.F. Denison. 2007. Human selection and the relaxation of legume defences against ineffective rhizobia. *Proceedings of the Royal Society B* 274:3119–3126.

Kim, Y., and L. Wicker. 2005. Soybean cultivars impact quality and function of soymilk and tofu. *Journal of the Science of Food and Agriculture* 85:2514–2518.

Lammerts van Bueren, E.T., P.C. Struik, M. Tiemens-Hulscher, and E. Jacobsen. 2003. Concepts of intrinsic value and integrity of plants in organic plant breeding and propagation. *Crop Science* 43:1922–1929.

Leffel, R.C. 1992. Registration of high-protein soybean germplasm lines BARC-6, BARC-7, BARC-8, and BARC-9. *Crop Science* 32:502.

Leffel, R.C., P.B. Cregan, A.P. Bolgiano, and D.J. Thibeau. 1992. Nitrogen metabolism of normal and high-seed-protein soybean. *Crop Science* 32:747–750.

Luna, R., and C. Planchon. 1995. Genotype × *Bradyrhizobium japonicum* strain interactions in dinitrogen fixation and agronomic traits of soybean (*Glycine max* L. Merr.). *Euphytica* 86:127–134.

Massantini, F., P. Caporali, and G. Zellini. 1977. Evidence for allelopathic control of weeds in lines of soybean. *In*: Proceedings: Symposium on the different methods of weed control and their integration. pp. 23–28. Uppsala, Sweden: European Weed Research Society.

McWhorter, C.G., and W.L. Barrantine. 1975. Cocklebur control in soybeans as affected by cultivars, seeding rates, and methods of weed control. *Weed Science* 23:386–390.

McWhorter, C.G., and E.E. Hartwig. 1972. Competition of johnsongrass and cocklebur with six soybean varieties. *Weed Science* 20:56–59.

Mebrahtu, T., T.E. Devine, P. Donald, and T.S. Abney. 2005. Registration of 'Asmara' vegetable soybean. *Crop Science* 45: 408–409.

Messina, M. 2010. A brief historical overview of the past two decades of soy and isoflavone research. *Journal of Nutrition* 140:1350S–1354S.

Mimura, M., C.J. Coyne, M.W. Bambuck, and T.A. Lumpkin. 2007. SSR diversity of vegetable soybean [*Glycine max* (L.) Merr.]. *Genetic Resources and Crop Evolution* 54:497–508.

Morrison, M.J., H.D. Voldeng, and E.R. Cober. 1999. Physiological changes from 58 years of genetic improvement of short-season soybean cultivars in Canada. *Agronomy Journal* 91:685–689.

Motoki, M., and K. Seguro. 1994. Trends in Japanese soy protein research. *INFORM* 5:308–313.

Murphy, K.M., K.G. Campbell, S.R. Lyon, and S.S. Jones. 2007. Evidence of varietal adaptation to organic farming systems. *Field Crops Research* 102:172–177.

National Organic Program. 2002. U.S. Code of Federal Regulations, Title 7, Part 205: National Organic Program, §205.206 Crop Pest, Weed, and Disease Management Practice Standard, Part (e). Accessed October 1, 2010. http://ecfr.gpoaccess.gov/cgi/t/text/text-idx?c=ecfr&tpl=/ecfrbrowse/Title07/7cfr205_main_02.tpl.

Nicolas, M.F., C.A. Arrabal Arias, and M. Hungria. 2002. Genetics of nodulation and nitrogen fixation in Brazilian soybean cultivars. *Biology and Fertility of Soils* 36:109–117.

Nielsen, N.C. 1996. Soybean seed composition. *In*: Verma, D.P.S., and R.C. Shoemaker (eds.) Soybean, genetics, molecular biology and biotechnology. pp. 127–163. Wallingford, UK: CAB International.

Orf, J.H., B.W. Diers, and H.R. Boerma. 2004. Genetic improvement: Conventional and molecular-based strategies. *In*: Boerma, H.R., and J.E. Specht (eds.) Soybeans: Improvement, production, and uses, 3rd Edition. Agronomy, no. 16. pp. 417–450. Madison, WI: American Society of Agronomy.

Palomeque, L., L. Li-Jun, W. Lin, B. Hedges, E.R. Cober, and I. Rajcan. 2009. QTL in mega-environments: I. Universal and specific seed yield QTL detected in a population derived from a cross of high-yielding adapted × high-yielding exotic soybean lines. *Theoretical and Applied Genetics* 119:417–427.

Pazdernik, D.L., L.L. Hardman, and J.H. Orf. 1997. Agronomic performance and stability of soybean varieties grown in three maturity zones in Minnesota. *Journal of Production Agriculture* 10:425–430.

Pazdernik, D.L., L.L. Hardman, J.H. Orf, and F. Clotaire. 1996a. Comparison of field methods for selection of protein and oil content in soybean. *Canadian Journal of Plant Science* 76:721–725.

Pazdernik, D.L., S.J. Plehn, J.L. Halgerson, and J.H. Orf. 1996b. Effect of temperature and genotype on the crude glycinin fraction (11S) of soybean and its analysis by near-infrared reflectance spectroscopy (Near-IRS). *Journal of Agriculture and Food Chemistry* 44:2278–2281.

Pester, T.A., O.C. Burnside, and J.H. Orf. 1999. Increasing crop competitiveness to weeds through crop breeding. *In*: Buhler, D.D. (ed.) Expanding the context of weed management. pp. 59–76. New York: Haworth Press.

Pimentel, D., P. Hepperly, J. Hanson, D. Douds, and R. Seidel. 2005. Environmental, energetic, and economic comparisons of organic and conventional farming systems. *BioScience* 55:573–582.

Poysa, V., L. Woodrow, and K. Yu. 2006. Effect of soy protein subunit composition on tofu quality. *Food Research International* 39:309–317.

Qian, D., F.L. Allen, G. Stacey, and P.M. Gresshoff. 1996. Plant genetic study of restricted nodulation in soybean. *Crop Science* 36:243–249.

Ragsdale, D.W., D.A. Landis, J. Brodeur, G.E. Hempel, and N. Desneux. 2011. Ecology and management of the soybean aphid in North America. *Annual Review of Entomology* 56:375–399.

Rose, S.J., O.C. Burnside, J.E. Specht, and B.A. Swisher. 1983. Competition and allelopathy between soybeans and weeds. *Agronomy Journal* 76:523–528.

Salvagiotti, F., K.G. Cassman, J.E. Specht, D.T. Waters, A. Weiss, and A. Dobermann. 2008. Nitrogen uptake, fixation and response to fertilizer N in soybeans: A review. *Field Crops Research* 108:1–13.

Specht, J.E., D.J. Hume, and S.V. Kumudini. 1999. Soybean yield potential – a genetic and physiological perspective. *Crop Science* 39:1560–1570.

Sudarić, A., D. Šimić, and M. Vratarić. 2006. Characterization of genotype by environment interactions in soybean breeding programmes of southeast Europe. *Plant Breeding* 125:191–194.

Vlachostergios, D.N, and D.G. Roupakias. 2008. Response to conventional and organic environment of thirty-six lentil (*Lens culinaris* Medik.) varieties. *Euphytica* 163:449–457.

Voldeng, H.D., E.R. Cober, D.J. Hume, C. Gillard, and M.J. Morrison. 1997. Fifty-eight years of genetic improvement of short-season soybean cultivars in Canada. *Crop Science* 37:428–431.

Vollmann, J., C.N. Fritz, H. Wagentristl, and P. Ruckenbauer. 2000. Environmental and genetic variation of soybean seed protein content under Central European growing conditions. *Journal of the Science of Food and Agriculture* 80:1300–1306.

Vollmann, J., H. Wagentristl, and W. Hartl. 2010. The effects of simulated weed pressure on early maturity soybeans. *European Journal of Agronomy* 32:243–248.

Wehrmann, V.K., W.R. Fehr, S.R. Cianzio, and J.F. Cavins. 1987. Transfer of high seed protein to high-yielding soybean cultivars. *Crop Science* 27:927–931.

Wilcox, J.R. 1983. Breeding soybeans resistant to diseases. *Plant Breeding Reviews* 1:183–235.

Wilcox, J.R., and J.F. Cavins. 1995. Backcrossing high seed protein to a soybean cultivar. *Crop Science* 35:1036–1041.

Xiao, C.W. 2008. Health effects of soy protein and isoflavones in humans. *Journal of Nutrition* 138:1244S–1249S.

Zhang, B., P. Chen, S.L. Florez-Palacios, A. Shi, A. Hou, and T. Ishibashi. 2010. Seed quality attributes of food-grade soybeans from the U.S. and Asia. *Euphytica* 173:387–396.

Zhe, Y., J.G. Lauer, R. Borges, and N. de Leon. 2010. Effects of genotype × environment interaction on agronomic traits in soybean. *Crop Science* 50:696–702.

13 Faba Bean: Breeding for Organic Farming Systems

Wolfgang Link and Lamiae Ghaouti

Purposes of Breeding and Growing Faba Bean

Faba bean (*Vicia faba* L.; Ackerbohne, field bean, féverole) is a traditional and important nitrogen-fixing crop with protein-rich seeds (about 30%). In Europe, dry faba bean seeds are mainly used as a component of feed for ruminants (cattle, sheep), monogastrics (pigs), and birds (chicken, pigeon). Like pea and sweet lupine, faba bean is used as substitute of soybean. Faba bean performs very well under neutral pH, fertile, and humid conditions, and is very tolerant to summertime episodes of cool temperatures. In such environmental conditions, faba beans are very appropriate to achieve self-sufficiency with respect to the protein feed component that is demanded by organic livestock standards (Hancock et al., 2005).

Another use of faba bean is food production for human consumption. Faba bean is a traditional and important staple food crop around the Mediterranean Basin, in Ethiopia, in the Hindu Kush region, in parts of China, and in the Andean Community.

The world faba bean area in 2009 was 2.51 Mha, and average yield was about 1.6 t ha^{-1}. Main growers were China (945,000 ha) and Ethiopia (521,000 ha). North Africa and Australia produce faba beans in significant amounts. In Europe, France and the UK are the main producers with 100,000 to 200,000 ha per year. Average yield in France from 2007 to 2009 was 4.9 t ha^{-1}. Bolivia, Algeria, and Morocco were the main producers of faba bean green pods and seeds. No precise figures are known for organic production of faba beans in Europe. A total of 1.3 Mha of pulses were grown in 25 European countries, with 4.4% produced under organic conditions; in Germany, this figure was 26% (i.e., 21,900 ha) in organic production. The German faba bean hectarage (12,000 ha) in 2009 was 50% organic (AMI, 2011; Eurostat, 2011; FAOSTAT, 2010).

Farmers in the UK and France export combine-harvested beans for a premium price to food markets in the Nile Valley. The main traits to satisfy this market are: White hilum; an even, light seed color; absence of symptoms from insect or fungus attack; and absence of bean beetles. In the UK and northern parts of Germany, faba bean seed is a component of some traditional dishes (e.g., *dicke Bohnen mit Speck*). The vegetable types of beans used for these purposes are quite different from beans used for animal feeds in that they are large-seeded. Breeding objectives depend on the type of use, which is either consumption as fresh, frozen, or canned product, very young pods, or mature dry seed. Cooking quality, color and taste are major features (AEP, 2010).

Organic Crop Breeding, First Edition. Edited by Edith T. Lammerts van Bueren and James R. Myers.
© 2012 John Wiley & Sons, Inc. Published 2012 by John Wiley & Sons, Inc.

Additional uses of faba beans are as rolled and cracked grain, which is used as nitrogen-rich fertilizer in allotment or community gardens (Raupp, 2010). In certain field rotations, faba bean may serve as a catch crop, with the sole purpose of fixing additional nitrogen, or as a mixed crop, to prevent nitrogen from leaching into ground water. It may also serve as a break in crop rotations to interrupt pest, weed, and disease infection or infestation cycles. Winter-hardy types of faba bean (similar to winter pea) may also serve as a catch crop in a biogas rotation, producing biomass for silage and being followed by corn for the same purpose (Backhaus, 2009; Roth, 2010). In this chapter, we focus on faba beans used for animal feed and grown as arable crop for combine harvesting.

Genetic and Botanical Basics of Breeding Faba Bean

Faba bean is diploid, $2n = 2x = 12$. It is closely related to the $2n = 2x = 14$ *Vicia narbonensis* or *Vicia peregrine* species. Two chromosomes in these related species were united in the evolution of *Vicia faba* to give rise to one very large faba bean chromosome (Fuchs et al., 1998). The genome of *Vicia faba* is huge and consists of about 13,000 Mbp, which is about 30 times the content of barrel medic (*Medicago truncatula*) and about 100 times the content of mouse-ear cress (*Arabidopsis thaliana*) – the latter two being known as model crops for dicot genomic research because of their small genomes. Faba bean's large genome size causes difficulty in molecular and genomics research (Ellwood et al., 2008).

In temperate regions, both spring and winter bean cultivars are grown. In the Mediterranean Basin and similar climates (e.g., parts of China, Australia, and South Africa), the crop is sown in late autumn, grows during mild winters, and matures very early in spring, thereby escaping summer drought and heat. Winter types show only a limited need of vernalization. In order to speed up breeding schemes, breeders may have recourse to two cycles per year, using greenhouse facilities or shuffling seed across both hemispheres.

Unlike grain legumes such as soybean or *Phaseolus* bean, faba bean cannot be crossed with related species. No wild progenitor species is known. Yet, a number of important features wait to be transferred from related vetches, such as tolerance to frost and drought, and resistance to fungal diseases and pests such as aphids (Link, 2006). The genetic diversity within the species is large, reflecting its wide distribution across its agro-ecological niches on all continents. It is only in the tropics that faba beans are not grown. Generally, *Vicia faba* var. *paucijuga*, *V.f.* var. *minor*, *V.f.* var. *equina* and *V.f.* var. *major* are identified as subspecies, classified mainly based on seed size (Mansfield, 2010).

Faba bean seeds are large. Thousand seed weight is approximately 200 to 500 g regarding *V.f.* var. *paucijuga* and *minor* types, 600 to 900 g regarding *V.f. equina* types, and 1,000 to more than 2,500 g regarding *V.f. major* types ("broad bean"). Thousand seed weight of typical feedstuff beans like Fuego or Sultan is 500 to 600 g. With a sowing density of at least 30 seeds per m^2, yield of one faba bean plant only allows the seeding of about one m^2 of field. Thus, faba bean has a low reproduction coefficient.

Mature grain legume seeds do not contain endosperm. Seed coat is maternal tissue with seed coat color, hilum color and tannin content being determined by maternal genotype. The embryo is inside the seed coat. Embryo traits such as cotyledon color (green vs. yellow) are determined by the embryo's genotype. Seed injuries, caused for example by combining or transporting, frequently injure the embryo. Small faba bean seeds are globular in shape but become less so with increasing size. Large broad bean (*major*) seeds are flat and therefore sensitive to mechanical impacts.

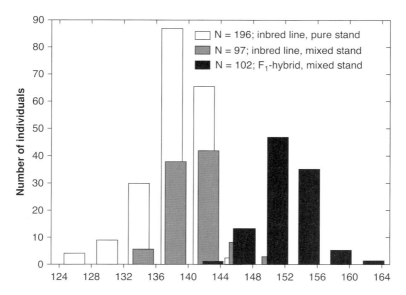

Figure 13.1 Distribution of individual plant height in a topcross-progeny plot (mixed stand) of faba bean, compared with its pure-line seed parent (pure stand). F_1-hybrid plants were identified in the mixed stand based on flower color. Data taken from Leineweber and Link (unpublished).

The anthracnose-causing fungus *Ascochyta faba* is partly seed transmitted. This is true as well for the stem nematode (*Ditylenchus dipsaci*), the broad bean true mosaic virus, and the broad bean stain virus (Rohloff, 1980). Thus, actual provenience, health, and quality of seed are important topics in faba bean breeding and production (Hebblethwaite, 1983).

Faba bean, unlike most other grain legumes, is partly self- and partly cross-fertilized (mixed mating). Both self-pollen and foreign pollen are fully compatible and fertile. The degree of cross-fertilization is about 40 to 50% (see fig. 13.1), and highly variable, with variation depending on the supply of pollinators, the genotype, and the inbreeding status of the plant (Link, 1990; Palmer et al., 2009). Honey bees and solitary bees such as *Osmia rufa* and bumble bees (*Bombus terrestris* and *Bombus hortorum*) are frequent and efficient pollinators. Other insects such as flies, aphids, butterflies, and thrips do not contribute to pollination.

A typical visit of a pollinating insect to a flower causes a mechanical action and stimulation of the stigma (called a "tripping" effect; Zaleski, 1956), thus allowing pollen to germinate and fertilize the ovules. Tripping happens every time the flower is visited by a pollinator. Even if an insect carries no pollen, its visit still stimulates fertilization with self-pollen in the flower. Typically, pollinators are dusted heavily with foreign pollen (pollen from other individuals), thus inducing the germination of a mixture of self-pollen and cross-pollen. The mixed mating system of faba bean is quite different from other grain legumes (such as pea, common bean, lentil, chickpea, groundnut, and soybean), where fertilization is nearly exclusive via selfing, which is caused by anther dehiscence prior to anthesis.

When pollinators are absent, only self-fertilization occurs. European faba beans are variable but generally have a marked dependency of tripping. A low tripping requirement (autofertile genotype) is correlated with a high level of selfing. Without being tripped, seed set may vary from zero to 100%, depending on the genotype, its inbreeding status, and partly on the environmental conditions. High levels of inbreeding favor low autofertility and a high degree of cross-fertilization. Late flowers on a plant tend to need less tripping for seed set. Faba bean is one of those crops that depend on the currently threatened pollinator fauna (UNEP, 2011).

The tripping effect can be triggered and imitated manually. One holds the banner petal of a flower with two fingers of one hand and the wing petals with the other hand and opens and closes the flower two or three times, thus making the stigma, anthers, and pollen extrude past the tip of the keel. Tripping manually in pollinator-free conditions is a standard job of faba bean breeders that ensures full seed set while avoiding contamination with foreign pollen (controlled, tripping-assisted selfing; Link, 1990).

Methodological Considerations

General Considerations for Breeding Diploid, Mixed Mating Crops

Faba bean populations consist of a mixture of more or less inbred individuals. If cross-fertilization was 50% in the previous generation, then 50% of plants in the current generation are hybrid (i.e., non-inbred). If the same was true in the penultimate generation, then there would be, in addition, 25% of F_2-individuals (half-inbred) in the current generation; and so on. For heterotic traits like vigor, plant height, or yield, mixed mating causes the better performing plants to be, on average, less inbred than the weaker ones (see fig. 13.1). Heterosis for grain yield may vary from 40 to 119% (Zeid et al., 2004). The most vigorous individuals are superior mostly because of their higher level of heterozygosity and to a lesser extent because of superior genes. Unfortunately, it is the genes and not the inbreeding statuses that are transmitted to offspring. The individual differences in inbreeding status partly mask the "true" genetic value. Thus, mass-selection of superior individuals is less promising in a mixed mating crop than in a self-fertilizing or a cross-fertilizing crop (where genotypes are not different in inbreeding status).

As is obvious from the example of oilseed rape, sorghum, or cotton, conventional breeders of mixed mating crops prefer to breed hybrid cultivars. As an attractive alternative, synthetic (population) cultivars can be bred.

In the breeding of synthetic cultivars, testing pure lines used to assemble the synthetics may not be appropriate because complete homozygosity is not representative of the inbreeding status of a synthetic population. Inbred line ranking for agronomic value per se is not necessarily identical to their ranking for their "breeding value" (i.e., the "value as parental components" of a synthetic cultivar). A useful solution is to produce topcross- or polycross-progeny using the pure lines. Progeny are tested and superior lines selected based on these results, instead of testing the parental lines themselves (see fig. 13.2). These polycross-derived offspring are very similar in heterozygosity and heterogeneity to that realized in a synthetic cultivar. The progeny effectively reflect their parental lines' differences in (1) per se performance, (2) combining ability and (3) degree of cross-fertilization (Becker, 1982; Link and Ederer, 1993). By maintaining pure lines as parents, cultivars can be resynthesized at will.

A standard initiation of a synthetic cultivar is to allow between (about) four and eight lines to reproduce under natural (partial allogamous) conditions, the mixture being called Syn-0 (the subsequent generations being Syn-1, Syn-2, etc.). As long as cross-fertilization is less than 100%, maximum heterosis is not achieved in Syn-1 but will be realized in later generations. For instance, with 67% self-fertilization, it will take until generation Syn-4 to realize (nearly) the maximum achievable heterosis. Only 50% of possible heterosis can be realized in this case (minimum

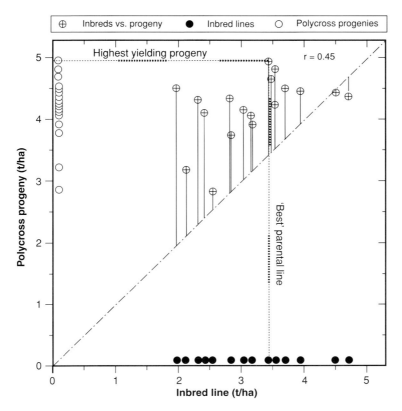

Figure 13.2 Correlation between grain yield of inbred lines and their polycross progenies. Data on y-axis reflect the inbred lines' value as parental components of synthetic cultivars. Results were taken from participatory faba bean experiments (Ghaouti, 2007).

inbreeding coefficient is F = [s/(2-s)] = 0.5). With only 33% self-fertilization, only two generations (Syn-2) are required, and the ultimate minimum inbreeding coefficient is F = [s/(2-s)] = 0.2, hence 80% of heterosis can be realized.

These considerations are based on an assumption of "constant degree of cross-fertilization" (Busbice, 1970). If there is environmental variation for outcrossing, then it is of utmost importance to allocate the highest possible degree of cross-fertilization to the last step of seed production, thus offering the lowest inbreeding and the highest share of heterosis to the farmer (in favor of high vigor, yield, and yield stability in farmers' fields; Geiger, 1982).

General Considerations of Breeding Faba Bean

Hybrid breeding in faba bean is not yet feasible, due to lack of a stable cytoplasmic male sterility (CMS) system and lack of any other tool to control pollination on a large scale (Link, 2006).

To breed and multiply pure lines, for line cultivars, or as components of synthetic cultivars, faba bean breeders use isolation cages that prevent access by pollinators. This allows controlled, pure selfing but requires manual tripping. In conventional breeding, the production of doubled haploid lines would be an attractive means to bypass this procedure. For this and other tissue culture techniques (callus and protoplast regeneration, and genetic transformation), *Vicia faba* is recalcitrant. While genetic transformation has been successful, it is very demanding with low efficiency; there has been very little activity in this area of research. No method of doubled haploid line production is known (Link et al., 2008; Ochatt et al., 2009). Due to the absence of such techniques, conventional and organic breeding differ less in the technologies that are available from each other in faba bean than in many other crops.

Local versus Formal Breeding, Heterogeneity and Heterosis, Intercropping, and Participatory Breeding

With the focus on organic breeding, the following interrelated items must be considered: (1) use of genotype × environment interaction, (2) deployment and exploitation of genotypic heterogeneity and of heterosis in farmers' fields, (3) breeding for intercropping, (4) farmers' participation in the breeding process.

In organic farming, genotype × location interactions are, on average, larger than in conventional agriculture; the purpose of agrochemical inputs in conventional agriculture is largely the mitigation of and compensation for environmental stress. Large genotype × location interaction favors regional or local breeding rather than breeding for geographically wide adaptation. Cultivars are selected to be specifically adapted to the agro-ecological features of their regions. In the local breeding approach, genotype × location interactions are transformed from a nuisance to a heritable part of variation (Ghaouti and Link, 2009). Nevertheless, cultivars have to be well buffered against variability for factors such as weed competition and fungal infection, because part of this variability is not specific to regions but, rather, is associated with specific years.

Important questions are whether (1) the entire breeding process has to be organic; (2) conventional (formal) breeding is fine enough as long as the resulting genotypes are tested in organic conditions; or (3) the ultimate results from conventional breeding are good enough for organic farmers (Desclaux et al., 2008; Sperling et al., 2001). These questions cannot be solved in general but need to be addressed on a crop by crop basis (Schmidt, 2009). From a series of faba bean trials in organic farms, Ghaouti and Link (2009) found that the locally realized gain from selection based on a formal breeding approach was 75.5% of that gain, which resulted from testing and selecting locally, underlining the superiority of local (and regional) breeding (see table 13.1 and see Chapters 2 and 6 of this book).

As with many crops, increasing genotypic heterogeneity and heterosis of faba bean cultivars increases yield stability (Link et al., 1996; Stelling et al., 1994). Both factors seem to be of similar importance in faba bean. The stabilizing impact of heterogeneity is mainly a statistical phenomenon. First, each individual existing within a heterogeneous cultivar shows its own type of interaction with environmental features; thus, the cultivar's interaction is near to the mean of its individuals'

Table 13.1 Ratio of responses to selection for faba bean grain yield: Response realized locally and based on formal breeding results versus response realized locally and based on local results; data are from a series of participatory, organic trials (Ghaouti and Link, 2009)

Expected heritability from testing at an organic location (Local)	$h^2_{(Org.)} = 0.803$
Expected heritability from multi-location testing (Formal)	$h^2_{(Form.)} = 0.765$
Genetic correlation coefficient (Local vs. Formal)	$r_{G(Org.; Form.)} = 0.773$
Expected ratio of responses	Ratio $= 0.755$

interactions, and variances of means are smaller than variances of single values. The stabilizing impact of heterosis reflects the finding that under stress, heterosis tends to be larger (Abdelmula et al., 1999). Thus, the higher the expected level of stress and the expected level of genotype × environment interaction, the more strongly increased levels of heterosis and more heterogeneous cultivars are favored. Assuming that stress is a more frequent feature of organic farming, more parental lines (6 < n) should be used to construct an organic synthetic cultivar compared to a conventional synthetic cultivar (Link and Ederer, 1993). The same reasoning applies if faba bean cultivars are specifically bred to show improved frost or drought stress tolerance. However, some traits such as days to maturity should be homogenous. In support of this approach to cultivar development, regulations of cultivar release (distinct-uniform-stable or DUS) should be relaxed (see Chapter 8). Alternatively, farmers could experiment with cultivar blends.

The reasoning given above leads as well to the topic of intercropping. Positive experience has been observed from intercropping winter wheat with winter faba beans (Hof-Kautz et al., 2008), and spring oats with spring beans is a well-known mixture (Köpke and Nemecek, 2010). Testing faba bean in a mixture with a cereal would not only allow us to breed cultivars that are better adapted to this situation, but would also favor genotypes with high symbiotic performance (Hof and Rauber, 2003). The intercrop niche is small and there is no one specific breeding for it.

In the UK, where faba bean was never abandoned and knowledge of its production, marketing, and use is actively pursued, participatory breeding is attractive. In Germany fewer farmers have experience with the crop. Ghaouti et al. (2008) reported that organic farmers in Germany favored bean types that best fit their local needs, such as tall types in dry and weedy locations, and short types in sites with fewer weeds and greater lodging risk. Paradoxically, farmers appreciated neat, homogenous inbred lines in spite of their lower grain yield, compared to shaggy, heterogeneous polycross-progenies (see fig. 13.1). This was in contradiction to the praise of heterogeneity and diversity that is commonly espoused by the organic community. We must be aware that faba bean is in peril of becoming too minor of a crop, and it may be necessary to allow for a mutual learning phase before embarking on participatory breeding (see Chapter 6).

Traits To Be Improved in Faba Bean Breeding

General

Faba bean seed production has a unique set of production problems. The production of pre-basic seed for synthetics (like Syn-1, Syn-2) and line cultivars should be conducted in environments with as little infestation of *Ascochyta*, *Sitona* beetles and aphids, as possible to keep seed free from fungi and viruses. When producing certified seed of synthetics, pollinator activity must be maximized and combine harvesting and conditioning of seed must be carefully and gently done to keep seed intact and clean and the embryo vigorous.

Traits ranked from most important to least important for spring bean improvement are: (1) grain yield and yield stability, (2) resistance to lodging, (3) *Botrytis, Ascochyta*, and downy mildew resistance, and (4) in specific locations resistance to rust and the soil-borne *Fusarium*-complex. Traits that are adequate and need (at least) to be kept at the current levels are: tolerance to viruses, non-shattering, synchronous maturity of pods and straw, date of maturity directly after combine harvest of winter wheat, thousand seed weight between 400 and 600 g, and round seed shape.

Resistance to drought is important; there is limited genetic variation for this trait, and the genetic correlation for grain yield under drought and under non-stressed conditions is high (Link et al.,

1999). Resistance breeding is most promising in the case of *Ascochyta faba* (anthracnose) and *Uromyces fabae* (rust), since major genes exist (Link, 2006). Yet, even in these promising cases, breeding activities are very limited.

Additional traits for winter beans are winter hardiness and resistance to fungal attack in early spring. There is some ongoing work on frost resistance and winter-hardiness, and with the advent of climate change, winter faba beans will probably extend their adaptation beyond the UK to more continental climates such as Germany (Arbaoui et al., 2008; Roth and Link, 2010).

For Organic Conditions

Following Schmidtke (2010), the main constraints to a successful organic faba bean production include weeds, aphids, and *Sitona* beetles and larvae. Breeding for insect resistance is not very promising because no marked genetic differences seem to exist within the primary gene pool. Thus other areas of research have to contribute to a solution. However, faba bean does possess favorable genetic variation for weed suppression. Based on a two-year experiment at Göttingen, Ghaouti (2007) reported on the impact of genotypic differences, heterosis, and heterogeneity on grain yield of spring beans with and without *Camelina sativa* (serving as artificial "weed"). *Camelina* caused, on average, 25% yield reduction of the bean. Bean yield was highly correlated with and without *Camelina* ($r = 0.84^{**}$; see fig. 13.3). Nevertheless, genotypes were significantly different for yield reduction. The most important factor was inbreeding status. Without weed stress, blends of hybrids yielded 35% higher than pure lines, and even 95% higher with stress. Hybrids lost 6% of their yield and inbred

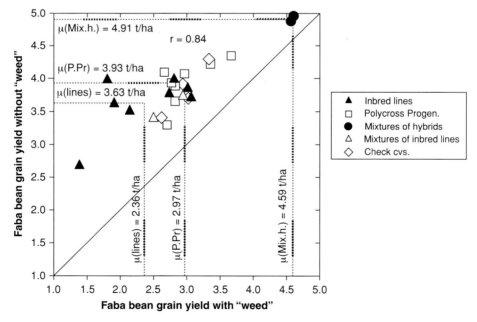

Figure 13.3 Correlation of grain yield with and without artificial weed competition (*Camelina sativa*) for different faba bean genotypes (based on data from Ghaouti, 2007).

lines lost 35% of their yield due to the *Camelina* competition. Yield of *Camelina* was positively correlated with yield loss of the beans (r = 0.55*): Competing with hybrid bean mixtures, *Camelina* yield was 0.29 t ha^{-1}, whereas in inbred bean lines it yielded on average 0.59 t ha^{-1}. Plant height had no major influence on weed tolerance as long as genotypes were of the same inbreeding status.

Tolerance to late sowing (spring beans) may be of specific importance to organic agriculture, because it offers time to mechanically control weed before sowing. This trait is connected to drought and heat tolerance and to the speed of juvenile growth. Prospects of genetic progress in the near future are uncertain.

High quality seed has high (>30%) protein, relatively high methionine content, and low burden of anti-nutritive compounds. Quality is especially important if beans are used on-farm as animal feed. Protein content could be improved, but it is a low priority because buyers are not willing to pay a premium for high protein beans. Genetic variation for methionine content is low, and prospects for improvement are poor unless new genetic variation is discovered, e.g., following mutagenesis (Schumacher et al., 2009). Tannins and vicine are anti-nutritive compounds for pigs and chickens, respectively (Crépon et al., 2010). Hence, zero tannin cultivars like Tangenta and low vicine cultivars like Divine and Mélodie are especially attractive. Since these are monogenic traits, breeding is easy and straight-forward. In the case of zero tannin, caution is advised because seed coat tannins mimic fungicide seed dressings, controlling soil-borne fungi during germination and emergence. Based on past experience, harvesting seed under favorable and dry conditions, employing a cautious transport and proper storage will avoid moldy seed in store, and this is often more important for feed quality than genetic differences of seed quality.

Maximum efficiency of symbioses (to feed the rotation with air-derived nitrogen) is a further specific topic. Current cultivars are able to adequately fix high amounts (>100 kg N per ha, even in high-N soils; Köpke and Nemecek, 2010). Yet, there is currently not much applied research on the interaction between environment, *Rhizobium*, and faba bean (Roskothen, 1989).

Open Questions, Need for Action

In France, faba bean average yield in 2008 was 5.18 t ha^{-1}, compared to 4.46 t ha^{-1} of dry pea. Hence, aiming at a premium market (export to Egypt), conventional farmers are able to produce quite high faba bean yields with current cultivars.

Yet, there are unsolved problems for breeding and growing faba bean as a feed component, especially under organic conditions: (1) susceptibility of the bean to aphids (mainly *Aphis fabae*) and to the *Sitona* leaf weevil, and (2) difficulties in production of high quality seed without chemical insecticides and fungicides. Very likely, breeding alone will not be able to overcome these obstacles for the short term.

Furthermore, with no interspecific crosses available, more care must be devoted to the available intraspecific genetic variation. Hence, more gene bank accession must be phenotyped for agronomic traits, and in certain regions new collections must be acquired, e.g., in the Hindu Kush area and in parts of China (Maharajan et al., 1990).

In Germany, farmers are not allowed to use farm-saved seed if it is a synthetic or hybrid cultivar. For such cultivars, there is not even an option to pay the so-called "fee for farm-saved seed." In case of synthetic cultivars, this makes no sense, because a core feature of a synthetic (population) cultivar is that (unlike a hybrid) its grain yield is adequate to be used as seed without genetic yield penalty (see Chapter 8 of this book for more on this issue).

References

Abdelmula, A.A., W. Link, E. von Kittlitz, and D. Stelling. 1999. Heterosis and inheritance of drought tolerance in faba bean (*Vicia faba* L.). *Plant Breeding* 118:485–490.

AEP Grain Legumes Portal. Accessed September 6, 2010. http://www.grainlegumes.com.

AMI. 2011. Accessed March 21, 2011. http://www.ami-informiert.de.

Arbaoui, M., C. Balko, and W. Link. 2008. Study of faba bean (*Vicia faba* L.) winter hardiness and development of screening methods. *Field Crops Research* 106:60–67.

Backhaus, G.F. 2009. Anbau und Züchtung von Leguminosen in Deutschland – Sachstand und Perspektiven. Fachgespräch im Julius Kühn-Institut. April 21–22, 2009, Braunschweig. *Journal für Kulturpflanzen* 61:301–364.

Becker, H.C. 1982. Züchtung synthetischer Sorten. II. Leistungsvorhersage und Selektion der Eltern. *Vorträge für Pflanzenzüchtung* 1:23–40.

Busbice, T.H. 1970. Predicting yield of synthetic varieties. *Crop Science* 10:265–269.

Crépon, K., P. Marget, C. Peyronnet, B. Carrouée, P. Arese, and G. Duc. 2010. Nutritional value of faba bean (*Vicia faba* L.) seeds for feed and food (review). *Field Crops Research* 115:329–339.

Desclaux, D., J.M. Nolot, Y. Chiffoleau, E. Gozé, and C. Leclerc. 2008. Changes in the concept of genotype × environment interactions to fit agriculture diversification and decentralized participatory plant breeding: Pluridisciplinary point of view. *Euphytica* 163:533–546.

Ellwood, S.R., H.T.T. Phan, M. Jordan, A.M. Torres, C.M. Avila, S. Cruz-Izquierdo, and R.P. Oliver. 2008. Construction of a comparative genetic map in faba bean (*Vicia faba* L.); conservation of genome structure with *Lens culinaris*. *BMC Genomics* 9:380.

Eurostat. 2011. Accessed March 21, 2011. http://www.eds-destatis.de.

FAOSTAT. 2010. Accessed September 6, 2010. http://faostat.fao.org.

Fuchs, J., Strehl, S., Brandes, A., Schweizer, D., and Schubert, I. 1998. Molecular-cytogenetic characterization of the *Vicia faba* genome – heterochromatin differentiation, replication patterns and sequence localization. *Chromosome Research* 6:219–230.

Geiger, H.H. 1982. Züchtung synthetischer Sorten III. Einfluss der Vermehrungsgeneration und des Selbstungsanteils. *Vorträge für Pflanzenzüchtung* 1:41–72.

Ghaouti, L. 2007. Comparison of pure line cultivars with synthetic cultivars in local breeding of faba bean (*Vicia faba* l.) for organic farming. PhD thesis, University of Göttingen.

Ghaouti, L., W. Vogt-Kaute, and W. Link. 2008: Development of locally-adapted faba bean cultivars for organic conditions in Germany through a participatory breeding approach. *Euphytica* 162:257–268.

Ghaouti, L., and W. Link. 2009. Local vs. formal breeding and inbred line vs. synthetic cultivar for organic farming: Case of *Vicia faba* L. *Field Crops Research* 110:167–172.

Hancock, J., R. Weller, and H. McCalman. 2005. 100% Organic Livestock Feeds in UK – preparing for 2005. Report for Organic Centre Wales, March 2003. Accessed September 6, 2010. http://orgprints.org.

Hebblethwaited, P.D. (ed.) 1983. The faba bean. A basis for improvement. London: Butterworths.

Hof, C., and R. Rauber. 2003. Anbau von Gemengen im ökologischen Landbau. Booklet in German language. Accessed March 17, http://www.uni-goettingen.de/en/44360.html.

Hof-Kautz, C., K. Schmidtke, and R. Rauber. 2008. Backweizen im Gemenge. *Lebendige Erde* 59:14–15.

Köpke, U., and T. Nemecek. 2010. Ecological services of faba bean. *Field Crops Research* 2010:217–233.

Link, W. 1990. Autofertility and rate of cross-fertilization: Crucial characters for breeding synthetic varieties in faba beans (*Vicia faba* L.). *Theoretical and Applied Genetics* 79:713-717.

Link, W., and W. Ederer. 1993. The concept of varietal ability for partially allogamous crops. *Plant Breeding* 110:1–8.

Link, W., W. Ederer, and E. von Kittlitz. 1994. Zuchtmethodische Entwicklungen: Nutzung von Heterosis bei Ackerbohnen. *Vorträge für Pflanzenzüchtg.* 30:201–230.

Link, W., B. Schill, and E. von Kittlitz. 1996. Breeding for wide adaptation in faba bean. *Euphytica* 92:185–190.

Link, W., A.A. Abdelmula, E. von Kittlitz, S. Bruns, H. Riemer, and D. Stelling. 1999. Genotypic variation for drought tolerance in *Vicia faba*. *Plant Breeding* 118:477–483.

Link, W. 2006. Methods and objectives in faba bean breeding. International Workshop on faba bean breeding and agronomy. pp. 35–40. October 25–27, Córdoba, Junta de Andalucia, Spain.

Link, W., M. Hanafy, N. Malenica, H-J. Jacobsen, and S. Jelenić. 2008. Broad bean. *In*: Kole C., and T.C. Hall (eds). Compendium of transgenic crop plants: Transgenic legume grains and forages. pp. 71–88. Oxford: Blackwell Publishing.

Maharajan, P.L., B. Bhadra, P. Roy, R.P. Yadav, and Z. Ronsu. 1990. Environmental diversity and its influence on farming systems in the Hindu Kush-Himalayas. *In*: International workshop on Mountain Agricultural and Crop Genetic Resources. February 1987, Kathmandu, Nepal. New Delhi: IDRC and ICIMOD, Rajbandhu Industrial Co.

Mansfeld's Word Database of Agricultural and Horticultural Crops. Accessed September 6, 2010. http://mansfeld.ipk-gatersleben.de.

Ochatt, S., C. Pech, T. Grewal, C. Conreux, M. Lulsdorf, and L. Jacas. 2009. Abiotic stress enhances androgenesis from isolated microspores of some legume species (*Fabaceae*). *Journal of Plant Physiology* 166:1314–1328.

Palmer R.G., P. Perez, E. Ortiz-Perez, F. Maalouf, M.J. Suso. 2009. The role of crop-pollinator relationships in breeding for pollinator-friendly legumes: From a breeding perspective. *Euphytica* 170:35–52.

Raupp, J. 2010. Leguminosenschrote als Düngemittel. Ist die Nährstoff – und Humusversorgung damit gesichert? *Lebendige Erde* 4:16–17.

Rohloff, H. 1980. Die Bedeutung der Viruskrankheiten bei der Ackerbohne (*Vicia faba* L.). für die Resistenzzüchtung. Mitt. aus der *Biol. Bundesanstalt für Land – und Forstwirtschaft* 197:31–38.

Roskothen, P. 1989. Genetic effects on host × strain interaction in the symbiosis of *Vicia faba* and *Rhizobium leguminosarum*. *Plant Breeding* 102:122–132.

Roth, F. 2010. Evaluierung von Winterackerbohnen als Zwischenfrucht für eine ökologische Biogas-Produktion. PhD thesis, University of Göttingen.

Roth, F., and W. Link. 2010. Selektion auf Frosttoleranz von Winterackerbohnen (*Vicia faba* L.): Methodenoptimierung und Ergebnisse. 60. *In*: Tagung der Vereinigung der Pflanzenzüchter und Saatgutkaufleute Österreichs. pp. 31–37.

Schmidt, W. 2009. Fünf Jahre Öko-Mais-Züchtung. Erfahrungen und Perspektiven. Accessed September 6, 2010. http://www.kws.de.

Schmidtke, K. 2010. Masterplan Körnerleguminosen – vom Anbaufrust zur Anbaulust. Alnatura Bauerntag, March 17, 2010. Accessed September 6, 2010. http://www.bodenfruchtbarkeit.org/167.

Schumacher, H., H.M. Paulsen, and A.E. Gau. 2009. Phenotypical indicators for the selection of methionine enriched local legumes in plant breeding. *Landbauforschung – vTI Agriculture and Forestry Research* 59:339–344.

Sperling, L., J.A. Ashby, M.E. Smith, E. Weltzien, and S. McGuire. 2001. A framework for analyzing participatory plant breeding approaches and results. *Euphytica* 122:439–450.

Stelling, D., E. Ebmeyer, and W. Ly. 1994. Effects of heterozygosity and heterogeneity. *Plant Breeding* 112:30–39.

UNEP. 2011. Gobal honey bee disorders and other threats to insect pollinators. Accessed March 22, 2011. http://www.unep.org.

Zaleski, A. 1956. Tripping, crossing and selfing in lucerne strains. *Nature* 177:334–335.

Zeid, M., C-C. Schön, and W. Link. 2004. Hybrid performance and AFLP-based genetic similarity in faba bean. *Euphytica* 139:207–216.

14 Potato: Perspectives to Breed for an Organic Crop Ideotype

Marjolein Tiemens-Hulscher, Edith T. Lammerts van Bueren, and Ronald C.B. Hutten

Introduction

Potato (*Solanum tuberosum* L.) is the world's number one non-cereal food crop. In organic agriculture, potato is an important crop, but also one of the most problematic to produce, due to its susceptibility to many pests and diseases. Organic potato yields are not only relatively low but also highly variable, ranging from 12 to 35 Mg ha^{-1} in Europe, whereas the average yields for conventional potato production varies from 36–43 Mg ha^{-1} (Tamm et al., 2004). The main reason for year-to-year variation is the deviation in timing and severity of late blight attacks, caused by the oomycete *Phytophthora infestans* and low nitrogen input (Möller et al., 2006). Resources in the agronomic toolbox to reduce the susceptibility of a potato crop to late blight are limited (Finckh, 2008; Finckh et al., 2006; Speiser et al., 2006). This certainly applies to countries with the humid, maritime conditions conducive to late blight, and the condition is exacerbated with national organic rules, such as those in The Netherlands and Denmark, preventing use of copper sprays. Moreover, in the Netherlands, there is a national regulation that requires vine killing when approximately 7% of the leaf surface is visually affected by late blight, in order to ultimately reduce sporulation. Due to lack of late blight resistant cultivars, organic potato area in the Netherlands has decreased by 20% in the past decade (Lammerts van Bueren et al., 2008), so improved cultivars are urgently needed.

Most organic potato research is focused on field trials to identify the best suitable cultivars derived from conventional breeding programs for organic agriculture (e.g., in United States at Cornell University; Halseth et al., 2010) or on improvement of organic seed potato production (e.g., in the U.S. Midwest; Genger and Charkowski, 2010). Several breeding companies have selected certain genotypes with multiple resistances from their conventional breeding program and have offered them for organic seed potato production (e.g., Bradshaw et al., 2007). New projects have been initiated for specific organic breeding programs, e.g., Scrabule (2010). In this chapter we will discuss the specific cultivar requirements and the breeding perspectives for cultivars better adapted to organic potato production. We will also illustrate the participatory potato breeding approach that has taken off in the Netherlands (Lammerts van Bueren et al., 2009).

Organic Crop Breeding, First Edition. Edited by Edith T. Lammerts van Bueren and James R. Myers.
© 2012 John Wiley & Sons, Inc. Published 2012 by John Wiley & Sons, Inc.

Required Cultivar Characteristics

All required quality and appearance traits should be also present in organic cultivars because the market demands are the same for organic and conventional. Growers not only select their cultivars based on market requirements, but also as a tool for managing pests and diseases to achieve yield stability rather than high yields. For profitable potato production, e.g., under Dutch growing conditions (fertile soils with high land rent), organic growers strive for a yield of 30–40 Mg ha^{-1}, whereas conventional growers strive for 50–60 Mg ha^{-1} due to lower prices. However, average potato yields in the organic sector often do not reach the break-even point due to two main yield-limiting factors: Low nutrient availability and lack of cultivars adapted to low-input conditions (Van Delden, 2001) and late blight infestation and lack of late blight resistant cultivars. The requirements for resistance to other diseases can differ compared to conventional farming systems. Some potato pests, e.g., cyst nematodes, can be well managed by a long crop rotation and are therefore not a problem in organic farming systems. However, other pests and diseases with a wider host range are hard to control, and resistant varieties are the only option for the organic potato sector. Therefore, certain disease resistances require higher priority in organic breeding programs than in conventional programs (see table 14.1). When resistance to late blight is not available in cultivars, other traits can reduce the risk of yield loss by escaping late blight infection with the choice of early to mid-early cultivars that mature in 90 to 100 days, including early tuber setting and tuber filling. However, this is only a successful strategy when late blight does not appear too early in the season.

There are also diseases, such as early blight (*Alternaria solani*), that do not show up under high inputs of nutrient and water conditions but do cause yield losses under low-input and stress conditions; they have not yet received much attention in breeding.

How to Achieve Required Traits

Potato breeding is complicated by the auto-tetraploid composition of the potato genome, which makes the inheritance of traits more complex. Potatoes are an outcrossing species, and four different alleles may reside at a locus creating high levels of heterozygosity. In crosses this will create an abundance of diversity (Bradshaw, 2007). On the other hand, selection in potato is simplified because of clonal multiplication of the tubers after creating the F_1 thus avoiding further segregation. This means that each of the genotypes of the F_1 is fixed in the next generation. Each seedling from a cross is potentially a new cultivar. The breeder has "merely" to find the best one by testing them in several environments over several years.

Further on we will discuss our perspective in breeding for organic farming systems.

Late Blight
Late blight is one of the world's most devastating diseases in potato. The disease is caused by the oomycete *Phytophthora infestans* (Mont.) de Bary and infects foliage as well as tubers. In years with an early and severe infestation of late blight, yield losses can rise to more than 50%. Although there are at least 11 *R*-genes (race-specific resistance conditioned by single genes) originating from the wild species *S. demissum* that have been incorporated into various potato cultivars, resistant cultivars are scarce. The dominant *R*-genes confer resistance based on a race-specific hypersensitive reaction. This type of resistance is not durable because new virulent races of the oomycete can easily occur (Wastie, 1991).

Table 14.1 Crop ideotype of organic potato cultivars in the Netherlands with an indication of priorities for the characteristics

Characteristics	Minimal	Strive for	Priority[1]
Yield			
Net yield	30 ton/ha	30–40 ton/ha in 90–95 DAP[2]	++++
Grade	35–65 mm	35–65 mm	+++
Crop			
Rooting	Rapid under cold conditions	Rapid and deeper under cold conditions	++++
Nutrient efficiency	Sufficient under cold conditions	Good under cold conditions	+++
Drought tolerance	Good	Excellent	+++
(Early) growth vigor	Powerful	Powerful	+++
Plant structure	Open	Open	++
Maturity[3]	6	6–8	+++
Tuber bulking	Early	Early	++++
Diseases			
Resistance/tolerance[4]			
Phytophthora infestans (foliage)	6	8–9	++++
Phytophthora infestans (tuber)	6	8–9	++++
Alternaria solani	6	8	++++
Rhizoctonia solani	6	8	++++
Helminthosporium solani	6	8	++++
Verticillium dahlieae	6	8	+++
Streptomyces spp.	6	8	+++
Erwinia spp.	6	8	+++
Viruses, PVX, PVY, PLRV	6	8	+++
Globodera spp	5	7	+
Meloidogyne spp	5	7	+
Trichodorus spp	5	7	++
Leptinotarsa decemlineata (Colorado potato beetle)	6	8	+
Weed control			
Canopy development	55 DAP 100% ground cover	45 DAP 100% ground cover	++
Storage			
Sprouting	Late	Late	++++
Quality			
Skin	Smooth	Smooth	++
Skin color	Red or yellow	Red or yellow	+
Flesh color	Yellow	Yellow	+++
Eyes	Shallow	Shallow	+++
Shape	Oval	Elongated oval	+++
Cooking type[5]	A–B	A–B	+++
Taste	good	good	++
Sensitivity to blue light	Low sensitive	Not sensitive	+++
Specific gravity after 90–95 DAP	300	325 or higher for frying and chipping	++++

[1] ++++ = very high priority, + = very low priority
[2] DAP = Days After Planting
[3] 6 = mid early, 8 = early
[4] 1 = very susceptible, 9 = complete resistant
[5] A = firm cooking, B = rather firm cooking, C = mealy

Knowing that potato *R*-genes are not very durable, a desirable option would be to search for more durable, quantitative resistance (race-nonspecific foliage resistance), but progress has been slow and has not yet contributed significant levels of resistance. Breeding for a high level of field resistance in potato is technically more difficult because of the quantitative character of this trait and the fact that quantitative resistance can easily be masked by *R*-genes (Solomon-Blackburn et al., 2007). Another complicating factor in breeding for quantitative resistance is the association between resistance and maturity type. Although some variation in resistance was found within the same maturity class, the effect of quantitative trait loci (QTL) was not sufficient to obtain satisfactory levels of resistance in early maturing genotypes (Visker et al., 2005).

Currently, new cultivars with genes from *S. bulbocastanum* are entering the market, 35 years after Hermsen and Ramanna (1973) made the initial bridge crosses with *S. acaule* and *S. phureja* to introduce these resistance genes into *S. tuberosum*. These genes confer a broad spectrum resistance to late blight and the resistance appears to be more durable and effective in the field (Song et al., 2003; Van der Vossen et al., 2002). Other more recent sources of late blight resistance have been discovered in *S. berthaultii, S. edinense, S. venturii* and *S. avilesi*, of which some are less difficult to introgress into *S. tuberosum* compared to *S. bulbocastanum* (e.g., Pel et al., 2009; Rauscher et al., 2006).

An approach that is in agreement with organic principles and exploits diversity as a key management tool to reduce disease pressure, is combining (pyramiding) resistance genes from different sources to prolong the durability of the resistance in cultivars (Tan et al., 2010). Molecular markers are indispensable to be able to select for those genotypes that have combined genes due to potential epistatic interactions among resistance genes.

Early Blight
Early blight disease, caused by the fungus *Alternaria solani* Sonauer, is a severe disease in most potato-growing regions worldwide and is expected to increase under changing climate conditions. Cultivars grown under stress conditions (e.g., low nitrogen input or in dry and warm regions) are more vulnerable to the pathogen. It infects both foliage and tubers, and yield losses can rise to 20 to 30% (Johnson et al., 1986). Sources of resistance are rarely found in *S. tuberosum*. Some moderate resistant cultivars and advanced breeding selections were identified by Platt and Reddin (1994). Several authors found strong correlations between early blight resistance and late maturity (e.g., Douglas and Pavek, 1972). Boiteux et al. (1995) however, found some clones out of 934 to be both resistant and early maturing. In the literature, resistance to early blight always refers to field resistance and is related to the progress of the disease infection. Haga et al. (2010) and Webber et al. (2010) identified higher levels of resistance in the diploid wild species *S. berthaultii* and *S. raphanifolium*. Jansky et al. (2008) identified species and individual accessions with a high percentage of resistant plants. They introgressed the resistance of the diploid wild species *S. tarijense* and *S. raphanifolium* into *S. tuberosum* and are investigating if this resistance can be combined with early maturity.

Rhizoctonia (Stem Canker and Black Scurf)
Rhizoctonia stem canker and black scurf are caused by the soil and tuber born fungus *Rhizoctonia solani* Kühn. It reduces quality (black scurf on the tubers) and productivity (stem canker and dieback and distortion of newly emerging underground stems and stolons). Resistance to R*hizoctonia* is not common. Some organic growers have experienced *Rhizoctonia* decline in their soil when producing their own seed potatoes over several years. This phenomenon is based on a biological nature, because it disappears after sterilization of the soil (Wiseman et al., 1996; see also for take-all decline in wheat, Cook, 2009). More research is needed to clarify which organisms are playing a role and under which conditions the disease decline has the best response (Hospers et al., 2004). Therefore, there is

still a need for improvement of resistance and/or tolerance to R*hizoctonia* through breeding. Bains et al. (2002), Dowley (1972), and Naz et al. (2008) screened several cultivars and breeding selections for susceptibility to *Rhizoctonia*. They found a range of susceptibility reactions, but none of the cultivars or breeding lines showed a complete resistance to the disease. This indicates that resistance to *Rhizoctonia* in potato is polygenically inherited. However, Bains et al. (2002) concluded that differences in the genetic variation for susceptibility to *Rhizoctonia* may be associated with both inheritance and maturity level of the cultivars. Another complicating factor is that the different isolates of *Rhizoctonia* vary in virulence to the various parts of the plant being attacked (Wastie, 1994).

Some resistance to stem canker was found in wild species *S. phureja* and *S. vernei* and crossed into cultivars of *S. tuberosum* (Gadzhiev, 1986), but it is not clear whether the resistance was polygenically inherited or based on a few recessive genes (Wastie, 1994).

A practical method for screening *Rhizoctonia* for breeders has been lacking, but recently Naz et al. (2008) developed a greenhouse screen that uses six disease parameters to identify resistance.

Silver Scurf

Helminthosporium solani Durieu and Mont. is the causal agent of silver scurf disease on the periderm of potato tuber and emerged in the 1990s as an economically important disease of fresh table and processing potatoes (Errampalli et al., 2001). As a greater number of supermarkets offer organically produced tubers as washed potatoes, any infection on the skin diminishes the appearance. Tuber infections can occur in the field and during storage. Silver scurf is primarily a seed-borne disease but low levels of conidia can survive from one season to the other in the soil (Secor and Gudmestad, 1999). Screening for partial resistance or susceptibility is quite easy. Until present, no resistant cultivars have been available for the market. However, many authors have mentioned genetic variability in susceptibility (e.g., Rodriquez et al., 1996). Similar to other diseases there is a correlation between maturity and silver scurf. Mérida et al. (1994) found early maturing cultivars to be more susceptible than late maturing cultivars. Rodriquez et al. (1995) found partial resistance in wild species with some accessions of *S. demissum, S. chacoense, S. acaule, S. stoloniferum, S. oxycarpum,* and *S. hondelmannii* consistently showed less sporulation in laboratory experiments and in storage. So far, this has not yet resulted in resistant cultivars.

Verticilium Wilt

Verticilium wilt, mainly caused by the soil-born fungus *Verticillium dahliae* Kleb., is a widespread disease in potato in production areas with average daily high temperatures exceeding 27°C. It causes substantial yield losses from premature senescence of the crop (Secor and Gudmestad, 1999). Although *V. dahliae* has a broad host range, crop rotation with non-host crops such as cereals can help to reduce the disease (Mulder and Turkensteen, 2008). Jansky (2009) evaluated 14 advanced clones from U.S. potato breeding programs and 11 cultivars. She developed a three-tiered approach to identify resistant clones including, (1) collecting maturity and symptom scores on a single date that was not too early in the season, (2) measuring levels of conidia in stem sap, and (3) evaluating levels of microsclerotia in dried stems. There was variation in resistance among cultivars but complete resistance was not found.

Colorado Potato Beetle

The Colorado potato beetle (*Leptinotarsa decemlineata* Say) is a devastating insect pest in potato and may become even more of a pest when summers become warmer as influenced by climate change. Adults as well as larvae feed from the leaves and plants can be defoliated within a few days. Resistance to the beetle is a promising management tool. But selection for resistance is quite difficult because of the narrow genetic base within the modern potato cultivars (Jansky et al., 1999).

Two sources of host plant resistance are known. The first reported by Sinden et al. (1986) concerned the production of high leptine glycoalkaloid levels found in the wild species *S. chacoense*. The second resistance mechanism documented by Gibson (1976) was glandular trichomes studied in the wild species *S. berthaultii*. Jansky et al. (1999) combined these two sources of resistance by somatic fusion of *S. chacoense* and *S. berthaultii*, with the expectation that combining resistance mechanisms might produce more stable and durable clones than those with only one resistance factor. However, for organic breeding, another approach is needed because somatic hybridization is not compatible with the values of organic agriculture (see Chapter 7).

Nitrogen Efficiency
Empirical observations have revealed that there is genotypic variation in the need for nitrogen in potatoes. Often with the release and marketing of a new cultivar, advice is provided on optimal nitrogen fertilization levels. Rarely has the reverse information on what genotypes can cope with limited nitrogen been addressed in potato breeding. Breeding nitrogen efficient cultivars is very complex because of the interaction among all of the factors that can influence nitrogen efficiency in potato. In the Netherlands, two breeding research projects have focused on the question of how to develop selection tools for nitrogen efficiency in potato (Tiemens-Hulscher et al., 2009; Ospina et al., 2010). Preliminary results showed that genetic variation for nitrogen uptake, nitrogen use efficiency, and harvest index was found under high and low nitrogen conditions. To a large extent, genetic variation was explained by maturity type. Late maturing cultivars had higher nitrogen uptake efficiency, whereas early maturing cultivars had higher nitrogen use efficiency. Whether early or late maturing cultivars will yield better under low nitrogen conditions depends on the length of the growing season, which depends strongly on the time of infestation with late blight. In short seasons (less than 90 days) early cultivars have a higher yield than late cultivars. In more prolonged seasons (more than 95 days) it will be the other way round. Probably one of the most important indicators for nitrogen uptake efficiency is the speed and amount of canopy cover in the developing crop. However, there is a large genotype × environment interaction that requires further exploration.

Other traits can be beneficial to nutrient efficiency, such as colonization with mycorrhizal fungi or an improved root system that explores a larger (deeper) volume of soil. Little is known about the ability of potato cultivars to profit from symbiosis with mycorrhizal fungi. Genetic variation in rate and degree of root colonization was found by Bhattarai and Mishra (1984) in an experiment with three potato genotypes.

Maturity
Maturity in variety lists generally refer to foliage-, not tuber maturity. Compared to late-maturing cultivars, early cultivars transport relatively more photo-assimilates to the tubers early in the season. Under conventional conditions this results in not only shorter crop seasons but also lower yields. Under organic conditions, the growing period is often the limiting factor due to low input or late blight and therefore the potential higher yield of late cultivars is hardly guaranteed.

Breeding for late or early cultivars is not difficult, as maturity has a high heritability. The midparent values are highly predictive of offspring performance (Visker et al., 2003).

Introgression Breeding and Applied Techniques

Wild *Solanum* species contain many desirable traits, especially resistance genes for many diseases. However, crossing with wild species to introgress these traits is tedious work that requires many years of backcrossing and selection. Problems can occur in wide crosses such that plants abort flowers or fruits due to male sterility (common), ovule sterility (rare), self-incompatibility factors, or differences in Endosperm Balance Number (EBN; Gopal, 2006). To overcome such problems several techniques have been used, some of which are not permitted or are discouraged in the organic sector. Non-permissible techniques would include somatic hybridization, and discouraged practices would be in vitro techniques and colchicine treatment (for chromosome doubling; Lammerts van Bueren et al., 2003; also see Chapter 7, this book). Crossing techniques that can be conducted at whole plant level such as bridge crosses and the use of natural occurring 2n-gametes are preferred for organic breeding. The later approach can be used as an alternative for somatic hybridization to manipulate EBN and to cross diploid wild relatives with the cultivated tetraploid species (see Gopal, 2006). Watanabe and Peloquin (1991) provide an excellent review of the natural production of 2n-gametes in many tuber-bearing *Solanum* species.

Participatory Approach: An Example from the Netherlands

The need for improved cultivars for organic agriculture was the basis for a publicly funded potato breeding program in the Netherlands. This program (BioImpuls) was led by two research institutes (Louis Bolk Institute and Wageningen University), which were in close cooperation with several breeding companies (Lammerts van Bueren et al., 2008, 2009). BioImpuls was started in 2008 and was able to further build on pre-breeding material bred at Wageningen University in order to develop late blight resistant materials based on *S. demissum, S. bulbocastanum, S. edinense, S. brachycarpum,* and *S. iopetalum.* The aim of the program was to develop potato germplasm for breeding companies that are well adapted to organic growing conditions with special emphasis on late blight resistance, early blight, silver scurf, *Rhizoctonia*, Potato Virus Y, early maturity, increased tuber dormancy, and nitrogen efficiency (see table 14.1). The project purpose was to conduct all breeding steps from introgression to cultivar testing under organic conditions (figure 14.1).

Based on an existing and successful conventional farmer breeder system in the Netherlands, the BioImpuls program included participation of organic potato growers as farmer breeders for selection in more advanced crossings. The program supported the farmers with courses, field visits, and a practical guide book. In 2011, the program included 11 organic farm breeders. Farmers contributed mainly to the first three years of field selection (see table 14.2). Farmer breeders annually received a number of seedlings or seedling tubers from their affiliated breeding company and/or the BioImpuls program. The number of seedlings varied among farmer breeders, and was on average

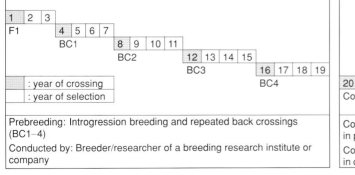

Figure 14.1 Breeding scheme to introgress genes from wild species into commercial cultivars.

Table 14.2 The collaborative approach of farmer-involved potato breeding in the Netherlands (adapted from Lammerts van Bueren et al., 2008)

Year	Activity	Company breeder	Farmer breeder
0	Choice of crossing parents	x	x
0–1	Crossing and harvesting seeds		
	Providing first-year seedlings	x	
1–3	Visual field selection for basic agronomic characteristics		x
4–12	Advanced trials for production, resistances, quality, and adaptation	x	
	Research for potential market, registration on national cultivar list, and obtaining plant breeders' rights		
	Market introduction and maintenance		

300–1500 per year. The farmer breeders selected for basic agronomic traits, such as even tuber size distribution, skin properties, limited stolon length, and special traits such as late blight resistance. After three years of field selection, the farmer breeder returned the selected clones (approx 0.1–1.0% of the original material) to one of the participating breeding companies for further evaluation and selection, and eventual market introduction (see table 14.2).

Outlook

Currently, most organic potato growers depend on cultivars derived from conventional breeding programs, and in many countries, companies successfully produce an assortment of organic seed potatoes of the most suitable cultivars. However, there is still a need for improved cultivars to optimize organic farming systems and to improve yield stability. Conventional breeders find that the traits that organic potato growers indicate as important are also important for conventional breeding programs. The difference is the lower weight given to characters important for organic production in the selection process, which reduces the chances that optimal cultivars for the organic sector will be developed. The increased number of organic farmer breeders in the BioImpuls program should stimulate conventional breeders to place higher priority in breeding for those traits that are essential for organic growers. By involving extra farmer breeders, more clones can be evaluated annually and this will increase the chances of selecting useful cultivars. Nevertheless, companies indicate that the market is too small to sell the selected cultivars only for the organic market. The selected cultivars should also serve additional markets (e.g., in low-input countries) to enlarge the volume of these extra cultivars. As the conventional sector shifts toward sustainable management, many traits that have high priority in organic agriculture will also meet the future needs of the conventional sector.

References

Bains, P.S., H.S. Bennypaul, D.R. Lynch, L.M. Kawchuk, and C.A. Schaupmeyer. 2002. Rhizoctonia disease of potatoes (*Rhizoctonia solani*): Fungicidal efficacy and cultivar susceptibility. *American Journal of Potato Research* 79:99–106.
Bhattarai, I.D., and R.R. Mishra. 1984. Study on the vesicular-arbuscular mycorrhiza of three cultivars of potato (*Solanum tuberosum* L.). *Plant and Soil* 79:299–303.

Boiteux, L.S., F.J.B. Reifschneider, M.E.N. Fonseca, and J.A. Buso. 1995. Search for sources of early blight (*Alternaria solani*) field resistance not associated with vegetative late maturity in tetraploid potato germplasm. *Euphytica* 83: 63–70.

Bradshaw, J.E. 2007. Potato-Breeding Strategy. *In*: Vreugdenhil, D., with J. Bradshaw, G. Gebhardt, F. Govers, D.K.L. MacKerron, M.A. Taylor, and H.A. Ross (eds.) Potato biology and biotechnology, advances and perspectives. pp. 157–174. Amsterdam, The Netherlands: Elsevier.

Bradshaw, J., C. Coelman, and F. Dale. 2007. Potato cultivar Lady Balfour, and example of breeding for organic production. *In*: Lammerts van Bueren E.T., H. Østergård, I. Goldringer, and O. Scholten. 2007. Abstract book of Eucarpia Symposium Plant breeding for organic and sustainable, low-input agriculture: Dealing with genotype-environment interactions. November 7–9, 2007, Wageningen, The Netherlands. Wageningen, The Netherlands: Wageningen University.

Cook, R.J. 2009. Take-all of wheat. *Physiological and Molecular Plant Pathology* 62:73–86.

Douglas, D.R., and K.K. Pavek. 1972. Screening potatoes for field resistance to early blight. *American Potato Journal* 49:1–6.

Dowley, L.J. 1972. Varietal susceptibility of potato tubers to *Rhizoctionia solani* in Ireland. *Irish Journal of Agricultural Research* 11:281–285.

Errampalli, D., J.M. Saunders, and J.D. Holley. 2001. Emergence of silver scurf (*Helminthosporium solani*) as an economically important disease of potato. *Plant Pathology* 50:141–153.

Finckh, M.R., E. Schulte-Geldermann, and C. Bruns. 2006. Challenges to organic potato farming: Disease and nutrient management. *Potato Research* 49:27–42.

Finckh. M.R. 2008. Integration of breeding and technology into diversification strategies for disease control in modern agriculture. *European Journal of Plant Pathology* 121:399–409.

Gadzhiev, N.M. 1986. The nature of the inheritance of black scurf resistance in interspecific potato hybrids. *Selektsiya I Semenovodstvo, USSR* 1:23–24.

Genger, R., and A. Charkowski. 2010. Production of healthy seed potatoes on organic farms (poster). 21st Annual MOSES Organic Farming Conference. February 25–27, 2010, La Crosse, WI. Accessed April 13, 2011. http://www.mosesorganic.org/forumposters2010.html.

Gibson, R.W. 1976. Trapping of the spider mite *Tetranychus urticae* by glandular hairs on the wild potato *Solanum berthaultii*. *Potato Research* 19:179–182.

Gopal, J. 2006. Considerations for successful breeding. *In*: Gopal, J., and S.M.P. Khurana (eds.) Handbook of potato production, improvement, and postharvest management. pp. 77–108. New York: Food Products Press, an imprint of the Haworth Press, Inc.

Haga, E., S. Jansky, and D. Halterman. 2010. Characterisation of early blight resistance in interspecific potato hybrids. Accessed January 6, 2011. www.ars.usda.gov/research/publications/publications.htm?SEQ_NO_115=257020.

Halseth, D., E. Sandsted, R. MacLaury, J. Kelly, and D. Hoy. 2004–2010. Organic potato variety trials 2004–2010. Ithaca, NY: Cornell University. Accessed March 18, 2011. http://www.vegetables.cornell.edu/crops/organic%20potato%20results.htm.

Hermsen, J.G.T., and M.S. Ramanna. 1973. Double-bridge hybrids of *Solanum bulbocastanum* and cultivars of *Solanum tuberosum*. *Euphytica* 22:457–466.

Hospers, M., J. Postma, and L. Colon. 2004. *Rhizoctonia* in organic potato production. *In*: Lammerts van Bueren, E.T., R. Ranganathan, and N. Sorensen (eds.) Proceedings of the 1st World IFOAM/ISF/FAO Conference on Organic Seed – Challenges and opportunities for organic agriculture and the seed industry. pp. 157–159. July 5–7, 2004, Rome, Italy. Bonn: IFOAM.

Jansky, S.H., 2008. Identification of *verticillium* wilt resistance in U.S. potato breeding programs. *American Journal Potato Research* 86:504–512.

Jansky, S.H., S. Austin-Phillips, C. McCarthy. 1999. Colorado potato beetle resistance in somatic hybrids of diploid interspecific *Solanum* clones. *HortScience* 34:922–927.

Jansky, S.H., R. Simon, and D.M. Spooner. 2008. A test of taxonomic predictivity: Resistance to early blight in wild relatives of cultivated potato. *Phytopathology* 98:680–687.

Johnson, K.B., E.B. Radcliffe, and P.S. Teng. 1986. Effects of interacting populations of *Alternaria solani, Verticillium dahliae,* and potato leafhopper (*Empoasca fabae*) on potato yield. *Phytopathology* 76:1046–1052.

Lammerts van Bueren, E.T., Tiemens-Hulscher, M., and P.C. Struik. 2008. Cisgenesis does not solve the late blight problem of organic potato production: Alternative breeding strategies. *Potato Research* 51:89–99.

Lammerts van Bueren, E.T., R.C.B. Hutten, M. Tiemens-Hulscher, and N. Vos. 2009. Developing collaborative strategies for breeding for organic potatoes in the Netherlands. *In*: Hoeschkel, Z. (ed.) Breeding diversity: Proceedings of the First International IFOAM Conference on Organic Animal and Plant Breeding. pp. 176–181. August 25–28, 2009, Santa Fe, New Mexico.

Lammerts van Bueren, E.T., P.C. Struik, M. Tiemens-Hulscher, and E. Jacobsen. 2003. The concepts of intrinsic value and integrity of plants in organic plant breeding and propagation. *Crop Science* 43:1922–1929.

Mérida, C.L., R. Laria, and D.E. Halseth. 1994. Effects of potato cultivar and time of harvest on the severity of silver scurf. *Plant Disease* 78:146–149.

Möller, K., J. Habermeyer, V. Zinkernagel, and H. Reents. 2006. Impact and interaction of nitrogen and *Phytophthora infestans* as yield-limiting and yield-reducing factors in organic potato (*Solanum tuberosum* L.) crops. *Potato Research* 49: 281–301.

Mulder, A., and L.J. Turkensteen. 2008. Aardappelziektenboek. Aardappelwereld BV., The Hague, The Netherlands.

Naz, F., C.A. Rauf, N.A. Abbasi, I. Ul-Haque, and I. Ahmad. 2008. Influence of inoculum levels of *Rhizoctionia solani* and susceptibility on new potato germplasm. *Pakistan Journal of Botany* 40:2199–2209.

Ospina, C.A., E.T. Lammerts van Bueren, S. Allefs, P. van der Putten, G. van der Linden, and P.C. Struik. 2010. Phenotyping of an extensive potato germplasm set for nitrogen use efficiency: Nitrogen effect on physiological mode parameters for canopy development. *In*: Goldringer, I., J. Dawson, A. Vettoretti, and F. Rey (eds.) Breeding for resilience: A strategy for organic and low-input farming systems? Proceedings of EUCARPIA conference Section Organic and Low-Input Agriculture. p. 86. December 1–3, 2010, Paris. France: INRA and ITAB, France.

Pel, M.A., S.J. Foster, T.H. Park, H. Rietman, G. Van Arkel, J.D. Jones, H.J. Van Eck, E. Jacobsen, R.G.F. Visser, and E.A. Van der Vossen. 2009. Mapping and cloning of late blight resistance genes from *Solanum venturii* using an interspecific candidate gene approach. *Molecular and Plant Microbe Interactions* 22:601–15.

Platt, H.W., and R. Reddin. 1994. Potato cultivar and accession responses to late blight, early blight, and grey mold. *Annals of Applied Biology* 124:118–119.

Rauscher, G.M., C.D. Smart, I. Simko, M. Bonierbale, H. Mayton, A. Greenland, and W.E. Fry. 2006. Characterization and mapping of $R_{Pi\text{-}ber}$, a resistance gene from *Solanum berthaultii*. *Theoretical and Applied Genetics* 112:674–687.

Rodriquez, D.A., G.A. Secor, N.C. Gudmestad, and K. Grafton. 1995. Screening tuber-bearing *Solanum* species for resistance to *Helminthosporium solani*. *American Potato Journal* 72:669–679.

Rodriguez, D.A., G.A. Secor, N.C. Gudmestad, and L.J. Franci. 1996. Sporulation of *Helminthosporium solani* and infection of potato tubers in seed and commercial storages. *Plant Disease* 80:1063–1070.

Scrabule, I. 2010. Evaluation of potato clones in organic and conventional growing conditions. *In*: Goldringer, I., J. Dawson, A. Vettoretti, and F. Rey (eds.) Breeding for Resilience: A strategy for organic and low-input farming systems? Proceedings of EUCARPIA conference Section Organic and Low-Input Agriculture. pp 105–108. December 1–3, 2010, Paris. France: INRA and ITAB.

Secor, G.A. 1994. Management strategies for fungal disases of tubers. *In*: Zhender, G.W., M.L. Powelson, R.K. Jansson, and K.K.V. Raman (eds.) Advances in potato pest biology and management. pp. 155–157. St. Paul: American Phytopathological Society Press.

Secor, G.A., and N.C. Gudmestad. 1999. Managing fungal diseases of potato. *Canadian Journal of Plant Pathology* 21:213–221.

Sinden, S.L., L.L. Sanford, and K.L. Deahl. 1986. Segregation of leptine glycoalkaloids in *Solanum chacoense* Bitter. *Journal of Agricultural and Food Chemistry* 34:372–377.

Speiser, B., L. Tamm, T. Amsler, J. Lambion, C. Bertrand, C.A. Hermansen, M.A. Ruissen et al. 2006. Improvement of late blight management in organic potato production systems in Europe: Field tests with more resistant potato varieties and copper based fungicides. *Biology Agriculture and Horticulture* 23:393–412.

Song, J., J.M. Bradeen, S.K. Naess, J.A. Raasch, S.M. Wilgus, G.R. Haberlach, J. Liu et al. 2003. Gene RB cloned from *S. bulbocastanum* confers broad spectrum resistance to potato late blight. *Proceedings of the National Academy of Sciences* 100:9128–9133.

Solomon-Blackburn, R.M., H.E. Stewart, and J.E. Bradshaw. 2007. Distinguishing major-gene from field resistance to late blight (*Phytophthora infestans*) of potato (*Solanum tuberosum*) and selecting for high levels of field resistance. *Theoretical and Applied Genetics* 115:141–149.

Tamm, L., A.B. Smit, M. Hospers, S.R.M. Janssens, J.S. Buurma, J-P. Mølgaard, P.E. Laerke et al. 2004. Assessment of the socio-economic impact of late blight and state-of-the-art management in European organic potato production systems. Research Institute of Organic Agriculture FiBL, Frick-CH.

Tan, M.Y.A., R.C.B. Hutten, R.G.F. Visser, and H.J. van Eck, 2011. The effect of pyramiding *Phytophthora infestans* resistance genes $R_{Pi\text{-}mcd1}$ and $R_{Pi\text{-}ber}$ in potato. Theoretical and Applied Genetics 121(1):117–125.

Tiemens-Hulscher, M., E.T. Lammerts van Bueren, and P.C. Struik. 2009. Identification of genotypic variation for nitrogen response in potato (*Solanum tuberosum*) under low nitrogen input circumstances. *In*: Hoeschkel, Z. (ed.) Proceedings of the First International IFOAM Conference on Organic Animal and Plant Breeding. pp. 354–361. August 25–28, 2009 Santa Fe, New Mexico.

Van Delden, A. 2001. Yield and growth of potato and wheat under organic N-management. *Agronomy Journal* 93:1370–1385.

Van der Vossen, E., A. Sikkema, B.T.L. Hekkert, J. Gros, M. Musken, W.J. Stiekema, and S. Alefs. 2002. Cloning of an R gene from *Solanum bulbocastanum* conferring complete resistance to *Phytophthora infestans*, Late blight: Managing the global threat. Proceedings Global Initiative on Late Blight Conference. p. 155. Hamburg, Germany: GILB.

Visker, M.H.P.W., L.C.P. Keizer, H.J. Van Eck, E. Jacobsen, L.T. Colon, and P.C. Struik. 2003. Can the QTL for late blight resistance on potato chromosome 5 be attributed to foliage maturity type? *Theoretical and Applied Genetics* 106: 317–325.

Visker, M.H.P.W., H.J.B. Heilersig, L.P. Kodde, W.E. van de Weg, R.E. Voorrips, P.C. Struik, and L.T. Colon. 2005. Genetic linkage of QTL's for late blight resistance and foliage maturity type in six related potato progenies. *Euphytica* 143:189–199.

Wastie, R.L. 1991. Breeding for resistance. *In*: Ingram, D.S., and P.H. Williams (eds.) Phytophthora infestans: The cause of late blight of potato. pp. 193–224. San Diego, CA: Academic Press.

Wastie, R.L. 1994. Inheritance of resistance to fungal diseases of tubers. *In*: Bradshaw, J.E., and G.R. Mackay (eds.) Potato genetics. pp. 411–427. Wallingford, UK: CAB International.

Watanabe, K., and S.J. Peloquin. 1991. The occurrence and frequency of 2n pollen in 2x, 4x, and 6x wild, tuber bearing, *Solanum* species from Mexico and Central and South America. *Theoretical and Applied Genetics* 82:621–626.

Webber, B., S. Jansky, and D. Halterman. 2010. Breeding for early blight resistance in potato using the wild species *Solanum raphanifolium*. Accessed January 6, 2011. www.ars.usda.gov/research/publications/publications.htm?seq_no_115=257022&.

Wiseman, B.M., S.M. Neate, K.O. Keller, and S.E. Smith. 1996. Suppression of *Rhizoctonia solani* anatomosis group 8 in Australia and its biological nature. *Soil Biology and Biochemistry* 28:727–732.

15 Tomato: Breeding for Improved Disease Resistance in Fresh Market and Home Garden Varieties

Bernd Horneburg and James R. Myers

Introduction

Tomato is a frost-tender warm season annual crop in temperate environments that tolerates a range of temperatures. In glasshouse environments and in the tropics, tomato plants may live for more than a year.

The domestication of tomato (*Lycopersicon esculentum* Mill. syn. *Solanum lycopersicum* L.) has not been fully understood due to scarce archeological remains and missing wild progenitors. The present name derives from "tomatl" in the Nahua language of Mexico where most of the genetic diversity for the cultivated crop is found. The distribution after the Columbian exchange was rather slow in most parts of the world. The ambivalent attitude of Europeans toward the strange new plants was reflected by the scientific name *Lycopersicon,* translated as "wolf's peach." Tomatoes became highly esteemed for their culinary properties in Italy by the seventeenth century, but remained rare plants used for witchcraft, medicinal, or ornamental purposes in most other parts of Europe until the twentieth century. J. Metzger (1841) wrote: "The fruits were mainly used for sauces, for preserves, and in pastry where they are particularly essential to the Italian [translation BH]." This quote hints at nineteenth century diversity of uses from fresh fruits, in salads, and in various processed products.

Today the tomato is the most important vegetable in the world, with over 141 million Mg produced on 4.98 million ha in 2009 (FAOSTAT, 2011). They are a significant source of vitamin C, carotenoids, flavonols, and phenolics in the human diet. For those individuals with nutritionally limited diets, tomatoes may be the single most important source of pro-vitamin A carotenoids and vitamin C. While the phenolics and flavonoids in tomatoes are found in low amounts relative to other vegetable species (Hertog et al., 1992; Stewart et al., 2000; Vinson et al., 1998), tomato is the leading source of phenolic compounds in the U.S. diet because tomatoes are so heavily consumed.

Most organic market gardeners in Europe and the United States value tomato as one of their most important crops. Certified organic tomato production has been increasing at a rate of about 16% per year over the past decade and was at 3,738 ha in 2009 in the United States. Certified organic tomato hectarage grew more rapidly than the 12% overall growth of certified organic area under production. Despite the relatively rapid growth, only 2.13% of all U.S. tomato production was certified organic in 2009 (ERS, 2011).

Organic Crop Breeding, First Edition. Edited by Edith T. Lammerts van Bueren and James R. Myers.
© 2012 John Wiley & Sons, Inc. Published 2012 by John Wiley & Sons, Inc.

Botanical and Genetic Characteristics of Tomato

All nine species of the genus Lycopersicon have 2n = 24 chromosomes. *L. pimpinellifolium* (L.) Mill. (syn. *Solanum pimpinellifolium* L.), *L. hirsutum* Dunal (syn. *S. habrochaites* S. Knapp & D. Spooner), *L. cheesmanii* Riley (syn. *S. cheesmaniae* [L. Riley] Fosberg), *L. chmielewskii* Rick, Kes., Fob. and Holle (syn. *S. chmielewskii* [C. M. Rick et al.] D. M. Spooner et al.), *L. parviflorum* Rick, Kes., Fob., and Holle (syn. *S. neorickii* D. M. Spooner et al.), and *L. pennellii* (Corr.) D'Arcy (syn. *S. pennellii* Correll) can be hybridized with *L. esculentum*. Crosses with *L. peruvianum* (L.) Mill. (syn. *S. peruvianum* L.) and *L. chilense* Dun (syn. *S. chilense* [Dunal] Reiche) are only successful with embryo rescue (Kalloo, 1991).

In cultivated species intraspecific genetic variation is low, so tomato breeders have relied heavily on introgression from the previously mentioned wild species for many traits. Breeding using species from the secondary gene pool is an efficient means of mitigating the effects of diseases and pests and introgressing novel quality traits.

Tomato is normally highly self-pollinated although some cultivated varieties and *L. esculentum* var. *cerasiforme* types may have higher levels of outcrossing. Outcrossing rates observed in temperate regions frequently covered the range from zero to a few percent as shown in the example in figure 15.1. In the tropical region of tomatoes' origin, carpenter bees (*Xylocopa* spp.) are present, and outcrossing can reach much higher levels (Rick, 1950). The enclosure of the stigma and style by the anther cone enhances self-fertilization and reduces chances of cross pollination. Higher outcrossing rates are sometimes associated with a style and stigma that extends beyond the anther cone (see fig. 15.1). To maintain varietal purity, isolation and roguing of off-types is required for such genotypes.

Hand crosses are performed by emasculating buds in the developmental stage shown in figure 15.2. Pollen is collected with the aid of a tuning fork that simulates the buzzing of a bumblebee (Gladis et al., 1996).

Some monogenetic dominant traits are useful tools to detect successful hybridization or undesired cross pollination. Red fruit color is dominant to yellow, indeterminate growth is dominant to determinate, and tomato leaf is dominant to potato leaf. The latter trait can already be observed at the first true leaf.

Rationale for Breeding Tomatoes within Organic Systems

Crops grown in organic systems need specific adaptation for certain traits. It is possible to find some varieties developed in conventional systems that perform well in organic systems, but this is not always the case and some studies have shown that the top performing breeding lines observed in conventional versus organic environments are not necessarily the same (Murphy et al., 2007). Most examples come from other crops, as no studies of this nature have been published for tomato.

The first pure line varieties bred within organic systems are available in Europe. Kultursaat (2011) released Oldenrot and Ruthje for low input greenhouse and polytunnel production, and Pilu and Tica for intensive cropping. The Organic Outdoor Tomato Project of the University of Göttingen released three varieties (Clou, Dorada, and Primavera). Genotypes used in organic breeding include heritage varieties stored in genebanks or maintained on farm by seed savers and NGOs, as well as pure line and hybrid varieties.

Landraces, heirlooms, and some varieties from organic or conventional public breeding programs are open pollinated (OP), but most commercially bred varieties released since the early 1980s are

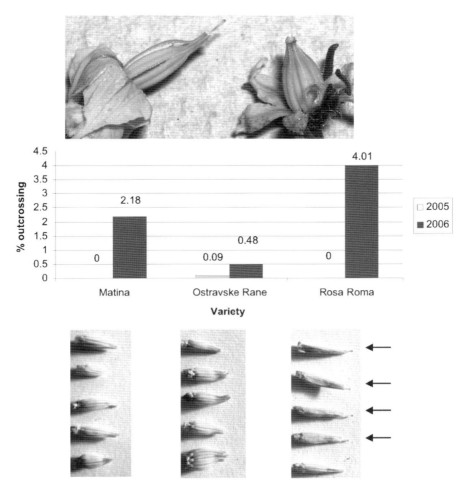

Figure 15.1 Outcrossing in tomato. Protruding (top left) and recessed stigma (top right). Outcrossing in three outdoor tomato varieties in two years in Central Germany (centre) and their respective anther cone morphology (bottom). (□ indicates protruding stigma). Top photo by J.R. Myers, lower photo by S. Tröster.

F_1 hybrids. Tomatoes do not show severe inbreeding depression when selfed to homozygosity, but they may exhibit heterosis for a number of traits when two inbred lines are crossed (Kalloo, 2000). Some organic growers would prefer pure lines over hybrids because a goal of organic farming is to minimize off-farm inputs, including growing seeds for crops on the farm. OP and pure line varieties allow the farmer to save their own seed whereas hybrids will not breed true, so the farmer must purchase new seed every year. In addition, contemporary hybrids bred in conventional systems may not possess some traits (quality, low input production, robustness) desired by farmers and valued by consumers. Evidence of this comes from the revival of heirloom tomatoes in farmers' markets and home gardens (Jordan, 2007).

Seed companies breeding for conventional markets commonly produce F_1 hybrid tomato seed by hand crosses, frequently in areas of the world with abundant cheap labor, such as East Asia. There is a growing demand for regional production chains including breeding and seed production, which

   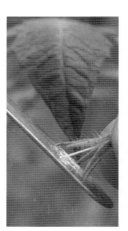

Figure 15.2 Tomato manual hybridization. From left to right, tomato buds at proper stage for crossing (female); removal of the anther cone; collection of pollen from an open flower to be used as the male using a tuning fork; and application of pollen to the emasculated flower. Photos by B. Horneburg.

would favor the use of OP and pure line varieties. However, there will always be a need for all types of varieties (hybrid, pure line, and OP) depending on market demands and needs of the grower.

Another reason for organic breeding programs to focus on the development of pure line varieties is to develop future genetic resources by maintaining and increasing genetic diversity. Hybrid breeding has contributed to a reduction in genetic diversity because F_1 hybrid varieties are only rarely stored in genebanks or taken into dynamic development on farm, because of the segregation in the F_2 and subsequent generations (Horneburg, 2010).

Developing tools for small breeding programs helps to encourage the breeding of genotypes adapted to the wide range of pedoclimatic conditions found in organic production sites. Nutrient use efficiency and tolerance to biotic and abiotic stresses are more important for varietal performance in organic systems as compared to conventional systems.

Types of Production Systems

In North America, processed production is outdoors, whereas a small portion of the fresh market hectarage is grown in polytunnels or heated greenhouses. Most of the fresh market crop is produced in Florida, California, and various states in Mexico. Most processing production takes place in California, with small areas of processed production in Indiana and Ohio. Much of the organic production takes place in small fresh market–based farms outdoors, and unheated polytunnels are used to extend the spring and fall seasons. Glasshouse production is rare, but can be found in areas where winter sunlight is plentiful (southwestern United States) or where electricity is cheap (British Columbia, Canada, where the majority of electricity is provided by hydropower).

Europe has much the same stratification with outdoor production in Spain, Italy, and the Balkans and indoor production in colder areas, such as the Netherlands. Outdoor production in large parts of Europe is mostly restricted to home gardeners, small scale organic, or traditional market garden businesses. In southern and some central European regions, both field and protected tomato production systems are used. The processing market is supplied exclusively from field production, while protected production is used mainly for the fresh markets (Lammerts van Bueren et al., 2010). Non-commercial production can play an important role, particularly in regions of the world with limited hectarage for processing tomatoes. For instance, in Germany, amateur gardeners account for approximately 50% of total production (Siebold, 2006).

In both organic and conventional production, pest control is based on an integrated pest management approach that utilizes biological controls. Foliar disease is controlled by lowering humidity, and when necessary, with sulfur treatment. Late blight is the restricting factor in outdoor production as well as in unheated protected cropping when the temperature drops below the dew point for several hours (Becktell et al., 2005). The main differences between protected organic and conventional production systems in northern regions are that in organic production, plants are grown in soil and fertilized with certified organic fertilizers, while conventional production is based on mineral fertilizers and is increasingly reliant on hydroponic systems. The potential for soil borne fungal pathogens and nematodes is greater in organic than in hydroponic systems, because in the latter case, the growth substrate can be replaced when pathogen pressure becomes too great. As a result, steam soil treatments, disease suppressive composts, and grafting onto soil borne disease resistant rootstocks are widely used for organic production (Lammerts van Bueren et al., 2010).

Breeding Needs with Focus on Organic Production

Organic and conventional production systems differ in soil management practices and subsequent effects on the rhizosphere environment. Organic systems mostly utilize organic matter–based fertilizer with nitrogen (N) and phosphorus (P) derived by mineralization. The availability of N and P will differ significantly from that available in conventional systems. Organic crops may experience limited N and P availability when soil temperatures and moisture are suboptimal for mineralization by soil biota. In fields that have been under organic management for some time and with regular organic matter applications, increases in soil biological activity and biodiversity and associated increases in mineralization capacity have been documented (Raviv, 2010). Organic fertilization regimes may also suppress pathogens and pests and stimulate systemic acquired resistance and induced systemic resistance (see also Chapter 3).

Crop varieties bred with fertilizer regimes that supply abundant N may have lost the ability to perform well in low-N environments. For example, wheat varieties bred before the advent of chemical fertilizers are better at N uptake in low-N environments compared to modern varieties (Murphy et al., 2008).

In tomatoes, adaptation to organic systems is likely to require genotypes that are able to form active symbiotic relationships with beneficial organisms in the rhizosphere. Mechanisms that may increase nutrient-use efficiency include vigorous root systems with modified architecture, ability to form active mycorrhizal associations, reduced root losses due to pathogens, and ability to maintain a high mineralization activity in the rhizosphere through root exudates.

When high levels of P are available, arbuscular mycorrhizal colonization is less important for plant nutrition and may be detrimental because the microorganisms extract photosynthates from the plant. In tomato, a major gene associated with mycorrhizal competence has been identified (Larkan et al., 2007; Cavagnaro et al., 2006), but germplasm has not been surveyed to determine whether this trait is present at different frequencies in modern varieties bred under high input conditions compared to heirloom varieties. Bryla and Koide's (1998) work suggests that it is possible to have different strategies for P acquisition and use in tomato. The genotype that responded to mycorrhizal infection also showed a strong response to P fertilization under non-mycorrhizal conditions. The other genotype was not mycorrhizal responsive, and it did not respond strongly to P deficit or to increasing P fertilization. It had a large root system regardless of the fertility level that explored the soil whereas the mycorrhizal responsive genotype relied on the fungus for soil exploration. Both P acquisition strategies have limitations, and without more research into the two mechanisms, it is not possible to predict which would provide better adaptation to organic production. Whether the two mechanisms are mutually exclusive is also not known.

Gene expression in tomatoes has been shown to be influenced by previous cover crop and associated microflora. In Mattoo and Teasdale's (2010) hairy vetch cover crop–tomato production system, soil microflora influenced the accumulation of secondary plant products in hairy vetch, which upon breakdown in the soil, induced higher root:shoot cytokinin levels in the tomato crop, which subsequently delayed senescence and up-regulated pathways involved in induced resistance. This area of the molecular and physiological level interactions among plant species and soil microbiota needs further exploration to determine whether there is a genetic component that could be exploited for improved adaptation to organic systems by breeding.

Many consumers expect extraordinary flavor and quality from organically grown tomatoes – especially from heirloom varieties purchased fresh from direct market operations. Soluble and total solids, and other compounds that influence taste and aroma have been reported to be higher in organic versus conventionally grown tomatoes where such factors as cultivar and production environment have been controlled (Chassy et al., 2006; Ordóñez-Santos et al., 2009). While breeders in both conventional and organic tomato breeding programs would be interested in preserving or improving flavor, the burden falls more heavily on organic tomato breeders because of the high expectations of consumers. The opportunity exists to improve the productivity, disease, and pest resistance of heirloom types with the best flavor characteristics while adapting them to organic systems.

The potential to produce fruit with a higher nutritional value depends on the genotype (Chassy et al., 2006), but no one has investigated whether significant genotype × system interaction for nutritional traits would bolster support for a dedicated organic breeding program. There is ample genetic variation in tomato to enhance the carotenoid (Levin et al., 2006) and flavonoid (Mes et al., 2008) levels.

Small, regionally based breeding programs can create advanced genotypes that serve the needs of the growers for yield and yield stability even in sub-optimal environments, and consumers' demands for improved taste and transparent production chains can be answered (Horneburg and Becker, 2008; Behrendt, 2009). A combination of regional breeding programs (exploiting genotype × environment interactions to optimize adaptation) with larger programs (to incorporate rare or difficult traits for improved field resistance, quality, and yield) is desirable for organic agriculture.

Resistances Needed for Organic Crop and Seed Production

Most contemporary tomato varieties are bred with resistance to one or more races of fusarium (*Fusarium oxysporum f.s. lycopersici*), verticillium (*Verticillium dahliae*), leaf mold (*Cladosporium fulvum*), and nematodes. These are diseases of intensive monoculture farming systems with short rotations, and may be of less concern in organic systems due to the disease suppressive effects of organic matter–based fertilization regimes (Mattoo and Teasdale, 2010; Stone et al., 2004) and wider crop rotations. Other diseases such as tomato mosaic virus (ToMV), tomato spotted wilt virus (TSWV), early blight (*Alternaria solani*), and late blight (*Phytophthora infestans*) are more widespread and occur independently of production system. Similar to potatoes (see Chapter 14), late blight in tomatoes is more of a concern in organic compared to conventional because of the lack of tools effective in controlling this disease (see also Chapter 3). It will be discussed in more detail in the following case study.

Seed-borne diseases are another area where control strategies differ between conventional and organic agriculture, with organic agriculture needing to rely more heavily on genetic control. The diseases of concern include ToMV, tobacco mosaic virus (TMV), bacterial speck and bacterial spot (caused by *Pseudomonas syringae* pv. *tomato* and *Xanthomonas campestris* pv. *vesicatoria*, respectively), and *Clavibacter michiganense*.

ToMV affects conventional and organic tomato seed production in a similar fashion and can be controlled by the use of resistant varieties. One complication with the most widely used form of resistance (Tm-2^2) is that it was derived from *L. peruvianum* using embryo rescue techniques (Hall, 1980). Such in vitro techniques are not compatible with the principles of organic agriculture, but since this form of resistance has been so widely used and is in much of the elite tomato germplasm, its use for organic agriculture is currently under debate (Lammerts van Bueren et al., 2003). While not as effective as Tm-2^2, Tm-1 provides resistance to the predominant strains of ToMV, and was transferred from *L. hirsutum*, which does not require special crossing techniques (Pelham, 1966). This illustrates a strategy (in this case choosing an alternative species from the secondary gene pool) that breeders may employ to obtain traits from exotic germplasm in ways that are compatible with the principles of organic agriculture.

Bacterial and fungal seedborne diseases may be controlled by antibiotics and fungicides in conventional agriculture, but only certified organic sulfur and copper-based products are allowed in organic agriculture, and copper compounds are being curtailed and already have been eliminated in some EU countries due to environment impact concerns (see Chapter 14). Alternative methods of cultural control include antagonistic micro-organisms, compost extracts, fermentation, acids and acidified nitrite, and hot water treatment; or the use of resistant varieties. Resistance to bacterial speck and bacterial spot is available and should be more widely incorporated into tomato varieties bred for organic systems.

Case Studies: Breeding for Late Blight Resistance in Europe and North America

Late blight is considered the most important disease in tomato grown in temperate humid conditions. Primary infections frequently derive from potato crops. Copper-fungicides currently used for the control often provide only partial control of the disease, and under growing conditions conducive to disease development (cool temperatures and long periods of leaf wetness/high humidity) late blight regularly causes significant yield losses (Becktell et al., 2005; Fry, 1998). Rapid evolution of *P. infestans* races is possible since both mating types of the fungus are present globally (Fry et al., 1993). As a consequence, cheap and resource-efficient outdoor production was almost abandoned in large parts of Europe and other regions of the world.

Genetic resistance to late blight is available within the tomato gene pool. Small-fruited varieties often exhibit higher levels of late blight field resistance. Resistance from *L. pimpinellifolium* has been characterized and transferred into fresh market cultivars. Four race specific genes (*Ph-1*, *Ph-2*, *Ph-3*, and *Ph-5*) have different levels of effectiveness (Foolad et al., 2008). *Ph-1* has been defeated by the races commonly found in North America whereas *Ph-2* provides delaying action to race US-8 but less so to US-11 (late blight isolates specific to potato and tomato, respectively). However, *Ph-2* was defeated by new strains of late blight that swept North America in 2009 and 2010. Currently, *Ph-3* provides the best available level of resistance to current races. *Ph-5* is a newly identified resistance gene from *L. pimpinellifolium* that is not yet widely available (Foolad et al., 2008).

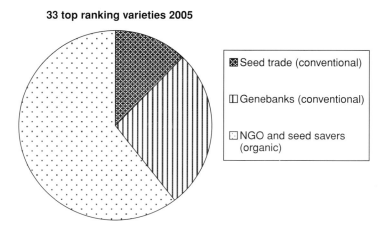

Figure 15.3 Origin of the best performing genotypes based on 3,500 accessions after three years of screening for outdoor performance at three locations in Central Europe.

Ph-1, *Ph-2*, and *Ph-3* have been incorporated into a limited number of commercial cultivars and breeding lines. The *Ph*-genes are dominant, qualitative, and confer a hyper sensitive response. Some quantitative forms of resistance are known (Brouwer et al., 2004) but have been less widely used. For organic agriculture, quantitative forms would be expected to offer race non-specific resistance, and therefore would be more durable.

The Organic Outdoor Tomato Project was started in 2003 by the University of Göttingen as a participatory selection and breeding program in Germany. A screen of 3,500 accessions identified cultivars with potential to improve amateur gardening (Horneburg, 2010). Eighty-eight percent of the 33 best performing (regarding yield, field resistance) genotypes tested in three years of evaluation came from non-commercial sources (i.e., genebanks, NGO, and private seed savers). More than 60% were originally maintained and distributed by seed savers and NGO within organic horticulture (see fig. 15.3).

The best parent cultivars were used as source material for the breeding program. Market gardeners, seed savers, advisors, and scientists worked together to safeguard, select, and disseminate suitable genotypes. The first pure line varieties resulting from the Organic Outdoor Tomato Project, the cocktail tomatoes 'Clou', 'Dorada', and 'Primavera', were released in 2010.

The Oregon State University (OSU) tomato breeding program focuses on varieties suitable for the maritime environments of the Pacific Northwest in the United States, where the challenges are low growing temperatures and late blight. Late blight most commonly affects early spring polytunnel and late summer or fall outdoor production when environmental conditions favor disease development. OSU released 'Legend' fresh market tomato with *Ph-2* resistance in 2000. The tomato breeding program has subsequently focused on combining *Ph-2* resistance with quantitatively inherited genes derived from *L. hirsutum* (TGRC accession LA3683) as well as with *Ph-3* resistance sources.

Both organic breeding programs select for productivity under suboptimal growing conditions. Improved genotypes are early maturing and less prone to pod and blossom drop at low temperatures. Selection for late blight field resistance was of major importance whereby leaf and fruit susceptibility was scored in field trials with natural infection, which was frequently enhanced by a neighboring potato plot. Field resistance is defined as a delaying action, whereby resistant plants eventually

Table 15.1 Performance of site-specific superior genotypes. The same populations of three crosses (F_2, F_3, and F_5, respectively) were selected for 1 to 2 years in three organic market gardens and subsequently tested at all three sites.

	Superior genotypes selected at:			
	Schönhagen	Ellingerode	Rhauderfehn	Mean
Test site	Yield per plant (g)			
Schönhagen	1,227	1,316	962	1,168
Ellingerode	1,718	2,115	1,491	1,775
Rhauderfehn	560	648	574	594
Mean	1,168	1,360	1,009	

succumb to disease but more slowly than susceptible varieties. For example, late blight resistance in Legend provides another two to four weeks of production (depending on environmental conditions) compared to susceptible varieties when US-8 is present. The selection of F_2 plants for *P. infestans* field resistance is a robust and efficient tool for organic breeding (Horneburg and Becker 2011).

Outlook

Several questions arise from the information compiled in this chapter, and these suggest avenues for further investigation. First, the effect of the selection environment on the breeding program is not fully understood. While the selection environment can have an effect, broad adaptation is generally better unless there are very specific needs of a niche environment (Atlin et al., 2001). In a study carried out by Horneburg and Becker (2008) the selection environment (Ellingerode) with the poorest water and nutrient supply was most efficient in selection for yield (see table 15.1). Second, no studies have been performed to determine if a genotype × production system interaction exists and whether it represents cross-over interaction, and for which traits is it most important. This is important to vegetable breeders because the answer to this question determines whether two separate breeding programs are needed. Third, below-ground components of growth, including nutrient acquisition and uptake and soil microbial interactions, are only minimally understood. In a number of crops, significant genetic variation for N-use efficiency has been demonstrated, making breeding feasible. To our knowledge, the genetic control of this trait has not been examined in tomato. Perhaps more is known about tomato than other crops with regard to soil microbial interactions with the pioneering work of Mattoo and Teasdale (2010) and the study of the genetic control of mycorrhizal associations (Larkan et al., 2007; Cavagnaro et al., 2006), which provides a framework for future breeding efforts. Fourth, the nature of induced resistance and its induction by organic matter or soil microflora needs to be studied, and the plant genetics of this response needs further clarification. Do breeding and selection need to be carried out under organic conditions in order not to lose the genetic ability to interact with soil microflora? One approach to investigating this question would be to compare old varieties (developed before widespread use of chemical fertilizers) with new varieties developed under high fertility regimes to see if there were differences in growth and interaction with soil microorganisms. Fifth, pyramiding of resistances remains a challenge, but pyramiding qualitative genes along with incorporation of quantitative resistance may be the best strategy to deal with rapidly evolving pathogens. Molecular markers linked to resistance QTL should be useful in selecting for quantitative resistance (Collard and McKill, 2008). Sixth, the organic breeding community is small and would benefit from greater international exchange of suitable genotypes and cooperative breeding improvement strategies. Because organic plant breeding in general and breeding of tomatoes for organic systems is relatively new, it is important to network with growers and consumers (see Chapter 6) to better understand the traits that are needed in adapted varieties.

References

Atlin, G.N., M. Cooper, and Å. Bjørnstad. 2001. A comparison of formal and participatory breeding approaches using selection theory. *Euphytica* 122:463–475.

Becktell, M.C., M.L. Daughtrey, and W.E. Fry. 2005. Temperature and leaf wetness requirements for pathogen establishment, incubation period, and sporulation of *Phytophthora infestans* on *Petunia* × *hybrida*. *Plant Disease* 89:975–979.

Behrendt, U. 2009: Tomato breeding for taste by Oldendorfer Saatzucht, *In*: Østergård, H., E.T. Lammerts van Bueren, L. Bouwman-Smits (eds.) Proceedings Eucarpia-Bioexploit workshop on The Role of Molecular Marker Assisted Selection in Breeding Varieties for Organic Agriculture. February 25–27, 2009, Wageningen, The Netherlands. Wageningen, The Netherlands: Bioexploit Project.

Brouwer, D.J., E.S. Jones, and D.A. St. Clair. 2004. QTL analysis of quantitative resistance to *Phytophthora infestans* (late blight) in tomato and comparisons with potato. *Genome* 47:475–492.

Bryla, D.R., and R.T. Koide. 1998. Mycorrhizal response of two tomato genotypes relates to their ability to acquire and utilize phosphorus. *Annals of Botany* 82:849–857.

Cavagnaro, T.R., L.E. Jackson, J. Six, H. Ferris, S. Goyal, D. Asami, and K.M. Scow. 2006. Arbuscular mycorrhizas, microbial communities, nutrient availability, and soil aggregates in organic tomato production. *Plant and Soil* 282:209–225.

Chassy, A.W., L. Bui, E.N.C. Renaud, M. Van Horn, and A.E. Mitchell. 2006. Three-year comparison of the content of antioxidant microconstituents and several quality characteristics in organic and conventionally managed tomatoes and bell peppers. *Journal of Agricultural and Food Chemistry* 54:8244–8252.

Collard, B.C.Y., and D.J. Mackill. 2008. Marker-assisted selection: An approach for precision plant breeding in the twenty-first century. *Philosophical Transactions of the Royal Society of London B Biological Sciences* 363:557–572.

ERS. 2011. Organic production. Accessed March 31, 2011. http://www.ers.usda.gov/Data/Organic/.

FAOSTAT. 2011. Accessed January 30, 2011. http://faostat.fao.org/site/567/DesktopDefault.aspx?#ancor.

Foolad, M.R., H.L. Merk, and H. Ashrafi. 2008. Genetics, genomics and breeding of late blight and early blight resistance in tomato. *Critical Reviews in Plant Sciences* 27:75–107.

Fry, W.E. 1998. Late blight of potatoes and tomatoes fact sheet. Cooperative Extension New York State Cornell University. Accessed March 31, 2011. http://vegetablemdonline.ppath.cornell.edu/factsheets/Potato_LateBlt.htm.

Fry W.E., S.B. Goodwin, A.T. Dyer, J.M. Matuszak, A. Drenth, P.W. Tooley, L.S. Sujkowski et al. 1993. Historical and recent migrations of *Phytophthora* infestans: Chronology, pathways, and implications. *Plant Disease* 77:653–661.

Gladis, T.H., K. Hammer, H.H. Dathe, and H. Pellmann. 1996. Insect pollination and isolation requirements in tomato collections (*Lycopersicon esculentum* Mill.). FAO/IPGRI PGR Newsletter 106:16–19; erratum FAO/IPGRI PGR Newsletter 107:68.

Hall, T.J. 1980. Resistance at the TM-s locus in the tomato to tomato mosaic-virus. *Euphytica* 29:189–197.

Hertog, M.G.L., P.C.H. Hollman, and M.B. Katan. 1992. The content of potentially anticarcinogenic flavonoids of 28 vegetables and 9 fruits commonly consumed in the Netherlands. *Journal of Agricultural and Food Chemistry* 40:2379–2383.

Horneburg B., and H.C. Becker. 2011. Selection for *Phytophthora* field resistance in the F_2 generation of organic outdoor tomatoes. *Euphytica* doi:10.1007/s10681-011-0384-3.

Horneburg, B. 2010. Participation, utilization and development of genetic resources in the Organic Outdoor Tomato Project. *In*: Goldringer, I., J. Dawson, F. Rey, and A. Vettoretti (eds.) Breeding for resilience: A strategy for organic and low-input farming systems? EUCARPIA 2nd Conference of the "Organic and Low-Input Agriculture" Section. pp. 139–142. Retreived from http://orgprints.org/18171/1/Breeding_for_resilience%2DBook_of_abstracts.pdf.

Horneburg, B., and H.C. Becker 2008. Does regional organic screening and breeding make sense? Experimental evidence from organic outdoor tomato breeding. *In*: Neuhoff, D., N. Halberg, and T. Alföldi (eds.) Cultivating the future based on science. Volume 1: Organic crop production. Proceedings of the Second Scientific Conference of the International Society of Organic Agriculture Research (ISOFAR). pp. 670–673. June 18–20, 2008, Modena, Italy. Bonn, Germany: ISOFAR.

Jordan, J. 2007. The heirloom tomato as cultural object: Investigating taste and space. *Sociologia Ruralis* 47:20–41.

Kalloo, G. (ed.). 1991. Monographs on theoretical and applied genetics 14: Genetic improvement of tomato. Berlin-Heidelberg: Springer.

Kalloo. 2000. Vegetable breeding, volume 1. Boca Raton, FL: CRC Press.

Kultursaat e.V. Accessed February 30, 2011. http://www.kultursaat.org/sorten.html.

Lammerts van Bueren, E.T., P.C. Struik, M. Tiemens-Hulscher, and E. Jacobsen. 2003. Concepts of intrinsic value and integrity of plants in organic plant breeding and propagation. *Crop Science* 43:1922–1929.

Lammerts van Bueren, E.T., S.S. Jones, L. Tamm, K.M. Murphy, J.R. Myers, C. Leifert, and M.M. Messmer. 2010. The need to breed crop varieties suitable for organic farming, using wheat, tomato and broccoli as examples: A review. *NJAS Wageningen Journal of Life Sciences*. 58:193–205.

Larkan, N.J., S.E. Smith, and S.J. Barker. 2007. Position of the reduced mycorrhizal colonization (*Rmc*) locus on the tomato genome map. *Mycorrhiza* 17:311–318.

Levin, I., C.H.R. De Vos, Y. Tadmor, A. Bovy, M. Lieberman, M. Oren-Shamir, O. Segev et al. 2006. High pigment tomato mutants-more lycopene (a review). *Israel Journal of Plant Sciences* 54:179–190.

Mattoo, A.K., and J.R. Teasdale. 2010. Ecological and genetic systems underlying sustainable horticulture. *Horticultural Reviews* 37:331–362.

Mes, P.J., P. Boches, R. Durst, and J.R. Myers. 2008. Characterization of tomatoes expressing anthocyanin in the fruit. *Journal of the American Society of Horticultural Science* 133:262–269.

Metzger, J. 1841. Landwirthschaftliche Pflanzenkunde oder praktische Anleitung zur Kenntniß und zum Anbau der für Oekonomie und Handel wichtigen Gewächse. Winter, Heidelberg. 2 Bände. pp. 558–559.

Murphy K., K. Campbell, S. Lyon, and S. Jones. 2007. Evidence of varietal adaptation to organic farming systems. *Field Crops Research* 102:172–177.

Murphy, K.M., P.R. Reeves, and S.S. Jones. 2008. Relationship between yield and mineral nutrient concentration in historical and modern spring wheat cultivars. *Euphytica* 163:381–390.

Ordóñez-Santos, L.E., E. Arbones-Maciñeira, J. Fernández-Perejón, M. Lombardero-Fernández, L. Vázquez-Odériz, and A. Romero-Rodríguez. 2009. Comparison of physicochemical, microscopic and sensory characteristics of ecologically and conventionally grown crops of two cultivars of tomato (*Lycopersicon esculentum* Mill.) *Journal of the Science of Food and Agriculture* 89:743–749.

Pelham, J. 1966. Resistance in tomato to tobacco mosaic virus. *Euphytica* 15:258–267.

Raviv, M. 2010. Sustainability of organic horticulture. *Horticultural Reviews* 36:289–333.

Rick, C.M. 1950. Pollination relations of *Lycopersicon esculentum* in native and foreign regions. *Evolution* 5:110–122.

Siebold, M. 2006. Gläserne Produktion am deutschen Tomatensaatgutmarkt? Aktuelle Saatgutflüsse und Bedarfskalkulationen. Bachelor-thesis, Faculty for Agriculture, Georg-August-University Göttingen.

Stewart, A.J., S. Bozonnet, W. Mullen, G.I. Jenkins, M.E. Lean, and A. Crozier. 2000. Occurrence of flavonols in tomatoes and tomato-based products. *Journal of Agricultural and Food Chemistry* 48:2663–2669.

Stone, A.G., S.J. Scheuerell, and H.M. Darby 2004. Suppression of soilborne diseases in field agricultural systems: Organic matter management, cover cropping, and other cultural practices, *In*: F. Magdoff, and R. Weil (eds.) Soil organic matter in sustainable agriculture. pp. 131–177. Boca Raton: CRC Press.

Vinson, J.A., Y. Hao, X. Su, and L. Zubik. 1998. Phenol antioxidant quantity and quality in foods: Vegetables. *Journal of Agricultural and Food Chemistry* 46:3630–3634.

16 Brassicas: Breeding Cole Crops for Organic Agriculture

James R. Myers, Laurie McKenzie, and Roeland E. Voorrips

Introduction

The edible crucifers represent a diverse group of species in the Brassicaceae family that are used as vegetables, condiments, and oilseeds. The majority of vegetable crops belong to *Brassica oleracea* and *B. rapa* (formerly *B. campestris*). Radish (*Raphanus sativus*) is the third largest group, in terms of diversity, with the major types being spring or winter radishes. A few other vegetable crops belong to allied species (i.e., rutabaga in *B. napus* and the *B. juncea* mustards). All of these species are relatively closely related and in many cases share genomes through allotetraploid relationships (Prakash et al., 2009).

We focus here on the vegetables that belong to *Brassica oleracea* (cabbage, broccoli, cauliflower, kale, collards, Brussels sprouts, and Kohlrabi) and *B. rapa* (turnip, Chinese or napa cabbage, pak choi, mizuna tat soi, and others). Cytogenetic and phylogenetic relationships of this family were recently reviewed by Prakash et al. (2009). Breeding reviews on the vegetable brassicas can be found for kale and cabbage (Ordás and Cartea, 2008) and cauliflower and broccoli (Branca, 2008). Additional valuable sources of information on the vegetable Brassicas include Dixon (2007) and Dickson and Wallace's (1986) cabbage breeding chapter.

In this chapter, we will concentrate on those species for which organic breeding is most advanced. We will contrast annual and biennial breeding systems as represented by broccoli and cabbage, with occasional reference to cauliflower. What we describe for these crops should be relevant to other vegetable crucifers.

Rationale for Breeding within Organic Systems

Organic agriculture (OA) differs from conventional agriculture primarily in soil attributes, biological diversity, nutrient delivery, weed control, and in some cases, pest and disease complexes. Organic growers place greater reliance on genetic solutions that provide adaptation to the organic system, and crop varieties adapted to this suite of differences are needed for optimal production in organic systems. With the exception of heirloom and landrace varieties developed more than 75 years ago,

---

Organic Crop Breeding, First Edition. Edited by Edith T. Lammerts van Bueren and James R. Myers.
© 2012 John Wiley & Sons, Inc. Published 2012 by John Wiley & Sons, Inc.

varieties have been bred in and for conventional agricultural systems. While some varieties work well in organic agriculture, others may require high nutrient inputs, or may lack essential disease resistances that provide adaptation. The Brassica vegetable crops are no exception. For broccoli, some older open-pollinated (OP) varieties work well in OA but they lack uniformity and many important quality traits required in contemporary markets. Even F_1 hybrids widely adapted and widely used in organic production may be only average performers in terms of yield and quality (Renaud et al., 2010). Their main advantage is probably their stability over a range of environments. Whether they be OPs or F_1 hybrids, it should be possible to breed varieties that are better adapted to OA, with improved performance over what is currently available. In an effort to close the loop and reduce reliance on off-farm inputs, some organic growers would like to have OP varieties in particular, with contemporary yield and quality characteristics that allow them to save their own seed.

Another aspect of organic agriculture is improving working conditions for workers. Production of the vegetable Brassicas is labor intensive, and developing traits that improve livelihood of field workers should be pursued in breeding for organic agriculture. Due to the proliferation of small organic fresh market farms with intensive hand harvesting, breeding for improved ergonomics is a worthy consideration (which would benefit conventional systems and mechanical harvesting as well). As an example in the Brassicas, nearly all modern broccoli cultivars are short with sessile heads. At Oregon State University (OSU), the broccoli program has focused on breeding varieties with exerted heads that are at or above leaf canopy height (Baggett et al., 2005). These require less stooping on the part of the person harvesting the heads, and fewer leaves are present on the heads requiring less trimming.

Studies in process that compare varietal performance in organic and conventional systems show that genotype × system interactions occur, indicating that best performers in conventional are not necessarily the best in organic, and vice versa (Renaud et al., 2010). Unlike crops such as wheat or maize where significant research in organic systems has been conducted, little is known about specific needs of Brassica vegetable crops for adaptation to organic systems.

Plant Biology

Understanding the life cycle and pollination system as well as growth characteristics is fundamental to determining how a Brassica breeding program should be established and implemented. Broccoli, cauliflower and cabbage are diploid (n = 9) and share a common ancestry with a progenitor most closely resembling kale. Both biennial and annual forms are found, with a biennial habit being ancestral. All cabbage varieties are biennial whereas most broccoli varieties are annuals and cauliflower varieties are a mix of annual and biennial types. Except for tropically adapted cauliflowers, *B. oleracea* crops require a minimum chilling period to induce flowering. The vernalization period for cabbage (approximately 4.5°C for about 8 weeks once stem diameter reaches 6–8 mm, depending on genotype) is generally much greater than that for broccoli (which will do fairly well in warm temperate zone climates but will fail to head under tropical conditions). These crucifers generally prefer a mild maritime climate where temperatures above 30°C cause abnormal growth and development (Björkman and Pearson, 1998). They are generally tolerant to freezing temperatures with cabbage being hardier than broccoli and overwintering forms of broccoli and cauliflower being hardier than annual forms. In addition, plants in vegetative stages of growth can withstand lower temperatures than when in reproductive stages. These crops have perfect flowers but are generally outcrossers with a sporophytic self-incompatibility (SI) system enforcing outcrossing. The SI system in *B. oleracea* is quite variable in enforcing outcrossing, with strong incompatibility occurring in cabbage, while SI in broccoli and cauliflower is generally weaker. Self-fertile individuals have been documented and self-compatibility is not absolute – especially in older plants, SI may break down and allow self-pollination to occur.

As a consequence of these life history and pollination characteristics, Brassica breeders generally use certain population structures and breeding techniques. Cabbage breeders, for example will grow the crop during the summer, then lift the plants and transfer to a cold frame that will protect them from extreme cold but which still allows vernalization. In the spring, plants flower and can be crossed and selfed (see following paragraph). A cycle in cabbage breeding is about 13 to 14 months from seed to seed. In broccoli, two generations per year can be achieved by using a combination of the field and greenhouse (or winter nursery) although broccoli has a relatively long life cycle and achieving a complete cycle of growth in six months can be problematic. An alternate one-year cycle in broccoli consists of planting trials midsummer with evaluation in the fall. At that time, cuttings are taken from selections and rooted in the greenhouse. These are used for crossing and selfing during the winter with seed set in the spring.

Breeding techniques that are relatively easy to implement include mass selection and various forms of recurrent selection for population improvement and development of open pollinated varieties. The cruciferers can be self-pollinated using techniques such as bud pollination and CO_2 treatment (Nakanishi and Hinata, 1973). The extent of inbreeding depression varies by crop and population with those that have weaker SI being less prone to inbreeding depression. As such, the development of inbreds to produce F_1 hybrids is an important aspect of many commercial breeding programs.

Traits Needed for Adaptation to Organic Production

The attributes needed to adapt Brassica crops to OA include adaptation to soil and fertility management in organic systems, ability to compete with weeds, and resistances to insects and diseases that would be controlled with chemical herbicides and pesticides in conventional agricultural systems.

Soils and Fertility

Key traits for adaptation of Brassica crops to organic systems include those that optimize growth in the soil and fertility systems commonly employed in organic production. Soils generally have substantial inputs of organic matter, which leads to increased microbial activity and a different complex of microfauna compared to conventionally managed soils. Soil nutrients may be similar in levels to conventional soils, but nutrients many not be as rapidly available. Arbuscular mycorrhizal (AM) fungi have been shown to increase uptake of phosphorus as well as certain micronutrients. With few exceptions, AM associations have not been found in the Brassicaceae family. Certain species have been shown to have low levels of colonization by AM (Nelson and Achar, 2001; Regvar et al., 2003), one of which was cabbage when inoculated with *Glomus* spp, but the relationship to phosphorus uptake has not been investigated. A larger question is why AM associations are not common in the Brassicaceae. In tomato, non-AM responsive types use different strategies to acquire phosphorus in low nutrient environments. One such strategy is to produce a large, highly branched root system that is better able to explore the soil (Bryla and Koide, 1998). Genetic variation for P-use efficiency has been described (Greenwood et al., 2005, 2006) and is under quantitative genetic control. Studies of cauliflower revealed that those with proportionally greater fine root systems are more N-use efficient (Kage et al., 2003). Studies are needed to evaluate genetic variability across a wide range of germplasm to explore the potential for incorporating favorable rhizosphere traits for organic systems.

Weeds

Weed management in Brassica crops is a concern, especially with direct seeding in the field (as is practiced in the United States sometimes with broccoli), because of the small size and relatively slow growth rate of the seedling. Because the Brassicaceae have C3 metabolism, planting them in warmer environments puts them at a particular disadvantage compared to C4 weeds. Weed pressure in crops grown from transplants – as many of the cole crops are grown – usually is less of a concern because plants at the transplant stage are entering a log phase in growth and are able to out-compete smaller weed plants that are still only in initial phases of growth. In addition, transplants are better adapted to cultivation for weed suppression. The decision to direct seed or use transplants may be based on several factors in addition to weed pressure, including cost of the seed (F_1 hybrid seed will be more expensive than OP seed), and labor and material costs for transplanting versus direct seeding. In direct-seeded cole crops, thinning may be required unless the grower has access to a precision planter. The Brassicas in general and cole crops in particular show an enormous range in variability in plant architecture. Lammerts van Bueren et al. (2004) found that cabbage varieties varied in leafiness and leaf erectness, and that these were traits related to weed suppression. The need for more leafy types with planophile leaves has been increasingly important as consumer demand in the Netherlands has shifted the market from traditional sized heads to smaller (800–1200 g) heads.

Brassicaceae is one of 16 families that produce glucosinolates (Fahey et al., 2001), with few cultivated species found within the other families. Among other effects, some glucosinolates are known to have weed, pathogen, and insect pest suppressive properties when converted to isothiocyanates and related compounds. When the plant tissues are macerated (such as in an insect or pathogen attack), myrosinase catalyzes glucosinolates to isothiocyanates and nitriles (Vaughn and Boydston, 1997). While these breakdown products have been demonstrated to have weed suppressive effects (Petersen et al., 2001), it is only when the Brassica crop is chopped and/or incorporated that the allelopathic effect is observed on the following generation of weeds. Where Brassicas have been studied to see if actively growing plants are allelopathic to weeds, no such effect has been discovered (Itulya and Aguyoh, 1998; Jimenezosornio and Gliessman, 1987). Thus, the weed suppressive effect of Brassica crops cannot be exploited directly for the benefit of that crop during the growing season, but must be considered as a long-term strategy, where incorporation of the crop residue after harvest can be used to reduce the weed seed bank size over time.

Diseases and Insect Resistances

As a consequence of intensified production (not only vegetable Brassicas in rotation, but also cruciferous green manure and oilseed crops) on limited hectarages, incidence of club root (*Plasmodiophora brassicae*) is increasing in Brassica crops among fresh market growers. This disease is found in every region where Brassica crops are grown and arguably represents the most widespread disease of Brassica vegetables. Once present in a field, resting spores will remain viable for at least 17 years (Diederichsen et al., 2009). Genetic resistance is available for most Brassica crop species, and in some cases resistant varieties have been developed. The genetics of resistance varies from species to species; dominant, race-specific resistance is found in *B. rapa* and *B. napus*, and predominantly recessive, race-nonspecific resistance is found in *B. oleracea*. Resistance has been difficult to study because of its quantitative nature, but molecular mapping studies with doubled haploids have revealed several major and minor QTL (Piao et al., 2009; Voorrips et al, 1997). Early attempts to deploy resistance resulted in resistance being overcome by more virulent races (Baggett, personal communication; Diederichsen et al., 2009). Because of its recessive nature, most commercial programs have not used the forms of resistance naturally occurring in *B. oleracea*. Resistance from *B. napus* and *B. rapa* has been transferred into *B. oleracea*, and the latter source in particular has been used commercially. With regard to breeding Brassica crops for OA, there are two aspects of clubroot resistance that are relevant. First is the interspecific transfer of resistance from related species. When this is done using in vitro techniques such as embryo rescue, it would be regarded as not compatible with the principles of organic agriculture (Lammerts van Bueren et al., 2003). Embryo rescue has been the approach that has been taken with transfer of clubroot resistance from *B. rapa* to *B. oleracea* (Bradshaw et al., 1997; Chiang et al., 1977; Diederichsen et al., 2009). Because it is possible to obtain interspecific hybrids from natural crosses with a success rate of about 0.42% when *B. rapa* is the maternal parent (Inomata, 1978, his table 5), this might be the approach that a breeder would take to transfer resistance to be used in organic agriculture. Second, commercial breeders have favored dominant resistance over recessive forms in F_1 hybrid production programs because the former only needs to be incorporated into one of the inbred parents, whereas for recessive resistance, both inbred parents must carry the appropriate alleles. This may not be the best strategy for obtaining durable resistance. Studies of disease resistance at the molecular level have led to the differentiation of resistance and susceptibility genes (Pavan et al., 2010). In their terminology, a resistance gene is a gene in which the dominant allele confers race-specific resistance, often of the hypersensitive response type. In contrast, a susceptibility gene is a gene where the dominant allele confers susceptibility, while a loss-of-function, recessive allele confers resistance. These recessive resistance alleles also appear to provide a more durable and robust form of resistance (Pavan et al., 2010). While most commercial breeding programs for Brassica crops are based on the production of F_1 hybrids, there remains a significant demand in OA for OP varieties, particularly those with contemporary horticultural and quality traits. For this market in *B. oleracea* in particular, the development of OP varieties that incorporate intraspecific recessive and quantitative resistances would be preferred over interspecific dominant race-specific forms of resistance because of their potentially higher durability. It should be noted that these forms of resistance are not effective against all clubroot isolates, so an additional strategy would be to pyramid *B. rapa* resistance on top of that present in *B. oleracea*.

Blackleg (*Leptosphaeria maculans*, formerly *Phoma lingam*) and Alternaria (caused by various *Alternaria* spp., but mainly *A. brassicicola*) are two diseases that affect Brassica production in Europe and eastern North America. While conventional growers can use fungicides to control these diseases, organic growers do not have similar tools. Hot water treatment can be used effectively to disinfect seed, but the technique may reduce germination if not applied with care (Dixon, 2007). Differences in genetic resistance to these diseases have been observed among various species, and this resistance needs to be transferred into a *B. oleracea* background.

Insect pests of Brassica crops are the principal biological constraint facing organic farmers. A number of dipteran, lepidopteran, coleopteran, and hemipteran pests attack Brassica crops, and in conventional agriculture, these pests are usually controlled by the use of pesticides. OA growers have fewer options for control. They may, for example, use floating row cover in early season to prevent infestation by cabbage root fly (*Delia radicum* and *D. brassicae*), flea beetle (*Phyllotreta* spp.), and lepidopteran pests (*Plutella xylostella*, *Pieris rapae*). While floating row cover is effective, it is not cost effective and sustainable to employ on large hectarages. Biological control products such as BT (based on *Bacillus thuringiensis*) and Spinosad (based on *Saccharopolyspora spinosa*) may also be used to control lepidopteran pests and aphids (*Brevicoryne brassicae, Lipaphis erysimi, Myzus persicae*). Insect pests are often seasonal with certain species more prevalent in the spring or fall. For example, flea beetle is more of a problem in spring than fall, whereas the opposite is true for aphids.

In general, genetic solutions to insect problems are difficult to find. Genetic variation does exist, but is often quantitatively inherited and therefore difficult to accumulate and introgress into new varieties. Selection for resistance may have unintended consequences on flavor and color as well as other plant traits (Voorrips et al., 2008).

Growers and researchers have observed that variation for leaf color and shine in *B. oleracea* crops is associated with differential feeding by insect pests. Dull, blue-green (glaucous) plants have relatively thick layers of epicuticular wax whereas plants with glossy, light green foliage lack or have reduced levels of wax. It has been shown that epicuticular wax may be positively or negatively associated with insect pest populations. In particular, glossy types show reduced damage from lepidopteran pests, lower whitefly populations (*Aleyrodes brassicae, Bemisia tabaci, B. argentifolii*), and have fewer cabbage root fly eggs laid near plants (Eigenbrode and Espelie, 1995; Farnham and Elsey, 1995). On the other hand thrips (*Thrips tabaci*) damage and flea beetle damage is greater on glossy plants, and both an increase and a decrease in aphid populations have been reported (Eigenbrode, 1996; Voorrips et al, 2008). One possible explanation as to why levels of certain pests are reduced is that insect predator species forage more efficiently on a glossy plant due to better traction on the leaf surfaces (Eigenbrode and Kabalo, 1999; Stoner, 1992). However, other factors such as wax composition and color may influence insect feeding and reproduction (Eigenbrode, 1996). A negative correlation between wax thickness and thrips damage has clearly been demonstrated but other factors including earliness and Brix also had an effect on resistance (Voorrips et al., 2008). The relationship between low wax and insect behavior is better characterized in comparison to the effect of higher than normal wax accumulation. Higher than normal wax variants might have greater resistance to flea beetle, for example. *B. oleracea* varieties with glossy leaf type have not been widely deployed (only two glossy collard accessions can be found in the USDA germplasm collection). Farnham (2010) released two isogenic pairs of broccoli inbreds that differ for leaf epicuticular wax that will allow plant breeders to investigate what insect pests and under what environmental conditions this trait may confer benefits to organic growers. Glossy types would be least useful in spring and early summer plantings when flea beetle pressure is highest, but might have a niche for fall production when flea beetle populations drop off and other pests such as aphids become more of a problem. The glossy trait might be particularly useful in overwintering cauliflower, Brussels sprouts, and cabbage. However, diseases must also be considered in any discussion about epicuticular wax. For example, the waxy cuticle apparently protects against autumn diseases such as *Alternaria brassicae* and *Mycosphaerella brassicicola*.

Glucosinolates and their breakdown products play a role in insect-plant relationships. High levels of glucosinolates serve as a feeding deterrent for generalist herbivores, but often act as a kairomone (feeding attractant) to specialist herbivores and as a synomone (mutually beneficial chemical signal) for predators and parasitoids of specialist herbivores (Hopkins et al., 2009). Herbivore specialists on Brassicas may sequester glucosinolate compounds and use them as a feeding deterrent to their predators. Because glucosinolates have an important influence on flavor characteristics of the Brassica vegetable crops, any manipulations in their content and composition to deter insect pests must be balanced with flavor requirements of humans. It may be possible to increase glucosinolate levels in particular plant organs of crops where the organ in question is not consumed. For example, higher levels of 2-phenylethylglucosinolate expressed in the roots appears to reduce cabbage root fly pupal mass and this glucosinolate is under separate genetic control in roots and shoots (Hopkins et al., 2009). Increasing levels of this compound only in the roots of *B. oleracea* vegetable crops is a possible strategy to achieving higher levels of cabbage root fly resistance. Otherwise, there are no clear-cut cases that plant breeders could exploit to increase insect resistance. Reducing glucosinolate levels might reduce attractiveness of the crop to certain specialist herbivores; however, it would make them more palatable to generalist herbivores including vertebrate species (Hopkins et al., 2009).

Consideration of Breeding Methods

We have alluded to the differences in types of cultivar (F_1 hybrid vs. OP) with regard to disease resistance, weed suppression, and compatibility with organic principles. Nearly all commercial seed companies with Brassica vegetable breeding programs are developing and releasing F_1 hybrids as their preferred cultivar type. The reasons include the advantages of uniformity, heterosis, and the easier combination of (dominant) traits. While there are many organic growers who prefer F_1 hybrid varieties for their uniformity and yield potential, others would like to have varieties they could maintain on their own farms. Since most of the Brassica vegetable crops are outcrossers, alternate types of cultivars to F_1 hybrids would be limited to OPs and synthetics. Historically, Brassica vegetable crops were maintained as OPs, and a number of heirloom and landrace varieties can still be found. While these varieties may possess traits that facilitate adaptation to organic systems, most of these lack the horticultural and quality traits found in contemporary F_1 hybrids because Brassica breeding has focused mostly on F_1 hybrids, and, consequently, improvement of OPs lagged behind. In theory, modern OPs could be developed that could compare favorably with F_1 hybrids. They may not have quite the same yield potential of an F_1 hybrid, but OPs could be developed that have equivalent quality, and would be superior in terms of providing genetic diversity to the farming system. The challenge with OP varieties is in their maintenance since reductions in population size below a critical threshold will result in inbreeding and genetic drift, and their inherently higher level of variation makes them less suitable for highly managed, mechanized production systems and more difficult to meet the uniformity requirements of variety registration.

Two technologies in particular in the vegetable Brassicas have promoted the development of F_1 hybrids. These are the availability of cytoplasmic male sterility (CMS) and the ability to produce dihaploids. Prior to the development of economically acceptable CMS, seedsmen used naturally occurring self-incompatibility (SI) to make F_1 hybrids. Although some hybrids are still produced using SI, its use requires the classification of SI alleles in inbreds. Only those inbreds with different SI alleles can be used effectively to produce hybrids in the field. SI will not produce 100% F_1 progeny because SI barriers may be relaxed toward the end of flowering. A level of 1 or 2% inbreds is considered acceptable for cabbage, and higher levels are acceptable in broccoli and cauliflower. CMS in *B. oleracea* has been achieved using somatic hybridization (Budar and Pelletier, 2001). The original cytoplasm donor was radish (Ogura, 1968) and more recently, *B. tournefortii* via *B. rapa* (Cardi and Earle, 1997). The original Ogura CMS was not economically useful because CMS lines exhibited cold temperature chlorosis. This defect was corrected by replacing radish chloroplasts with those of *B. oleracea* (Walters et al., 1992; Sigareva and Earle, 1997). To a large extent, *B. oleracea* F_1 hybrids currently on the market are produced on CMS maternal plants. While CMS is a convenient way to produce F_1 hybrids, if derived through cell-fusion, it is not considered to be compatible with organic principles (IFOAM, 2005). However, its use is widespread and its acceptability in OA is under discussion (Billmann, 2008; Lammerts van Bueren and Haring, 2009).

A second reason to avoid the use of CMS is its effect on the reduction of genetic diversity. Varieties derived through CMS are biological dead ends due to their cytoplasmic male sterility, and breeders cannot use these to develop new varieties.

The vegetable Brassicas are relatively easy to culture in vitro, and efficient doubled haploid techniques based on in vitro microspore culture are used by most commercial programs. The use of doubled haploids in a breeding program can dramatically reduce the time it takes to develop inbreds. This can be especially important in a biennial crop such as cabbage. However, as with cell-fusion derived CMS, microspore culture is not in agreement with the organic principles, but occupies more of a gray area in terms of acceptance in OA.

A Farmer Participatory Broccoli Breeding Program

The Vegetable Breeding Program (VBP) at OSU has taken on the task of creating an OP broccoli variety that is adapted to organic production using a farmer-participatory model similar to that described in Atlin et al. (2001). In particular, they suggest that one model that could be implemented in a farmer participatory plant breeding program would be shuttle breeding, whereby a network of farmers made selections in their own environments and exchanged or bulked seed to be redistributed to the farmer network for another round of selection in the next growing season (Atlin et al., 2001).

In the OP broccoli project, farmer participation has been utilized in multiple stages during this breeding effort, including the creating of breeding objectives, identifying important production traits, and facilitating population improvement and variety development. Traits used for selection included: heat tolerance, early and even maturity, medium head size (about 13 cm or 5 in), exserted heads, small bead (flower bud) size, vigorous growth in cool spring conditions, high side-shoot production, attractive (blue-green) color, clean leaf abscission, and good flavor.

Broccoli is an ideal crop for participatory plant breeding (PPB) because it is easy to breed using mass selection, which is a straightforward and accessible technique for farmers. Given sufficiently wide variation, gains from selections are rapid since evaluation and selection can be done prior to flowering. As long as a sufficient number of plants (~50) are maintained in each generation to avoid inbreeding depression, mass selection is an effective and easy means for broccoli improvement.

The broccoli population was created in 1997 from a random mating of six F_1 hybrid varieties (Arcadia, Barbados, Decathlon, Excelsior, San Miguel, and Shogun) with 17 inbreds that had been developed by the VBP. Four years (1997-2000) of random mating without selection were carried out within a conventional system at OSU to maximize recombination within the population. Population improvement was carried out over the next seven years (2001–2007) with farmer participation using a divergent-convergent scheme (see fig. 16.1). Each season, participating (organic) farmers and the VBP would select for the traits of head size, plant vigor, freedom from diseases, and heat tolerance, and allow random mating of the population and seed production on their farms. Farmers returned a portion of their seed to the VBP where seed from all sources was blended and aliquots sent back out; this process was repeated every year. In 2004, research production at OSU was moved to organic transitional ground to facilitate selection within an organic system. The performance of populations by year of selection from 2003 to 2007 are shown in table 16.1. In general, we saw no change in head size, canopy and head height, or bead size (not unexpected, given that the parental material was excellent for these traits). There was an increase in the number of over-mature heads and a decrease in young heads indicating that we were shifting toward earlier maturity over generations. We also noticed an increase in the overall variability of the population over time, and suspect that uncontrolled introgressions in farmers' plots was taking place. An obvious divergence

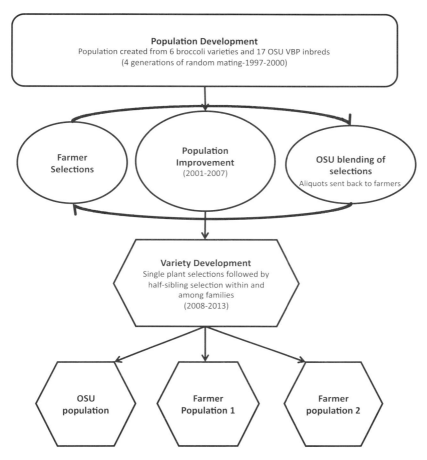

Figure 16.1 Breeding scheme for a project to develop an open pollinated broccoli variety for organic production that involves farmers in the breeding process.

Table 16.1 Performance of open pollinated broccoli subpopulations by original year of selection for plant and head characteristics when grown at the Lewis Brown Farm, Corvallis Oregon in 2008.

Year	Bloom	Flowering heads	Prime heads	Young heads[a]	Bead size[b]	Head compact[c]	Head dia.	Head ht.	Canopy ht.
	in days		%					cm	
2003	97.1	28.0	40.0	18.0	5.0	5.8	12.8	53.5	57.5
2004	110.8	5.0	65.0	29.0	6.5	6.0	12.8	52.9	57.7
2005	104.9	6.0	48.0	39.0	5.8	4.8	11.8	47.8	50.8
2006	95.3	2.3	52.0	16.0	4.5	4.7	13.4	48.1	55.5
2007	93.7	46.0	42.0	6.0	4.7	5.2	13.6	49.3	53.8
	NS	***	NS	***	NS	NS	NS	NS	NS

[a]Flowering + Prime + Young heads do not sum to 100% because non-heading plants were not counted.
[b]Beads are individual unopened flowers and are measured on a 1–9 scale where 1 = very coarse and 9 = very fine.
[c]Head compactness measured on a 1–9 scale where 1 = loose and 9 = compact.
NS = non significant; ***statistically significant at $p < 0.001$.

Figure 16.2 Divergence in the United States in broccoli types selected in the Pacific Northwest (left) and in Long Island and Ithaca, NY (right), from an open-pollinated population selected for organic production. The west coast material tends to be taller with darker, more compact heads and finer bead structure than the east coast material.

in broccoli morphology and phenotype occurred between selections being carried out on the west coast as compared to selections from the east coast, and this difference was probably due to regional preference, not environmental differences. The east coast selections were noticeably shorter, lighter in color, and presented a more open stature (see fig 16.2).

In 2008, the program shifted from population improvement to varietal selection with single plant selections from the populations. These selections were carried out by two participating farmers as well as by the VBP at the OSU Lewis Brown Research Farm. Each year, single-plant selections were planted to rows from which further single plant selections were made in order to achieve greater uniformity. At the time of this writing, three of four years (2008–2011) of this half-sibling mating and single plant selection approach had been completed. One participating farmer intended to develop a broccoli variety specifically adapted to her production needs (a community supported agriculture or CSA operation) and growing environment (low-input, low N, and cool maritime climate). The other participating farmer has begun distribution of the variety Oregon Longneck, which he developed from this population.

Outlook

The participatory plant breeding approach using a network of farmers was instrumental to the OSU VBP in transitioning into breeding for organic systems, and the choice of broccoli was fortuitous in that its lifecycle lends itself to a mass selection approach that is easy for farmer-breeders to use. The VBP had little knowledge of the production traits that were important for organic broccoli culture and treated the selection process as a "black box" by simplifying the selection procedure to selecting for productivity in the target environments. Farmers were engaged to define quality, handling, and harvest traits of highest priority.

A benefit to farmers was to provide them with resources and skills that allowed them to carry out plant breeding projects on their farms. Farmers have had "ownership" in the project, which has kept it relevant to the needs of farmers, and increased the potential for adoption and use of the variety upon future release. This project has seen farmers continue to breed without direct input of the OSU VBP (e.g., farmer development of the variety Oregon Longneck), providing the greater organic community with increased varietal options that are specifically bred for and adapted to organic production systems. There is still much to learn about the traits that are unconsciously selected for in an organic system, and this will require that the VBP pursue formal scientific studies. The convergent-divergent selection scheme employed appears to have led to increased stability over growing environments, but whether we have achieved greater robustness will only be determined as varieties are developed and evaluated from the population in multiple environments.

References

Atlin, G.N., M. Cooper, and Å. Bjørnstad. 2001. A comparison of formal and participatory breeding approaches using selection theory. *Euphytica* 122:463–475.

Baggett, J. R., D. Kean, and K. Kasimor 1995. Inheritance of internode length and its relation to head exsertion and head size in broccoli. *Journal of the American Society of Horticultural Science* 120:292–296.

Billmann, B. 2008. Impacts of banning protoplast fusion on the range of varieties available for organic arable cropping and vegetable production, FiBL Report as Part of the COOP Project on Module 1.3. Safeguarding organic seed and planting material – Impulses for organic plant breeding. Frick, Switzerland: FiBL.

Björkman, T., and K.J. Pearson. 1998. High temperature arrest of inflorescence development in broccoli (*Brassica oleracea* var. *italica* L.). *Journal of Experimental Botany* 49:101–106.

Bradshaw, J.E., D.J. Gemmell and R.N. Wilson. 1997. Transfer of resistance to clubroot (*Plasmodiophora brassicae*) to swedes (*Brassica napus* L. var. *napobrassica* Peterm) from *B. rapa*. *Annals of Applied Biology* 130:337–348.

Branca, F. 2008. Cauliflower and broccoli. *In*: Prohens, J., and F. Nuez (eds.) Vegetables I handbook of plant breeding, volume 1, part 2. pp. 151–186. New York: Springer.

Bryla, D.R., and R.T. Koide. 1998. Mycorrhizal response of two tomato genotypes relates to their ability to acquire and utilize phosphorus. *Annals of Botany* 82:849–857.

Budar, F., and G. Pelletier. 2001. Male sterility in plants: Occurrence, determinism, significance and use. *Comptes Rendus* de l'Académie des Sciences Series III 324:543–550.

Cardi, T., and E.D. Earle. 1997. Production of new CMS *Brassica oleracea* by transfer of "Anand" cytoplasm from *B. rapa* through protoplast fusion. *Theoretical and Applied Genetics* 94:204–212.

Chiang, M.S., B.Y. Chiang, and W.F. Grant. 1977. Transfer of resistance to race 2 of *Plasmodiophora brassicae* from *Brassica napus* to cabbage (*B. oleracea* var. *Capitata*). I. Interspecific hybridization between *B. napus* and *B. oleracea* var. *Capitata*. *Euphytica* 26:319–336.

Dickson M.H., and D.H. Wallace. 1986. Cabbage breeding. *In*: M.J. Bassett (ed.) Breeding vegetable crops, pp. 396-432. Westport, CT: AVI Publ. Co. Inc.

Diederichsen, E., M. Frauen, E.G.A. Linders, K. Hatakeyama, and M. Hirai. 2009. Status and perspectives of clubroot resistance breeding in crucifer crops. *Journal of Plant Growth Regulators* 28:265–281.

Dixon, G.R. 2007. Vegetable Brassicas and related crucifers. Crop production science in horticulture series, vol. 14. Wallingford, UK: CABI.

Eigenbrode, S.D. 1996. Plant surface waxes and insect behaviour. pp. 201–222. *In*: Kerstiens, G. (ed.) Plant cuticles: An integrated functional approach. Oxford, UK: BIOS Scientific Publishers.

Eigenbrode, S.D., and K.E. Espelie. 1995. Effects of plant epicuticular lipids on insect herbivores. *Annual Review of Entomology* 40:171–194.

Eigenbrode, S.D., and N.N. Kabalo. 1999. Effects of *Brassica oleracea* waxblooms on predation and attachment by *Hippodamia convergens*. *Entomologia Experimentalis et Applicata* 91:125–130.

Fahey, J.W., A.T. Zalcmann, and P. Talalay. 2001. The chemical diversity and distribution of glucosinolates and isothiocyanates among plants. *Phytochemistry* 56:5–51.

Farnham, M.W. 2010. Glossy and nonglossy near-isogenic lines USVL115-GL, USVL115-NG, USVL188-GL, and USVL188-NG of broccoli. *HortScience* 45:660–662.

Farnham, M.W., and K.D. Elsey. 1995. Recognition of *Brassica oleracea* L. resistance against the silverleaf whitefly. *HortScience* 30:343–347.

Greenwood, D.J., A.M. Stellacci, M.C. Meacham, M.R. Broadley, and P.J. White. 2005. Phosphorus response components of different *Brassica oleracea* genotypes are reproducible in different environments. *Crop Science* 45:1728–1735.

Greenwood, D.J., A.M. Stellacci, M.C. Meacham, P.J. White, and M.R. Broadley. 2006. Brassica cultivars: P response and fertilizer efficient cropping. *Acta Horticulturae* 700:91–96.

Hopkins, R.J., N.M. van Dam, and J.J.A. van Loon. 2009. Role of glucosinolates in insect-plant relationships and multitrophic interactions. *Annual Review of Entomolology* 54:57–83.

IFOAM. 2005. The IFOAM norms for organic production and processing version 2005. Bonn, Germany: IFOAM. Accessed March 3, 2011. http://www.ifoam.org/about ifoam/standards/norms.html.

Inomata, N. 1978. Production of interspecific hybrids between *Brassica campestris* and *Brassica oleracea* by culture *in vitro* of excised ovaries. III. Effects of coconut milk and casein hydrolysate on the development of excised ovaries. *Japanese Journal of Genetics* 53:1–11.

Itulya, F.M., and J.N. Aguyoh. 1998. The effects of intercropping kale with beans on yield and suppression of redroot pigweed under high altitude conditions in Kenya. *Experimental Agriculture* 34:171–176.

Jimenezosornio, J.J., and S.R. Gliessman. 1987. Allelopathic interference in a wild mustard (*Brassica campestris* L) and broccoli (*Brassica oleracea* L var *Italica*) intercrop agro-ecosystem. *ACS Symposium Series* 330:262–274.

Kage, H., C. Alt, and S. Stutzel. 2003. Aspects of nitrogen use efficiency of cauliflower I. A simulation modelling based analysis of nitrogen availability under field conditions. *Journal of Agricultural Science* 141:1–16.

Lammerts van Bueren, E.T., P.C. Struik, M. Tiemens-Hulscher, and E. Jacobsen. 2003. Concepts of intrinsic value and integrity of plants in organic plant breeding and propagation. *Crop Science* 43:1922–1929.

Lammerts van Bueren, E.T., C. ter Berg, and E.H.G. Bremer. 2004. Goede kool uit Noord-Holland: Vergelijkend rassenonderzoek voor Rode en Witte Kool voor Biologische Bedrijfssystemen 2001-2003. [Good cabbage from the North of The Netherlands. Variety trialling with white and red head cabbage cultivars for organic farming systems 2001-2003.] Report G42. Driebergen, The Netherlands: Louis Bolk Institute.

Lammerts van Bueren, E.T., and M. Haring. 2009. What is protoplast fusion and what are the objections against protoplast fusion from an organic point of view. *In*: Rey, F., and M. Gerber (eds.) Strategies for a future without cell fusion techniques in varieties applied in organic farming. pp. 5–8. Paris, France: ITAB and ECO-PB.

Nakanishi T., and K. Hinata. 1973. An effective time for CO_2 gas treatment in overcoming self-incompatibility in *Brassica*. *Plant and Cell Physiology* 14:873–879.

Nelson, R., and P.N. Achar. 2001. Stimulation of growth and nutrient uptake by VAM fungi in *Brassica oleracea* var. *capitata*. *Biologia Plantarum* 44:277–281.

Ogura, H. 1968. Studies on the new male sterility in Japanese radish, with special references to the utilization of this sterility towards practical raising of hybrid seed. *Memoirs of the faculty of agriculture, Kagoshima University* 6:39–78.

Ordás, A. and M.E. Cartea. 2008. Kale and cabbage. *In*: Prohens, J., and F. Nuez (eds.) Vegetables I handbook of plant breeding, volume 1, part 2. pp. 119–149. New York: Springer.

Pavan, S., E. Jacobsen, R.G.F. Visser, and Y. Bai. 2010. Loss of susceptibility as a novel breeding strategy for durable and broad-spectrum resistance. *Molecular Breeding* 25:1–12.

Petersen, J., R. Belz, F. Walker, and K. Hurle. 2001. Weed Suppression by release of isothiocyanates from turnip-rape mulch. *Agronomy Journal* 93:37–43.

Piao, Z., N. Ramchiary, and Y.P. Lim. 2009. Genetics of clubroot resistance in *Brassica* species. *Journal of Plant Growth Regulation* 28:252–264.

Prakash, S., S.R. Bhat, C.F. Quiros, P.B. Kirti, and V.L. Chopra. 2009. Brassica and its close allies: Cytogenetics and evolution. *Plant Breeding Reviews* 31:21–187.

Regvar, M., K. Vogel, N. Irgel, T. Wraber, U. Hildebrandt, P. Wilde, and H. Bothe. 2003. Colonization of pennycresses (*Thlaspi* spp.) of the Brassicaceae by arbuscular mycorrhizal fungi. *Journal of Plant Physiolology* 160:615–626.

Renaud, E.N.C., E.T. Lammerts van Bueren, J. Jiggins, C. Maliepaard, J. Paulo, J.A. Juvik, and J.R. Myers. 2010. Breeding for specific bioregions: A genotype by environment study of horticultural and nutritional traits integrating breeder and farmer priorities for organic broccoli cultivar improvement. pp. 127–130. *In*: Goldringer, I., J. Dawson, F. Rey, and A. Vettoretti (eds.) Breeding for resilience: A strategy for organic and low-input farming systems? EUCARPIA 2nd Conference of the Organic and Low-Input Agriculture Section. December 2010 1–3, Paris France. Paris France: ITAB and INRA.

Sigareva, M.A., and E.D. Earle. 1997. Direct transfer of a cold-tolerant Ogura male-sterile cytoplasm into cabbage (*Brassica oleracea* ssp. *capitata*) via protoplast fusion. *Theoretical and Applied Genetics* 94:213–220.

Stoner, K.A. 1992. Density of imported cabbageworms (Lepidoptera: Pieridae), cabbage aphids (Homoptera: Aphididae), and flea beetles (Coleoptera: Chrysomelidae) on glossy and trichome bearing lines of *Brassica oleracea*. *Journal of Economic Entomology* 85:1023–1030.

Vaughn, S.F., and R.A. Boydston. 1997. Volatile allelochemicals released by crucifer green manures. *Journal of Chemical Ecology* 23:2107–2116.

Voorrips, R.E., M.C. Jongerius, and H.J. Kanne. 1997. Mapping of two genes for resistance to clubroot (*Plasmodiophora brassicae*) in a population of doubled haploid lines of *Brassica oleracea* by means of RFLP and AFLP markers. *Theoretical and Applied Genetics* 94:75–82.

Voorrips, R.E., G. Steenhuis-Broers, M. Tiemens-Hulscher, and E.T. Lammerts van Bueren. 2008. Plant traits associated with resistance to *Thrips tabaci* in cabbage (*Brassica oleracea* var *capitata*). *Euphytica* 163:409–415.

Walters T.W., M.A. Mutschler, and E.D. Earle. 1992. Protoplast fusion-derived Ogura male-sterile cauliflower with cold tolerance. *Plant Cell Reports* 10:624–628.

17 Onions: Breeding Onions for Low-Input and Organic Agriculture

Olga E. Scholten and Thomas W. Kuyper

Introduction

Onion, *Allium cepa* L. (Alliaceae), is one of the main vegetable crops worldwide in terms of production value (FAOSTAT, 2010). The crop was domesticated about 3,000 to 4,000 years BCE (Shigyo and Kik, 2008). Plants are grown for their bulbs and mostly come from seed, but also sets (small bulbs) and transplants are used. Cultivars differ in adaptation to photoperiod and temperature, storage ability, dry matter content, flavor, and skin color (Brewster, 1994). Depending on geographic location, long-day, intermediate-day, or short-day onions are cultivated. Cultivars are available as open-pollinated or F_1-hybrid cultivars based on naturally derived cytoplasmic male sterile (CMS) mother lines.

Bulb onions are produced from the subarctic regions of northern Finland to the humid tropics, although they are best adapted to production in subtropical and temperate areas (Brewster, 1994). In 2008, the largest producers of dry onions were China and India, with production levels of 20.8 and 13.6 million Mg, respectively (FAOSTAT, 2010). These two countries together account for almost 60% of the onions produced by the top-20 onion-producing countries. The United States is third on the list with a production of 3.4 million Mg. The largest European producing country is the Netherlands with a production of almost 1.1 million Mg. These are long-storage, spring-sown, pungent onions of which large quantities are exported (Brewster, 1994; http://statline.cbs.nl/statweb/). Organically grown onions are an important commodity in the Netherlands, accounting for 2% of the total acreage of onion production.

Worldwide, there are many onion cultivars available to growers. In industrialized agricultural communities, breeding companies create their onion cultivars and conduct seed production mainly under high input conventional conditions. Organic onion seed production is costly due to the fact that onion is a biannual crop and weed management in onion seed production is labor intensive (Wolf et al., 2005). In the Netherlands, since 2007 organic growers are no longer granted derogations to use conventionally produced seed for organic yellow onion production. Although the market for organic onion seed is limited, Dutch seed companies successfully offer organically produced seed of several onion cultivars (open pollinated and F_1 hybrids) to Dutch growers. Cultivars that are available for organic growers are published in the Dutch National Annex (www.biodatabase.nl).

Organic Crop Breeding, First Edition. Edited by Edith T. Lammerts van Bueren and James R. Myers.
© 2012 John Wiley & Sons, Inc. Published 2012 by John Wiley & Sons, Inc.

Dutch breeding companies are interested in learning more about options to develop new cultivars for organic and low-input farming in order to contribute to a more sustainable agriculture. In addition, they collaborate in Dutch breeding research programs for organic agriculture. This review focuses on issues related to disease management and the opportunities for breeding for organic and low-input conditions. We focus particularly on the problem that onion is a very nutrient-demanding crop with low uptake efficiency. The need for much higher nutrient use efficiency in onion also arises from the fact that phosphorus (P) will become less available in the coming decades (Cordell et al., 2009).

Robust Onion Cultivars

During cultivation, onion is exposed to numerous biotic and abiotic stresses. Biotic stress includes diseases, such as downy mildew (caused by *Peronospora destructor*), leaf blight (caused by *Botrytis squamosa*), and Fusarium basal rot (caused by *Fusarium oxysporum,* as well as other species of that genus; e.g., Galván et al., 2008), and pests, such as the onion fly (*Delia antiqua*) and thrips (*Thrips tabaci*). Thrips may also transmit Iris yellow spot virus (IYSV; Gera et al., 1998), which may lead to yield losses of up to 60% when the crop is severely infested (Gent et al., 2006). During the growing season, organic growers have very little if any possibility of managing diseases compared to conventional growers (Osman et al., 2008). Therefore, breeding for disease resistance is highly desired. In onion, genetic variation for disease resistance is limited. Resistance has been described for Fusarium basal rot and several studies have been carried out to unravel the genetic basis of resistance (reviewed by Cramer, 2000). These studies showed that one to several genes are involved in reduced infection levels. Absolute resistance to Fusarium basal rot was not found, however. In species related to onion, high levels of resistance have been described for various diseases: In *A. roylei* Stearn and *A. fistulosum* L., resistance was found to *B. squamosa* (Bergquist and Lorbeer, 1971; Currah and Maude, 1984; De Vries et al., 1992) and to *P. destructor* (Kofoet et al., 1990; Van Heusden et al., 2000). In *A. fistulosum* resistance was also found to *F. oxysporum* (Galván et al., 2008; Holz and Knox Davies, 1974). The use of such related species in onion breeding programs is still limited, but serves as a potentially useful source of resistance that could be exploited. The first successful example of use of related species in onion breeding is the release of cultivars carrying an introgression from *A. roylei* against downy mildew (Scholten et al., 2007).

Breeding for organic cultivation not only focuses on breeding for disease resistance, but also more generally on breeding for robustness. Robust onion cultivars are reliable in yielding and able to adapt to varying and unfavorable growing conditions – adapted from a definition by Star et al. (2008) used in breeding for animal robustness. Variation in uniformity during the growing season, resistance to pests and diseases, and yield and quality during storage between onion cultivars grown under organically and conventionally managed conditions were studied in field studies in the Netherlands from 2001 to 2004. The aim of these studies was to investigate the need of testing onion cultivars in Official Variety Trials specifically under organic conditions (Osman et al., 2005). Experiments were carried out during four years with 16 cultivars grown every year. Although phenotypic variation was found for several of the traits examined, none of these traits was significantly influenced by the management system. The only trait in which cultivars differed phenotypically, and that was considered of importance for additional inclusion in the Official Cultivar Trials, was leaf erectness (Osman et al., 2005). Leaf erectness is an important trait for organic farmers, as leaves that are more erect enable farmers to mechanically weed between the rows without damaging the crop. Differential ranking of cultivars under both conditions demonstrated that those with the highest yield under conventional conditions were different from those with the highest yield under organic conditions. Specific selection for onion yield under organic agriculture apparently will be needed to develop cultivars optimally adapted to organic production.

Onion yields in conventional and organic farming systems in the Netherlands differ significantly. Yields range between 50 to 80 ton ha^{-1} under conventional conditions and between 25 to 40 ton ha^{-1} under organic conditions (Galván et al., 2009). In years with severe infestation by downy mildew, yields were even lower under organic conditions when resistant cultivars were not available. Disparities were partly caused by differences in planting density due to wider row spacing in organic onion compared to conventional systems, so as to avoid producing a canopy that was too dense and moist, with the aim of reducing disease pressure (e.g., downy mildew) and enabling mechanical weeding (Van den Broek, 2008). In addition, organic or low-input farming deals with substantially lower nutrient inputs. For example, in the Netherlands, 80 kg of N ha^{-1} is applied in organic onion cultivation, compared to 100 to 140 kg ha^{-1} in conventional cultivation systems. To build and sustain adequate inherent soil fertility, animal manure, plant composts and legume crops are used (Clark et al., 1998). Slow release of nutrients is typical for these organic amendments. Soil temperature and moisture are main determinants of nitrogen (N) mineralization, and therefore N availability during the growing season, especially in spring, can be less well regulated in organic systems. Phosphorus availability is another limiting issue for onion production, due to the common occurrence of poorly developed root systems.

Breeding for Improved Nutrient Acquisition

Lynch (2007) made a strong plea for breeding for improved nutrient acquisition through root traits as essential for a new Green Revolution. Improved nutrient acquisition is especially relevant for those farmers who have no fertilizers (such as in many developing countries) or who prefer limited use of fertilizers (as in organic agriculture). The impending global shortage of fertilizer, especially of P fertilizer (Cordell et al., 2009; Gilbert, 2009), makes an increased efficiency in nutrient acquisition and utilization even more imperative. We discuss two promising approaches for plant breeding, (1) improvement of root architecture and (2) enhanced interactions with beneficial root-inhabiting organisms such as arbuscular mycorrhizal (AM) fungi. At the end of this chapter we will reflect on potential trade-offs that such breeding efforts imply.

Breeding for Improved Root Architecture in Onion

A major characteristic of onion is its very poorly developed root system. The inevitable consequence of this poor root system is that for high onion productivity large amounts of fertilizer must be applied. The root system of onion is shallow and consists of short, thick stem-borne roots and poorly developed lateral roots without root hairs (Jones and Mann, 1963). Uptake of nutrients with low mobility in the soil solution (i.e., nutrients that are mainly delivered to the root surface by diffusion) is therefore very limited unless these nutrients are delivered in ample quantities. De Melo (2003) investigated variation for root traits with the ultimate aim of using this variation in a breeding program for onion cultivars with improved root systems. In total, 16 onion populations originating from Eastern Europe and the Netherlands and three accessions of *A. fistulosum* were studied in pot-experiments in the greenhouse. The three *A. fistulosum* accessions had a significantly larger number of stem-borne and lateral roots than did all onion cultivars. Among onion cultivars, virtually no significant variation was found. Therefore, De Melo suggested that introgression of genes from *A. fistulosum* into onion could improve the onion root system. To test the effect of N availability on the root system, an experiment was conducted with onion and *A. fistulosum* under two levels of N. In onion, low N levels resulted in much smaller root length density compared to high N levels, whereas N levels hardly affected root length density of *A. fistulosum*. Above-ground development was also affected by the lower N levels for onions, and again this was not the case in *A. fistulosum*. These results strengthened the hypothesis that *A. fistulosum* could be used to breed onions with an improved root system.

As direct crosses between *A. fistulosum* and onion do not result in fertile offspring, *A. roylei* needs to be used as a bridge species. Onion is diploid with $2n = 16$ chromosomes, and the other species have identical chromosome numbers (Kik, 2000). A tri-hybrid population was obtained from a cross between onion and the F_1 hybrid *A. roylei* × *A. fistulosum* (Khrustaleva and Kik, 1998). Genome in situ hybridization studies (GISH) with the tri-hybrid population and back-crosses between these plants and *A. cepa* showed that several recombination events randomly occurred between the chromosomes of the three species. These results indicated that introgression of traits from *A. fistulosum* into onion was feasible (Khrustaleva and Kik, 2000). Genotypes of the tri-hybrid population were phenotypically scored for root traits that were evaluated at full vegetative growth to assess the genetic basis of the root system of *A. fistulosum*. Quantitative Trait Loci (QTLs) were found for number of lateral roots per stem-borne root and relative length for fine roots (De Melo, 2003). Although population size in that study was too low to study genetics of root architecture (only 49 individuals were used), genotypes clearly differed for this trait. In a subsequent study with more individuals of the population (i.e., 68) a QTL was located for number of stem-borne roots (Galván et al., 2011). These results indicate that breeding for improved root systems in onion using the interspecific hybrid *A. roylei* × *A. fistulosum* is feasible.

Breeding for Enhanced Benefit by Arbuscular Mycorrhizal Symbioses in Onion

Roots of most agricultural crops, including all species of the genus *Allium*, are colonized by a specific group of fungi of phylum Glomeromycota. These fungi, collectively known as arbuscular mycorrhizal (AM) fungi enter into a mutualistic symbiosis with plant roots. The plant provides carbon to the fungus while the fungus provides the plants with nutrients, especially those of low mobility such as P and zinc. Carbon flow to the AM fungus usually ranges between 5 to 10% of photosynthesis, but there are important feedback loops between AM fungi and photosynthesis (Kaschuk et al., 2009). Consequently, mycorrhizal plants are larger and the mass fraction of P in leaves is also often higher. Plant species differ in the degree to which they benefit from these mycorrhizal associations. More importantly, there is also genetic variation within plant species in the benefit derived from mycorrhiza. Genetic variation for mycorrhizal benefit was first described for cereals (Hetrick et al., 1993; Kaeppler et al., 2000). Genetic variation has also been demonstrated for at least four species of *Allium*, i.e., *A. cepa* (Powell et al., 1982), *A. fistulosum* (Tawaraya et al., 2001), *A. porrum* (Sasa et al., 1987), and *A. vineale* (Ronsheim and Anderson, 2001). The presence of genetic variation indicates that natural or artificial selection can favor or select against traits associated with this symbiosis.

Because organic farming systems rely on lower levels of nutrient inputs than conventional agriculture, and because mycorrhizal benefit is larger under nutrient-poor conditions, the question has been raised of how breeding for mycorrhizal benefit can be integrated in breeding experiments for organic agriculture. Another question is whether the inherent genetic variation of onion is sufficient for such breeding programs, or whether introgression of genes from related species into onion is the better way forward. The latter approach was followed by Galván et al. (2011), who studied the above-mentioned tri-hybrid population *A. cepa* × (*A. roylei* × *A. fistulosum*) during two consecutive years for mycorrhizal benefit. In both years large variation in genetic benefit was observed. Ranking of genotypes in terms of benefit was significantly correlated between these years. However, a further analysis also suggested complications that demand reflection and these will be discussed below.

The classical method to assess mycorrhizal benefit (a relative measure, based on a comparison of mycorrhizal and non-mycorrhizal plants; see Janos, 2007) is problematic. The main problem is that high responsiveness can be caused either by good performance in the mycorrhizal conditions or by poor performance in the non-mycorrhizal condition (see fig. 17.1). Breeding for yield stability (which includes breeding for high yields under conditions of variable mycorrhizal inoculum potential) will likely result in an improved performance in the non-mycorrhizal conditions and consequently into a reduced mycorrhizal responsiveness. We consider it unlikely that poor performance in the non-mycorrhizal condition is a desirable trait. This problem has clearly hampered the evaluation of changes in mycorrhizal responsiveness in cereals. Contrary to the common knowledge that reduced mycorrhizal responsiveness of modern cereal cultivars (compared to landraces) was caused by

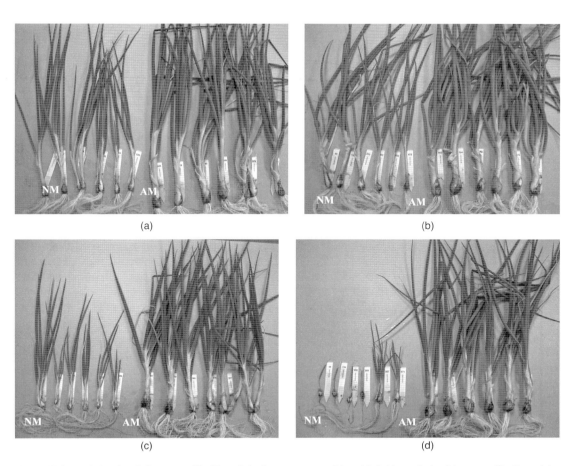

Figure 17.1 Variation in relative mycorrhizal benefit in four genotypes of the tri-hybrid population [*A. cepa* × F_1 (*A. roylei* × *A. fistulosum*)]. Six replicates of each genotype were grown without (left) or with (right) mycorrhizal fungi. Different genotypes varied in mycorrhizal benefit from low (a & b), through medium (c) to high (d); see also Galván et al. (2011).

selection against mycorrhiza in a P-rich environment, we propose that reduced responsiveness was a consequence of selection for yield stability (Hetrick et al., 1993; Kaeppler et al., 2000). For onion and its relatives the situation is somewhat different.

Species of *Allium* possess a very high dependency on arbuscular mycorrhiza, implying that these plants are unable to complete their life cycle in the absence of AM fungi, due to insufficient P uptake and hence insufficient growth. The extreme dependency on mycorrhiza translates into an equally high responsiveness, which was demonstrated for the *Allium* tri-hybrid population (Galván et al., 2011) and also for onion (Charron et al., 2001), leek (Sasa et al., 1987) and *A. fistulosum* (Tawaraya et al., 2001). The weak performance of onion and relatives under non-mycorrhizal conditions automatically translates into high responsiveness. Galván et al. (2011) therefore proposed to use the absolute benefit, i.e., the yield increment, as a measure to describe the genotypic response to AM fungi. For the tri-hybrid, plant weight increased under mycorrhizal conditions, suggesting that selection for plant weight (or yield) in the presence of mycorrhiza generally suffices for cultivar selection. Since all onion fields in the Netherlands on fertile clay soil were mycorrhizal independent of management systems applied (Galván et al., 2009), one could conclude that the selection environment to breed for mycorrhizal responsiveness is of minor importance. These results also provide evidence that modern onion breeding did not select against the response to mycorrhizal fungi. This is in contrast to wheat colonized by AM fungi, where the root length was 40% higher in organic farming systems than in conventional systems (Mäder et al., 2002).

We also investigated the response to mycorrhizal fungi in the field under different levels of P. In an organic field, two levels of P were compared (15 versus 31 mg P_2O_5 per 100 gr soil) for growth of genotypes of the tri-hybrid *Allium* population and the parental species. Results indicated that onion and *A. fistulosum* differed substantially in P uptake in similar ways as described above for N uptake, with onion performing better at higher P availability and *A. fistulosum* performing better at the lower P availability. Introgression of *A. fistulosum* genes into onion may therefore improve onion adaptation for growth under lower nutrient conditions.

Potential Trade-Offs

It is useful to elaborate on the effects of mycorrhizal symbiosis on plant productivity and root traits, both morphological and physiological, to predict the potential role of breeding for mycorrhizal symbiosis. Mycorrhizal symbiosis affects root architecture in various ways. In the study with the tri-hybrid *Allium* population, where both *A. fistulosum* and *A. roylei* have a larger root system than onion, a high mycorrhizal responsiveness was noted both in progeny plants with a small number of roots as in those with a large number of roots (Galván et al., 2011). Introgression of genes from *A. fistulosum* in onion to improve the root system while maintaining or even increasing its benefit from AM fungi could therefore contribute to improved plant performance. Increased branching of secondary roots of leek under the influence of the mycorrhizal symbiosis has also been reported (Berta et al., 1990); this was possibly mediated by hormonal changes. Mycorrhizal plants have shown altered levels of auxin, cytokinin, ethylene (either increased or decreased), and the recently reported plant hormone class strigolactones (Gómez-Roldán et al., 2008; Hodge et al., 2009). Increased root branching increases the surface that is available for mycorrhizal colonization and for uptake of immobile nutrients.

One could think that increased root length and surface would create more opportunities for soil pathogens to infest plant roots and thus may make a plant more vulnerable to root pathogens. However, increased susceptibility of onion as a result of development of a larger root system through mycorrhiza colonization of roots is not very likely. Results of a meta-analysis studying the effect of mycorrhizal fungi on plant-pathogen interactions demonstrated that in general, inoculation with AM fungi had a large negative effect on growth of pathogens, which may indicate increased host resistance (Borowicz, 2001). Alternatively, in some experiments AM colonization was reduced, suggesting competition between AM fungi and plant-pathogenic fungi. For plant-nematode interactions compared in the same analysis, AM fungi harmed sedentary nematodes and improved growth of migratory nematodes, but because only small numbers of experiments were included in the analysis, caution is necessary when interpreting these results. Potential mechanisms behind this bioprotection of mycorrhizal fungi against many soil-borne pathogens and some various nematodes were discussed by Harrier and Watson (2004). They concluded that mycorrhizal-bioprotection is the result of a complex interaction between plant, pathogen, and mycorrhizal fungus and may be the cumulative result of a number of possible mechanisms, such as enhanced crop nutrition, changes in root architecture, competition for colonization and infection sites, changes in the rhizosphere of microbial populations, and activation of plant defense responses. There is evidence for the accumulation of defensive plant compounds related to mycorrizal colonization, although to a much lower extent than in plant-pathogen interactions (Pozo and Azcón-Aguilar, 2007). These reactions are generally localized, suggesting a role in mycorrhizal establishment or control of the symbiosis. In addition, both mycorrhizal colonization and infection by pathogens change the plant's signaling defense pathways of salicylic acid and jasmonic acid (Pozo and Azcón-Aguilar, 2007). Upon germination, mycorrhizal fungi grow to the root and form appresoria at the root surface, leading to increased levels of salicylic acid, which decrease as the fungus colonizes the cortex. In cells that contain arbuscules (fungal bodies in plant cells that facilitate nutrient exchange), the levels of jasmonic acids increase. It is difficult to predict what the consequences would be of breeding programs that might take the mycorrhizal symbiosis into account. The association leads generally to a reduction of damage caused by soil-borne pathogens, but effects on above-ground organisms depend on the pathogenicity mechanism (Pozo and Azcón-Aguilar, 2007). The current consensus is that mycorrhizal fungi induce resistance to necrotrophic pathogens and generalist chewing insects but not against biotrophs.

The study by Galván et al. (2011) suggested another potential trade-off. Onion is grown for its development of large, single bulbs, whereas *A. fistulosum* develops a long, single, and only slightly enlarged pseudo-stem; *A. roylei* produces small bulbs. Genotypic segregation of the tri-hybrid population ranged from types with a high degree of bulbing to types that produced no more than a slightly enlarged pseudo-stem (Galvan et al., 2011; see also fig. 17.2). Genotypes with the highest degree of bulbing showed lower mycorrhizal benefit than genotypes with medium or low degrees of bulbing. This phenomenon was most apparent among genotypes that combined high bulbing with early leaf senescence. Whether there is a causal relationship between mycorrhizal responsiveness and early bulbing/early leaf senescence is not yet clear. Leaf production places a high demand for nutrients, whereas bulbs are carbohydrate-storage organs. The contribution of mycorrhizal fungi to the development of leaves and bulbs likely differs, and further research on this issue is needed. The finding, however, of several genotypes that combine a high degree of bulbing with large mycorrhizal responsiveness is promising.

Mycorrhizal Symbiosis and Product Quality

Mycorrhizal symbiosis not only enhances uptake of the macronutrient P, but it also has been shown to play a role in the uptake of other nutrients as well. A mycorrhizal role in the uptake of sulfur has been demonstrated for various *Allium* species. Increased sulfur uptake translates into higher levels of organosulfur compounds and hence into organoleptic qualities and health-promoting

Figure 17.2 Four genotypes of the tri-hybrid population [*A. cepa* × F_1 (*A. roylei* × *A. fistulosum*)] showing phenotypic variation for bulbing and number of stems.

effects. The mycorrhizal symbiosis also contributes to biofortification of crops with regard to micronutrients (zinc, copper, etc.). Various species of *Allium* derive their importance from micronutrients such as selenium, so the question is whether the mycorrhizal symbiosis can play a role in increasing selenium uptake and therefore selenium concentration within plant tissues. Enhanced uptake and thus enhanced product quality has been shown for garlic by Larsen et al. (2006). Breeding for mycorrhizal benefits could therefore contribute to increased food quality, an important area for further research.

Conclusion

Genetic variation in morphological and physiological root traits that allows plants to more efficiently acquire and utilize nutrients has always been of major interest to plant breeders. The need to further develop that knowledge in order to feed a growing world population that is simultaneously faced with declining supplies of fertilizers (Cordell et al., 2009) is evident. A specific focus on breeding (and redesigning crops in some cases) is required, as the large amount of genetic variation that allows plants to perform well under conditions of reduced nutrient supply has not yet been effectively used (Lynch, 2007). From a more holistic point of view, breeding should not only involve the portion of a plant's genome that is responsible for uptake and use of nutrients, but also the part of the plant's genome that is involved in beneficial interactions in the rhizosphere, which allows plants improved access to scarce nutrients. Onion, like other species of the same genus (e.g., leek) provides excellent opportunities for organic plant breeding in this respect. Compared to cereals, where a trade-off between improved root systems and enhanced benefit from mycorrhizal symbiosis has been noted (Gao et al., 2008; Kaeppler et al., 2000), onion and its allies provide opportunities to breed for better adaptation to low input and organic agriculture by improving both the root system as well as enhancing the mycorrhizal benefit. Such combined approaches may work out beneficially both for productivity and for product quality and also for health-promoting effects of onion and its relatives.

References

Bergquist, R.R., and J.W. Lorbeer. 1971. Reaction of *Alliums* pp. and *Allium cepa* to *Botryiotinia* (*Botrytis*) *squamosa*. *Plant Disease Reporter* 55:394–398.
Berta, G., A. Fusconi, A. Trotta, and S. Scannerini. 1990. Morphogenetic modification induced by the mycorrhizal fungus *Glomus* strain E_3 in the root system of *Allium porrum*. *New Phytologist* 114:207–215.
Borowicz, V. 2001. Do arbuscular mycorrhizal fungi alter plant-pathogen relations? *Ecology* 82:3057–3068.
Brewster, J. 1994. Onions and other vegetable *Alliums*. *In*: Atherton, J. (ed.) Crop production science horticulture, pp. 236. Wallingford, UK: CAB International.
Charron, G., V. Furlan, M. Bernier-Carou, and G. Doyon. 2001. Response of onion plants to arbuscular mycorrhizae. 1. Effects of inoculation method and phosphorus fertilization on biomass and bulb firmness. *Mycorrhiza* 11:187–197.
Clark, M.S., W.R. Horwath, C. Shennan, and K.M. Scow. 1998. Changes in soil chemical properties resulting from organic and low-input farming practices. *Agronomy Journal* 90:662–671.
Cordell, D., J.-O. Drangert, and S. White. 2009. The story of phosphorus: Global food security and food for thought. *Global and Environmental Change* 19:292–305.
Cramer, C.S. 2000. Breeding and genetics of Fusarium basal rot resistance in onion. *Euphytica* 115:159–166.
Currah, L., and R.B. Maude. 1984. Laboratory tests for leaf resistance to *Botrytis squamosa* in onions. *Annals of Applied Biology* 105:277–283.
De Melo, P.E. 2003. The root systems of onion and *Allium fistulosum* in the context of organic farming: A breeding approach, pp. 136. PhD Thesis. Wageningen Agricultural University, The Netherlands.
De Vries, J.N., W.A. Wietsma, and T. de Vries. 1992. Introgression of leaf blight resistance from *Allium roylei* Stearn into onion (*A. cepa* L.). *Euphytica* 62:127–133.
FAOSTAT. 2010. Food and agricultural commodities production. Accessed April 2011. http://faostat.fao.org/.
Galván, G.A., C.F.S. Koning-Boucoiran, W.J.M. Koopman, K. Burger, P.H. González, C. Kik, and O.E. Scholten. 2008. Genetic variation among *Fusarium* isolates from onion and resistance to Fusarium basal rot in related *Allium* species. *European Journal of Plant Pathology* 121:499–512.
Galván, G.A., T.W. Kuyper, K. Burger, L.C.P. Keizer, R.F. Hoekstra, C. Kik, and O.E. Scholten. 2011. Genetic analysis of the interaction between *Allium* species and arbuscular mycorrhizal fungi. *Theoretical and Applied Genetics* 122:947–960.
Galván, G.A., I. Parádi, K. Burger, J. Baar, T.W. Kuyper, O.E. Scholten, and C. Kik. 2009. Molecular diversity of arbuscular mycorrhizal fungi in onion roots from organic and conventional farming systems in the Netherlands. *Mycorrhiza* 19:317–328.
Gao, X., T.W. Kuyper, F. Zhang, C. Zou, and E. Hoffland. 2008. How does aerobic rice take up zinc from low zinc soil? Mechanisms, trade-offs, and implications for breeding. *In*: G.S. Banuelos and Z.-Q. Lin (eds.) Development and uses of biofortified agricultural products, pp. 153–170. Boca Raton, FL: CRC Press.
Gent, D.H., L.T. Du Toit, S.F. Fichtner, S. Krishna Mohan, H.R. Pappu, and H.F. Schwartz. 2006. Iris yellow spot virus: An emerging threat to onion bulb and seed production. *Plant Disease* 90:1468–1480.
Gera, A., J. Cohen, R. Salomon, and B. Raccah. 1998. Iris yellow spot tospovirus detected in onion (*Allium cepa*) in Israel. *Plant Disease* 82:127.
Gilbert, N. 2009. The disappearing nutrient. *Nature* 461:716–718.
Gómez-Roldán V., S. Fermas, P.B. Brewer, V. Puech-Pages, E.A. Dun, J.-P. Pillot, F. Letisse, R. Matusova et al. 2008. Strigolactone inhibition of shoot branching. *Nature* 455:189–194.

Harrier, L.A., and C.A. Watson. 2004. The potential role of arbuscular mycorrhizal (AM) fungi in the bioprotection of plants against soil-borne pathogens in organic and/or other sustainable farming systems. *Pest Management Science* 60:149–157.

Hetrick, B.A.D., G.W.T. Wilson, and T.S. Cox. 1993. Mycorrhizal dependence of modern wheat cultivars and ancestors: A synthesis. *Canadian Journal of Botany* 71:512–518.

Hodge, A., G. Berta, C. Doussan, F. Merchan, and M. Crespi. 2009. Plant root growth, architecture and function. *Plant and Soil* 321:153–187.

Holz, G. and P.S. Knox-Davies. 1974. Resistance of onion selections to *Fusarium oxysporum* f. sp. *cepea*. *Phytophylactica* 6:153–156.

Janos, D.P. 2007. Plant responsiveness to mycorrhizas differs from dependence upon mycorrhizas. *Mycorrhiza* 17:75–91.

Jones, H.A., and L.K. Mann. 1963. Onions and their allies. pp. 286. New York: Interscience Publishers, Inc.

Kaeppler, S.M., J.L. Parke, S.M. Mueller, L. Senior, C. Stuber, and W.F. Tracy. 2000. Variation among maize inbred lines and detection of quantitative trait loci for growth at low phosphorus and responsiveness to arbuscular mycorrhiza fungi. *Crop Science* 40:358–364.

Kaschuk, G., T.W. Kuyper, P.A. Leffelaar, M. Hungria, and K.E. Giller. 2009. Are the rates of photosynthesis stimulated by the carbon sink strength of rhizobial and arbuscular mycorrhizal symbioses? *Soil Biology and Biochemistry* 41:1233–1244.

Khrustaleva, L.I., and C. Kik. 1998. Cytogenetical studies in the bridge cross *Allium cepa* × *(A. fistulosum* × *A. roylei)*. *Theoretical and Applied Genetics* 96:8–14.

Khrustaleva, L.I., and C. Kik. 2000. Introgression of *Allium fistulosum* into *A. cepa* mediated by *A. roylei*. *Theoretical and Applied Genetics* 100:17–26.

Kik, C. 2002. Exploitation of wild relatives for the breeding of cultivated *Allium* species. *In*: H.D. Rabinowitch and L. Currah (eds.), *Allium* crop science: Recent advances. pp. 81–100. Oxford, UK: CABI Publishing.

Kofoet, A., C. Kik, W.A. Wietsma, and J.N. de Vries. 1990. Inheritance of resistance to downy mildew (*Peronospora destructor* [Berk.] Casp.) from *Allium roylei* Stearn in the backcross *Allium cepa* L. × *(A. roylei* × *A. cepa)*. *Plant Breeding* 105:144–149.

Lammerts van Bueren, E.T., P.C. Struik, and E. Jacobsen. 2002. Ecological aspects in organic farming and its consequences for an organic crop ideotype. *Netherlands Journal of Agricultural Science* 50:1–26.

Larsen, E.H., R. Lobinski, K. Burger-Meyer, M. Hansen, R. Ruzik, L. Mazurowska, P.H. Rasmussen, J.J. Sloth, O.E. Scholten, and C. Kik. 2006. Uptake and speciation of selenium in garlic cultivated in soil amended with symbiotic fungi (mycorrhiza) and selenate. *Analytical and Bioanalytical Chemistry* 385:1098–1108.

Lynch, J.P. 2007. Roots of the second green revolution. *Australian Journal of Botany* 55:493–512.

Mäder, P., A. Fließbach, D. Dubois, L. Gunst, P. Fried, and U. Niggli. 2002. Soil fertility and biodiversity in organic farming. *Science* 296:1694–1697.

Osman, A.M., L. van den Brink, R.C.F.M. van den Broek, W. van den Berg, and E.T. Lammerts van Bueren. 2005. Passende Rassen. Rassenonderzoek voor Biologische Bedrijfssystemen 2001–2004 – Zaaiuien & Zomertarwe. pp. 143. Report of Praktijkonderzoek Plant & Omgeving. The Netherlands: Wageningen UR and Louis Bolk Institute.

Powell, C.L., G.E. Clark, and N.J. Verberne. 1982. Growth response of four onion cultivars to several isolates of VA mycorrhizal fungi. *New Zealand Journal of Agricultural Research* 25:465–470.

Pozo, M.J., and C. Azcón-Aguilar. 2007. Unravelling mycorrhiza-induced resistance. *Current Opinion in Plant Biology* 10:393–398.

Ronsheim, M.L., and S.E. Anderson. 2001. Population-level specificity in the plant-mycorrhizae association alters intraspecific interactions among neighboring plants. *Oecologia* 128:77–84.

Sasa, M., G. Zahka, and I. Jakobsen. 1987. The effect of pretransplant inoculation with VA mycorrhizal fungi on the subsequent growth of leeks in the field. *Plant and Soil* 97:279–283.

Scholten, O.E., A.W. van Heusden, L.I. Khrustaleva, K. Burger-Meijer, R.A. Mank, R.G.C. Antonise, L.J. Harrewijn et al. 2007. The long and winding road leading to the successful introgression of downy mildew resistance into onion. *Euphytica* 156:345–353.

Shigyo, M., and C. Kik. 2008. Onion. *In*: J. Prohens and F. Nuez (eds.) Vegetables, vol. 2: Handbook of plant breeding. pp. 121–162. Verlag, Berlin: Springer.

Star, L., E.D. Ellen, K. Uitdehaag, and F.W.A. Abrom. 2008. A plea to implement robustness into a breeding goal: Poultry as an example. *Journal of Agricultural and Environmental Ethics* 21:109–125.

Tawaraya, K., K. Tokairin, and T. Wagatsuma. 2001. Dependence of *Allium fistulosum* cultivars on the arbuscular mycorrhizal fungus, *Glomus fasciculatum*. *Applied Soil Ecology* 17:119–124.

Van Heusden, A.W., M. Shigyo, Y. Tashiro, R. Vrielink-van Ginkel, and C. Kik. 2000. AFLP linkage group assignement to the chromosomes of *Allium cepa* L. via monosomic addition lines. *Theoretical and Applied Genetics* 100:480–486.

Van den Broek, R.C.F.M. 2008. Naar een optimal teeltsysteem voor biologische zaaiuien [Toward an optimal cropping system for organic onion]. Biokennisbericht March 2008. The Netherlands: Wageningen UR and Louis Bolk Institute.

Wolf, P.L. de, M.P.J. van der Voort, S.C. van Woerden, and F.J. Munneke. 2005. Concurrentieanalyse biologisch uitgangsmateriaal [Analysis of the competitve potential of organic propagation material]. Lelystad: Praktijkonderzoek Plant & Omgeving-AGV.

Index

Entries followed by *f* indicate figures. Entries followed by *t* indicate tables. Entries followed by *b* indicate boxes.

Accessions, (genebank), 52, 81, 88, 209, 210, 230, 231, 246, 246*t*, 256, 265
Adaptation
    Local, 31, 80, 81, 86, 90, 105, 193, 198
    Narrow (specific), 31, 49, 80, 154, 208, 240
    Wide, 31, 50, 205, 208, 220
Agro-biodiversity, 46, 120, 131, 131*t*, 149
Alfalfa (*Medicago sativa*), 71, 186
Allelopathy, 64, 64*t*, 68, 72, 205, 254
Antagonist, 44, 45, 48, 245
Autonomy
    of plants, 127, 128
    of farmers, 156

Barley
    composite cross, 90, 167
    mixtures, 45, 46, 86, 87
    nitrogen-use efficiency, of, 22, 28, 29
    participatory breeding in, 105–110
    resistance in, 53
    root system, 19
    weed competition, of, 62, 68–70
    Zn-efficiency, in, 24, 27
Basic attitude towards nature, 127
    stewardship, 3, 127, 130
    participant, 127, 129
    partner, 127, 129
Bacillus thuringiensis, 44*t*, 126, 255
Back-Up farms, 194, 198
Bean (*Phaseolus*)
    composite cross, 86
    farmer selection in, 87, 111, 113
    root density, 18
    Zn-efficiency, 24
    weed competition, 71

Biodiversity, 4, 5*t*, 9, 45, 46, 77, 79, 102, 108, 129–131, 136, 148, 149, 152, 155, 243
    above-ground biodiversity, 79
    below-ground biodiversity, 79
Biodynamic plant breeding, 8, 171
Bioethics, 127
Biofortification, 270
Biofumigation crops, 46, 54
Biotechnology, 11, 126, 152–154
Blending, *see* mixing
Brassica, 251–262
    allelopathic effect, 254
    alternaria, 255, 256
    Bacillus thuringiensis, 255
    biofumigation, 46
    blackleg (*Leptospaeria maculans*), 255
    breeding techniques, 253, 257
    cabbage root fly, 255, 257
    club root (*Plasmodiophora brassicae*), 254, 255
    cover crop, 53
    heirloom, 251, 257
    insect pests, 255, 256
    insect-plant relationship, 256
    genotype-system interaction, 252
    glossy leaf type, 256
    glucosinolates, 254–257
    mycorrhiza, 24, 253
    Mycospaerella, 256
    mass selection, 253, 258, 260
    N-use efficiency, 253
    plant architecture, 254
    recurrent selection, 253
    root system, 17, 253
    self-incompatibility, 252, 257
    waxy leaf type, 6, 256
    weed suppression, 253, 254

Breeders' exemption, 126, 127, 130, 153, 155
Breeders' rights, 126, 127, 130, 234
Breeding techniques, 125–138, 154–156, 159, 163, 253
  (*see also* Selection methods, strategies)
  cellular level, 127, 128, 131, 132
  cell fusion, 132, 134, 135, 257, 258
  cisgenesis, 134, 135
  cytoplasmic male sterility (cms), 12, 99, 134, 219, 257, 258
  DNA level, 127, 131, 132, 133*t*
  embryo culture (rescue), 12, 132, 133*t*, 240, 245, 255
  gene silencing, 134
  in-vitro, 131, 132, 133*t*, 233, 245, 255, 258
  mutagenesis, 134, 135, 223
  novel breeding techniques, 134, 136
  protoplast fusion, 131, 132, 134
  reverse breeding, 134
  self-incompatibility, 233, 252, 257
  somatic hybridization, 12, 131, 134, 232, 233, 257
  suitability for organic plant breeding, 133*t*
  tilling, 117, 134
  whole plant level, 131, 133*t*, 233
Broccoli, 251–254
  breeding scheme, 259*f*
  cytoplasmic male sterility, 134
  low-nutrient, 16
  mycorrhiza, 24
  participatory breeding, 258–260
  performance of subpopulations, 259*t*
  wax, 256
  weed management, 254
Buckwheat, 24
Buffering capacity, 79, 83

Cabbage, *see* Brassica
Canola, 23, 24
Carrot, 18, 134, 143
Centralized breeding, 99–123
Cereals, 11, 15, 26, 27, 42, 45, 47, 53, 61–76, 141, 201, 208, 231, 266, 267, 271
  weed suppression, 61–76
Chickpea (*Cicer arietinum*), 24, 71, 110, 217
Clover, 45, 68, 144, 186
Co-breeding for intercropping, 54
Co-evolution, 39
Colchicine, 133*t*, 233
Collaborative breeding, 19
Combinability
  general (GCA), 176, 182
  specific (SCA), 183
Competition
  below-ground, 68
  intra-cultivar, 83
  plant-plant, 45
  weed, *see* Weed competitiveness

Competitiveness against weeds, *see* Weed competitiveness
Composite cross populations, 87, 88, 88*f*, 90, 93, 147
  definition, 78*b*
Composite populations, definition, 78*b*
Conservation, 19, 45, 46, 48, 79, 103*t*, 129
  ex-situ, 80, 94
  in-situ, 81, 94
  on-farm, 80, 81
Conservation variety, (*see also* Heirloom/heritage variety), 148–151, 154–156
Cotton (*Gossipium*), 43, 72, 155, 218
Crop competitive ability, 61–76
Crop diversification, 40, 41, 43, 44
Crop growth
  branching, 17, 72, 204, 268
  early (season) vigor, 6, 65, 68, 69, 204
  early canopy closure, 65, 204
  early crop ground cover, 62, 63, 65, 65*t*, 67
  early (rapid) growth, 17, 64*t*, 67–71, 84, 229*t*
  erectophile leaf inclination, 63, 66*f*, 65, 67
  extensive leaf display, 62, 64*t*
  leaf length, size, 63, 64*t*
  light interception, 63–65, 67, 71
  planophile leaf inclination, 6, 63, 64, 64*t*, 65, 67, 69, 254
  plant height, 7*t*, 63–65, 64*t*, 67, 69, 71, 84, 85, 91, 106, 188, 196, 204, 207, 217*f*, 218, 223
  prostate growth, 67
  robust, 6, 69, 115, 128, 184, 208, 241, 247, 255, 261, 264
  shading ability, 63–65, 67, 69–72
  tallness, 62, 63, 67
  vigorous, 5, 114, 115, 184, 186, 218, 221, 258
Cropping systems
  intercropping, 31, 40, 54, 220, 221
  reduced tillage, 31, 169
Crop-weed interaction, 61, 62, 71
Cultivar requirements, 5, 227
  flexible, 6
  morphological, 6, 54, 62, 64, 132, 139, 142, 143, 145, 151, 268, 270
  (open) plant architecture, 6, 15, 115, 146, 254
  robust, 6, 69, 115, 128, 184, 208, 241, 247, 255, 261, 264
  wax layer, 6, 256
  weed competition, *see* Weed competitiveness
Cultivar (*see also* Variety)
  diversity, 3, 17, 48, 117
  mixtures, 8, 41, 44, 45, 48, 49, 51, 52, 77, 78, 79, 82–86, 93, 108, 109, 147, 150, 222, 223
  protocol (testing, evaluation), 6, 7*t*, 147
  testing, 6, 7*t*, 69, 70, 83, 84, 86, 87, 88, 92, 108, 110, 118, 142–148, 156, 175, 178, 180–182, 184–186, 205, 207, 208, 220, 220*t*, 221, 228, 233, 264

registration, 10, 12, 45, 77, 86, 92, 101, 117, 118, 139, 142–144, 147, 148, 150–152, 155, 156, 212, 234t, 257
  value for cultivation and use (VCU), 7t, 92, 118, 143, 144, 146–149
Cytoplasmic male sterility, *see* under Breeding techniques

Decentralized breeding (evaluation), 31, 87, 90, 99–123, 155
Derogations, 47, 146, 156, 263
Diallel crossing, 85, 88, 89
Direct/Indirect selection, 15–38, 69, 167, 168, 179, 180, 208, 209
Disease (resistance), *see* Resistance
Divers(ity)
  breeding for, 42, 54, 81, 83
  breeding programs, 131, 131t
  buffering capacity, 79, 83
  crop, 46, 50, 54, 108
  cultural, 120, 131t
  ecosystem-wide, 130
  farmers' needs, 99, 101, 200
  farmers' selection, 107
  farmers' knowledge, 119
  food, 47, 54, 239
  functional (beneficial), 4, 44, 83, 119, 131t
  genetic, 8, 28, 40, 41, 50, 51, 77, 78, 79, 80, 81, 87–94, 105, 106, 111, 119, 120, 129, 131, 131t, 147, 150, 192, 193, 197, 200, 201, 210, 211, 216, 239, 242, 257, 258
  interspecific (within species), 77, 79, 130, 216, 251
  intraspecific (within a population/variety), 130, 136
  levels of, 77, 80, 81
  maintaining, 110, 156
  management, 49, 131
  microbial, 52, 53
  pathogen, 48, 51
  plant, 39, 43, 44, 45, 54
  population, 77–98
  production (farming) systems, 3, 43, 54, 61, 99, 129, 150
  resistance, in, 52, 79, 82
  soil organisms, 5t, 130, 243
  spatial, 5t, 79, 108
  temporal, 5t, 46, 79, 108
  varietal (cultivar), 5t, 8, 12, 31, 48, 49, 51, 54, 77, 80, 117, 136, 156, 200
  within-species, 77
DNA markers, *see* Molecular markers
Drought stress, 18, 206, 221
DUS criteria, 83, 118, 142, 143, 147, 148, 150–152, 221
Dwarfing genes, 19

Eco-functional intensification, 30, 32
Ecological combining ability, 86

Ecosystem approach, 129
Ecosystem-wide diversity, 130
Ecosystem services, 44, 79, 130
Emmer wheat, 170
Empowerment of farmers, 12, 102, 105, 108, 110, 113, 163, 192, 193
Environments, heterogeneous, 86, 99
Environmental variability (variance), 50, 77, 80, 100, 113, 163, 170
Epigenetic mechanisms, 31, 178
Evolutionary
  breeding, 82, 87, 89, 91, 92, 164, 167
  fitness, 87, 89, 90
  populations, 77, 88, 91
  process, 130, 154
Extrinsic values, 128b

Faba bean (field bean), 23, 30, 45, 215–225
  anti-nutritive compounds, 223
  blends of hybrids, 221, 222
  disease resistance, 221, 222
  drought resistance, 221
  Ascochyta faba, 217, 222
  downy mildew, 221
  fusarium complex, 221
  nitrogen fixation, 215, 216, 223
  pest resistance
    aphids, 216, 217, 221–223
    sitona leaf weevil, 221–223
  plant height distribution in topcross progeny, 217f
  pollination, 217, 219
  polycross progenies, correlation grain yield with, 219f
  rust (*Uromyces fabae*), 221, 222
  rhizobium, 223
  quality, 215, 217, 223
  synthetic population, 218
  topcross breeding, 217, 218
  weed competition, 216, 221–223
  weed competition, correlation grain yield with, 222f
  winter hardiness, 222
  yield stability, 219–221
Farmers' rights (privilege), 110, 130, 153
Farmer participation, 93, 99–123, 193, 220, 233, 258
Farmer led/centered breeding, 105, 111, 191, 193, 199–202
Farmer selected spikes within diverse wheat varieties, 82f
Farmer variety, 78b
Fertility
  low, 5, 18, 22, 199t
  early season, 5, 16
Food safety, 126
Food security, 110, 111, 130, 145, 155, 156, 191, 201
Food sovereignty, 130
Food values, 130
Formal breeding, 87, 99, 120, 220, 220t

Genetic diversity, 8, 28, 40, 41, 50, 51, 77–94, 105, 106, 111, 119, 120, 129, 131, 131*t*, 147, 150, 193, 197, 200, 201, 210, 211, 216, 239, 242, 257, 258
Genetic
  base, 40, 79, 111, 131, 231
  engineering (GE), 4, 125, 126, 127, 129, 131, 132, 134–136
  erosion, 150
  heterogeneity, 114, 167
  variability, 24, 99, 105, 111, 113, 117, 168, 231, 253
Genetically heterogeneous populations, 79, 81, 84, 86, 89, 90, 93, 118, 130, 150
Genetically modified organism (GMO or GM) 7*t*, 126, 134, 183, 187, 188, 203, 208, 211, 212 (*see also* Genetic engineering (GE))
  contamination (outcrossing), 175, 183, 184, 187
General yielding ability (GYA), 85
General combining ability (GCA), 176, 182
General mixing ability (GMA), 85
Genotype-environment interaction (GxE), *see* Interaction
Genotype-system interaction, *see* Interaction
Genotype-tillage interaction, *see* interaction
Genotypic correlation (organic vs. conventional), 29, 30, 100, 180, 181, 181*t*
Green area index, 65
Green manure, 5*t*, 15, 16, 45, 49, 254
  grain legumes, 16, 45
Ground shading, 63, 64

Heirloom/heritage variety, 8, 9, 129, 155, 240, 241, 243, 244, 251, 257
  see also Conservation variety
Heritability, 29, 101, 108, 120, 164, 180, 181, 187, 206, 207, 209, 220*t*, 232
  broad sense, 100
  coefficient, 29
Heterogeneity
  of environments, 77, 94, 120
  genetic, 83, 114, 120, 155, 167, 168, 218, 220–222
  intra-varietal, 115
  population, 168
  of soils, 27
Heterogeneous populations, 79, 81, 84, 86, 89, 90, 93, 118, 130, 150
Heterosis, 115, 176, 218–222, 241, 257
Holistic, holism, 32, 39, 43, 99, 126–129, 271

IFOAM basic principles, 128, 201
Indigenous
  knowledge, 103*t*, 120, 129
  varieties, 130
  groups, 164
  farming systems, 192
Indirect breeding/selection, 15–32, 53, 69, 133*t*, 167, 179, 180, 209

Induced resistance, 7*t*, 51–53, 243, 244, 247
Inherent worth of organisms, 127
Insect-plant relationship, 256
Integrated pest management (IPM), 4, 5*t*, 42
Integrity
  genotypic, 135
  life/living systems, 3, 127, 131
  organic, 201
  organisms, 127
  plants, 7*t*, 127, 128, 132, 211
  plants, definition, 128*b*
  seed, 186
Intellectual property rights (IPR), 4, 11, 78, 129, 130, 139, 151–156
Interaction
  plant-plant, 83
  plant–pathogen, 83, 269
  plant-(soil)-microbe, 21, 22, 25, 26, 30
  genotype x environment (GxE), 5, 24, 62, 70, 103*t*, 129, 155, 208, 209, 202, 221, 232
  genotype x system (GxS), 208, 244, 247, 252
  genotype x tillage, 169
  genotype x year (GxY), 31
Intercropping, 31, 40, 45, 54, 111, 112*f*, 220, 221
Intra-cultivar competition, 83
Intra-varietal heterogeneity, 115
Intra-varietal erosion, 150
  Intrinsic value, *see* Value
Introgression breeding, 169, 170, 207, 209, 211, 232, 233*f*, 233, 240, 258, 264, 266, 268

Lactuca species, 17
Landrace, 28, 51, 77–81, 78*b*, 87, 89, 91, 93, 94, 149, 150, 151, 155
  barley, 68, 106–109, 109*t*
  brassicas, 251, 257
  cereals, 267
  definition, 78*b*
  lentil, 91
  maize, 181, 182, 184
  modern, 168
  tomato, 240
Laws, *see* regulation
Leaf
  angle, 65, 66*f*
  area index, 64, 65, 67, 207
  inclination, 63, 65, 67
  erectophile, 63, 65, 67
  planophile, 63, 64*t*, 64, 65, 67–69
  length, 63
  width, 67
Legumes, 15, 18, 42, 46, 49, 141, 208
  grain, 16, 45, 49, 216, 217
Lentil, 27, 91, 208, 217
Lettuce (*Lactuca*), 48, 149

Maize (corn), 9, 175–189
  corn borer, 26, 126
  energy maize, 186
  gametophytic incompatibility, 189
  general combining ability, 176, 181*t*, 182, 183
  genotypic correlation (org/conv), 180, 181
  heritability, 180*t*
  indirect selection gain, 179
  kernel color pigment, 186
  lysine, 185
  methionine, 185
  microbial endophytes, 186
  N-efficiency, 184
  nutritional value, 184, 186
  open-pollinated, 176, 177, 184
  participatory plant breeding, 87, 89, 93, 111, 141, 154
  P-efficiency, 18
  protein quality, 176, 184–186
  roots, 17, 18, 19, 21, 30
  specific combinability, 183
  sunflower, proxy for weed, 179*f*
  synthetic variety, 176, 177, 177*f*, 184, 185
  vigor, 140, 178, 184, 186
  weed competition, 176, 177, 177*f*, 168, 180, 182
Mapping studies, 22, 27, 255
Male sterility, cytoplasmic, *see* Breeding techniques
Mass selection, *see* Selection
Millet, 22, 205
Microbial endophytes, 186
Mixing ability, 84–86
  general, 85
  specific, 85
Mixtures (varietal), 8, 41, 44, 48, 49, 51, 77–79, 82, 83–87, 93, 147, 150
  apple, 48
  barley, 86, 108, 109
  bean, 221, 222*f*, 223
  definition, 78*b*
  grapevine, 48
  legume-grass, 49
  oat, 83
  potato, 83, 84
  rice, 83
  soy bean, 85
  wheat, 83, 85
Molecular (DNA) markers (assisted selection), *see* Selection methods
  mycorrhizal colonization, 24
  nutrient efficiency, 27
  neutral, 80, 81
  simple sequence repeat (SSR), 94
  stacking (pyramiding) genes, 230, 247
Multi-criteria mapping, 110
Multidisciplinary collaboration, 119

Multifunctional varieties, 119
Multilevel approach, 6
Multiline variety, 78*t*, 82, 83, 147
Multiline variety, definition, 78*b*
Multi-parental populations, 82
Multi Environment (local) trials, 108
Multiple race resistance, 83
Mycorrhizal fungi, 24, 25, 119, 169, 232, 243, 244, 247, 253, 265–271, 267*f*

Natural
  biological control, 54
  selection, 80, 86, 87, 89–92, 109, 164, 165, 167, 168, 171, 211
  crossing barriers, 131, 132
Nitrogen
  fixation/fixing, 205
  release, 63
  use efficiency, 22, 63, 68, 168, 169, 171, 232
Normative frame work of action, 127
  anthropocentric, 127
  bio-centric, 127, 128*t*
  zoocentric, 127
Nutrient
  availability, 16, 17, 22, 30, 40, 63, 113, 228
  (re)cycling, 15, 22, 46
  (use-) efficiency, 19, 27, 29, 68, 163, 168, 169, 171
  genotypic variation (e.g. in nutrient uptake, fixation, mycorrhizal colonization) for, 18, 28, 205, 232, 247, 253, 266
  management, 15, 42
  input (reduced), 15, 16
  release (slow), 16, 17, 27, 63
  uptake (efficiency, capacity), 15–19, 21, 63, 68
Nutritional
  value, 24, 88, 170, 176, 184, 186, 244
  quality, 170, 210
  traits, 210
    interaction between, 244
    taste (flavor), 7, 48, 51, 88, 117, 171, 210, 215, 229*t*, 244

Oats, 22, 45, 62, 221
On-farm
  breeding (development), 87, 92, 101, 114, 130
  conservation, 79, 80, 81, 149, 150
  evaluation (trails, testing), 91, 106, 107*t*, 108, 108
Onion, 263–272
  arbuscular mycorrhizal fungi, 265–271, 267*f*
  cytoplasmic male sterility, 134
  downy mildew (*Peronospora destructor*), 264, 265
  health promoting effects, 269, 271
  induced resistance, 269
  resistance, 264
  mycorrhizal benefit, 266, 267, 267*f*
  nematode, 269

Onion (*Continued*)
  nutrient efficiency, 264
    nitrogen, 265
    phosphorus (P), 264, 265
    sulfur, 269
  organic seed, 263
  onion fly, 264
  phenotypic variation, 270f
  quality, 264, 269–271
  related species, 264, 266
  robust, 264
  root architecture, 265–271
  thrips (*Thrips tabaci*), 264
  yield, 264, 265, 267, 268
Open pollinated variety (OP), 176, 177, 184, 240, 252, 253, 259t, 260f, 263
Organic
  agriculture, description, 4
  certified breeding program, 135
  cultivars (varieties), 72, 118, 132, 136, 146, 171, 180, 181, 228
  drivers, 3
  farming systems, 4, 5, 5t
  plant breeding, 3–14, 125–138
    history, 3–14
    funding, 10, 11
    key criteria, 131, 131t
    standards, 125, 132, 134, 135, 136
    time schedule for development, 7t
    type of programs, 8, 9, 10, 11
  plant protection, 39–59
  (basic) principles, 3, 125, 127, 128
    of health, 128
    of ecology, 129
    of fairness, 130
    of care, 129
  seed (production), 125–138
  seed regulation, 3–14, 125–138, 139–160
  (basic) values, 3, 125, 126, 127
  variety testing, 7, 146
  regulation, 4, 8, 130
    EU, 130, 204
    NOP (USDA), 4

Palatability, 49, 106, 110, 176
Patchiness of genotypes, 84
Patches, random, 84
Patents, 12, 126, 129, 151–155
Participatory plant breeding (approaches), 6, 10, 12, 31, 41, 91–93, 99–124, 131, 150, 154, 164, 167, 168, 171
  benefit-cost ratio, 109
  evolutionary, 109, 154, 167–172
  farmer-led, 105, 111, 191, 193, 195–197, 199–202
  farmer selection in sorghum, 111, 112f
  farmer selection in sorghum-bean, 111, 112f
  favorable context for, 103t
  functional approaches, 102
  gender dimension, 110
  scheme for zucchini, 116t
  project in Syria, type of selections, 107t
  project in Syria, performance of barley lines, 109t
  process approaches, 102
  northern countries, 99, 113
  southern countries, 102, 106
Participatory variety selection (PVS), 105, 106, 111
Peas (*Pisum sativum*), 23, 71, 215–217, 223
Pepper, 61
Phosphorus (P), 18, 19, 22–26, 28, 30
  efficiency, 18, 23, 24
  solubility, 23
  uptake, 18, 19, 24, 25, 28
  phosphatases, 23, 26
Phytochemical, 6, *see also* Secondary metabolites
Plant architecture, 6, 53, 115, 146, 254
  open, 6
  planophile growth habit, 6
Plant defense mechanism, 50–52
Plant ideotype, 7t, 65, 114, 115, 117, 227, 229t
Plant growth habit, 6, 65, 66f, 67, 69, 71
  erectophile, 65
  intermediate, 65
  planophile, 63, 65, 69
Plant height, 7t, 63, 64t, 65, 67, 69, 71
Population varieties, 81, 87
Population varieties, definition, 78b
Potassium (K), 18, 22
Potato (*Solanum tuberosum*), 40, 41, 61, 71, 149, 152, 227–237, 245, 246
  breeding scheme, 233f
  collaborative approach, scheme, 234t
  colorado beetle, 229t, 231
  crop ideotype, organic, 229t
  early blight (*Alternaria solani*), 228, 230, 233
  low nitrogen stress, 227, 230
  maturity, 229t, 230–233
  mixtures, 83
  mycorrhizal fungi, 232
  nitrogen efficiency, 232, 233
  late blight (*Phytophthora infestans*), 40–42, 50, 227, 228, 230, 232–234
  participatory breeding, 154, 227, 233, 234
  Rhizoctonia solani, 49, 229t, 230, 231, 233
  root system, 229, 232
  silver scurf (*Helminthosporium solani*), 231, 233
  2n-gametes, 233
  weed competition, 61, 229t
PPB, *see* participatory plant breeding
Producer-competitor model, 85
Product quality, 84, 143, 210, 269–271
Precautionary principle, 4, 43, 126, 129, 135
Public (plant) breeding, 11, 148, 154, 186, 212, 240

Push-pull system, 54
Pure line variety, 77, 84–86, 88*f*, 91, 92, 94, 117, 118, 163, 164, 168, 196, 203, 217*f*, 218, 219, 222, 240–242, 246

Quality, (product), 17, 43, 45, 47, 52, 79–81, 84, 87, 89, 91–94, 111, 115, 116, 118, 134, 136, 143, 144, 163, 168–170, 176, 180, 181, 184–187, 194, 195, 198, 203, 205, 207–210, 215, 223, 228, 229*t*, 230, 234*t*, 240, 241, 244, 252, 255, 257, 260, 264, 269, 270, 271
  appearance, 48, 83, 111, 117, 228, 231
  baking quality, 54, 68, 83, 93, 132, 144, 147, 169
  color, 84, 155, 178, 186, 191, 197, 200, 201, 215, 229*t*, 240, 256, 258, 260
  cooking quality, 215, 229*t*
  nutritional, 24, 51, 88, 163, 170, 176, 184, 186, 210, 239, 244
  organoleptic, 269
  taste, 7*t*, 48, 51, 88, 117, 171, 210, 215, 229*t*, 244
  texture, 7*t*, 117, 210
Quantitative Trait Loci (QTL), 23, 169, 207, 230, 248, 255, 266
  nodulation, 23
  N_2 fixation, 23
  number of stem-borne roots, 266

Reductionist approach, 126, 127, 129
Regulatory hurdles, 81, 92
Regulation, 3, 4, 8, 16, 42, 47, 82, 84, 92, 100, 101, 110, 119, 120, 130, 131*t*, 134, 139–156, 204, 209, 221, 227
  DUS, 83, 118, 142, 143, 147, 148, 150–152, 221
  VCU, 7*t*, 92, 118, 143, 144, 146–149
Resilience, resilient
  systems, 3, 4, 5*t*, 8, 50, 54, 79, 113, 114, 116*t*, 120, 128, 129, 131, 147
  crops, 128, 156
Resistance, 7*t*, 12, 39, 40, 41, 44*t*, 47–54, 80, 82–84, 86, 89, 90, 116, 125, 129, 141, 146, 163, 167–170, 180–183, 186, 191, 194, 195, 197, 198, 199*t*, 203, 216, 221, 222, 227, 228, 229*t*, 230–234, 234*t*, 244–248, 252–255, 264, 265, 269
  durability, 51, 197, 230, 255
  field resistance, 186, 230, 244–247
  horizontal, 128, 129
  induced resistance, 7*t*, 51, 52, 53, 244, 247
  management, 41
  major gene, 51, 222, 243
  non-specific, 39, 246
  pyramiding, 230, 247
  quantitative resistance, 51, 230, 247, 248, 255
  race specific, 41, 45, 50–52, 169, 228, 245, 254, 255
  R-gene, 228, 230
Resistance to disease
  Ascochyta, 217, 221, 222
  Alternaria solani, 228–230, 229*t*, 233, 244
  brown rust (*Puccinia triticina*), 42
  brown rot (*Molinia fructigena*), 48
  common bunt (*Tilletia tritici* syn. *T. caries*), 47, 169
  crown rust, 83
  downy mildew (*Pseudoperonospora cubensis*), 116
  downy mildew (*Plasmopara viticola*), 48
  downy mildew (*Peronospora destructor*), 264, 265
  early blight, *see Alternaria solani* (Resistance to disease)
  fusarium, 49, 221, 244, 264
  late blight (*Phytophthora infestans*), 40–42, 50, 227, 228, 230, 232–234, 243–247
  leaf rust, 51
  loose smut (*Ustilago tritici*), 47
  Mycosphaerella spp., 49, 256
  Phoma medicaginis, 49
  powdery mildew (*Blumeria graminis*), 53
  Rhizoctonia solani, 49, 229*t*, 230, 231, 233
  Scab (*Venturia inaequalis*), 48
  Sclerotinia, 49
  Seed borne, 47, 55, 168, 169, 231, 245
  Silver scurf (*Helminthosporium solani*), 231, 233
  Sooty blotch (apple), 48
  Stripe rust, 41, 83
  Xanthomonas oryza pv.oryzicola (rice bacterial leaf streak), 52
  Xanthomonas campestris, 245
  Yellow (stripe) rust (*Puccinia striiformis*), 41, 42, 50, 52, 147
Resistance to pests, 7*t*, 39–60, 86, 89, 125, 126, 147, 192, 194–198, 216, 227, 228, 240, 243, 244, 255, 256, 264
  aphids, 45, 46, 48, 205, 216, 217, 221–223, 256
  colorado potato beetle, 229*t*, 231
  corn borer, 126
  nematodes, 46, 54, 203, 217, 228, 243, 244, 269
  sitona, 221–223
Rice, 21, 22, 24, 28, 29, 32, 45, 52, 68, 83, 84, 117, 191–201
  adaptability of selection, 199*t*
  allelopathic, 68
  bacterial leaf streak (*Xanthomonas oryza pv.oryzicola*), 52
  blast, 45
  bulk breeding, 196, 197
  emasculation by farmer-breeder, 193, 193*f*, 195
  glutinous, 83, 199*t*, 200
  MASIPAG, 192–201
  mixtures, 84
  organic production in Philippines, 201*b*
  pedigree method, 196
  selection, panicle, 197*f*
  traditional rice variety (TRV), 28, 191, 192, 198, 199*t*, 200
  variety traits, 195, 196, 200

Rice (*Continued*)
  variety traits, farmer preferred, 199*t*
  yield, organic vs. conventional, 200*t*
Risks, environmental, 125–127, 129
Risks, health, 126, 149
Robust(ness), 6, 69, 115, 128, 184, 208, 241, 261, 264
Root
  competition, 68, 72
  density, 17, 18
  depth, 17, 19, 21, 47
  distribution, 17–19
  exploitation of inter-row soil, 17, 18
  exudates, 25, 26, 30, 205, 243
  functioning, 19
  growth, 17, 18, 19, 21, 26, 30, 63, 229*t*, 244, 265
    capacity, 2, 18
    duration, 18
    early root growth, 17, 18, 68
    inter-row, 2
    rate, 17, 18
    vigorous, 12, 68, 114, 117, 243
  improvement, 265, 271
  nodulating rhizobia, 205, 206*f*, 207
  penetration ability, 19
  pulling resistance, 19
Root architecture, root morphology, 7*t*, 15, 17, 18, 21, 24, 30, 265, 266, 268, 269
  adventitious-root formation, 17
  apical, 26
  axial root length, 17
  basal-root gravitropism, 17
  dauciform, 23
  genetic variation, 17–19, 21
  hair density, 18, 21, 265
  hair length, 18, 30
  horizontal root growth, 18
  lateral branching, 17, 253, 265, 268
  lateral root length, 17
  length, 19, 21, 232, 266, 268
  primary roots, 17
  root angling, 18
  seminal, 21
  taproot length, 17, 114
  tip, 26
  vertical, 18
  vigorous, 12, 68, 114, 117, 243
Root sampling/research methods, 19
  core break method, 19
  chelator-buffered nutrient solution, 26, 27
  electrical capacitance measurement, 19
  growth containers, 27
  greenhouse pots, 26, 27
  herbicide placement in deeper layers, 19
  high throughput phenotyping, 21
  hydroponic system, 21, 26
  Isotope labeled nutrients, 19, 21
  large scale screening, 21
  minirhizotron, 17, 20*f*
  Petri dish, 21
  root pouches, 21
  root pulling, 19
  root modeling, 31
  soil sampling, 19
  seedling test, 21
  transparent tubes, 21
  trench wall technique, 19
  washing roots, 19
Root-shoot ratio, 21
Rotation, crop, 5*t*, 15, 16, 18, 30, 32, 40, 42, 43, 45–47, 49, 53, 100, 117, 130, 176, 186, 203–205, 208, 216, 223, 228, 231, 244, 254
Rye, 8, 22, 26, 62, 68
Rye grass, 17, 68

Salt tolerance, 30, 53, 103*t*, 199*t*, 200
Secondary metabolites, 7*t*, 46, 50, 51, 116
  anthocyanin, 143, 186
  carotenoid, 186, 239, 244
  disease suppression, 46
  flavonoids, 6, 23, 239, 244
  glucosinolates, 6, 254, 256, 257
  phenolics, 25, 239
  provitamin-A, 239
  vitamin C, 239
Seed-borne diseases, 46, 47, 55, 168, 169, 231, 245
Seed embryo size, 65
Seedling growth rate, 65
Seed production, 4, 6, 7*t*, 8, 10, 11, 18, 47, 49, 104*t*, 108, 110, 125, 128, 130, 136, 139, 140, 143, 147, 148, 151, 156, 177*f*, 184, 187, 204, 219, 221, 241, 244, 245, 258, 263
Seed regulation, *see* Regulation
Seed size, 65, 210, 216
Seed system, 87, 88, 104*t*, 105, 110, 155, 156, 200
Seed testing, 140, 145, 148
Seed treatment, 5*t*, 7*t*, 27, 47, 125, 169, 183
Seed quality (control), 47, 92, 142, 144, 145, 147, 149, 203, 204, 208, 223
Selection environment, 27, 29, 90, 247, 268
Selection methods, strategies, 9, 11, 15–38, 40, 64, 69, 70, 72, 89, 90, 132, 140, 167, 168
  association mapping, 27
  bulk breeding, 90, 165–167
  decentralized, 31, 87, 90, 92, 99–123, 155
  direct(ed), 15, 21, 28–32, 50, 53, 69, 89, 90–91, 167, 168, 179, 180, 208
  ear-to-row, 89, 164, 165*f*
  evolutionary (population) breeding, 77–98, 164, 167, 168, 171
  hydroponics, 5*t*, 21, 26, 69
  indirect, 15, 19, 28–32, 53, 69, 133*t*, 167, 179, 180, 209
  mapping population (studies), 21, 22, 27, 255
  mass selection, 87, 89, 91–94, 116*t*, 133*t*, 140, 163–165, 218, 253, 258, 260

(molecular) marker assisted selection, 21, 24, 27, 30, 31, 80, 81, 94, 132, 183, 188, 207, 230, 248
  natural selection, 80, 86–92, 109, 164, 165, 167, 168, 171, 211
  pedigree, 86, 133*t*, 140, 143, 196, 211
  recombinant inbred lines, 29, 182
  recurrent selection, 87, 89, 93, 94, 114, 183, 209, 253
  regional selection/breeding, 8, 93, 94, 131, 132, 149, 192, 220, 241, 244, 260
  single seed descent, 211
  site-specific, 247*t*
  Quantitative Trait Loci (QTL), 23, 169, 207, 230, 248, 255, 266
Self incompatibility, 233, 252, 257
Selection intensity, 29, 180, 182
Self-organizing life, 127, 129, 131
Self-regulatory mechanisms (ability), 79, 128, 147
Simple sequence repeat (SSR) markers, 94
Soil
  carbon (C), 15, 22, 30, 266
  crop residue, 15, 22, 68, 254
  fertility, 5*t*, 15, 18, 28, 30, 49, 69, 200, 265
  microbial, biomass/activity, 15, 22, 26, 44*t*, 46, 49, 52, 53, 186, 247
  (N) mineralization, 5, 7*t*, 16, 31, 243, 265
  nitrogen, N, *see* Nitrogen
  nodules, 23, 206, 206*f*, 207
  organic matter, 4, 5, 15, 16, 26, 44, 47, 49, 170, 176, 183, 191, 243, 244, 247, 253
  N_2 fixation, 23
  tropical, 18
  potassium (K), 18
Soil-borne diseases, 40, 42, 49, 53, 221, 223, 243, 269
Soil-borne microbiota
  Arbuscular mycorrhizal fungi (AMF), 24, 25, 28, 119, 169, 232, 243, 244, 247, 253, 265–271
  Azospirillum, 169
  ectomycorrhizal, 24
  ericoid mycorrhizal, 24
  associative bacteria, 23
  associative diazotrophs, 23
  microorganisms, 22, 23, 25, 26, 52, 53, 169, 247
  microbial communities, 24, 25, 46
  microbial interaction, 15, 247
  mycorrhizae, 6, 7, 22, 46
  mycorrhizal,
    colonization, 24, 243, 268, 269
    associations, 24, 243, 247, 266
    competence, 243
    responsiveness, 24, 267–269
    symbiosis, 268–271
  nodules, 23, 206, 206*f*, 207
  non-mycorrhizal plants, 267
  plant growth promoting rhizosphere organisms (PGPR), 25
  rhizobium, 223
Sorghum, 19, 22, 61, 111, 112*f*, 113, 204, 218
Soybean, 203–214
  binary mixtures, 85
  food grade quality, 203, 207, 209, 210
  genetically diverse genotypes, 85
  genetic diversity, 210, 211
  herbicide resistance, 154
  iron (Fe) efficiency, 27
  mycorrhizae, 208
  nitrogen fixation, 205
  nodule formation, 205–207, 206*f*
  oil quality, 208–210
  protein content, 185, 209
  weed competitiveness, 61, 70, 71, 204, 205
  yield (stability), 207, 208
Spelt, 171
Squash (*Cucurbita pepo*), 114–116, 140, 149
Standards, organic plant breeding, *see* Organic
Stress
  abiotic, 22, 30, 31, 44*t*, 79, 83, 89, 94, 102, 109, 183, 186, 220, 221, 242, 264
  biotic, 22, 30, 31, 43, 48, 50, 79, 83, 89, 94, 109, 167, 242, 264
  drought, 18, 50, 206, 221
  nutrient deficiency, 26, 29, 30, 169, 222, 228, 230
Sulfur uptake efficiency, 269
Sugar beet (*Beta vulgaris*), 24
Sunflower (*Helianthus annuus*), 140, 178, 179*f*, 179
Synthetic populations, 78, 89, 111
  definition, 78*b*

Taste, 7*t*, 48, 51, 88, 117, 171, 210, 215, 229*t*, 244
Tillering ability (vigor, activity), 64*t*, 65, 67, 71, 167, 182, 194, 197, 198, 199*t*
Tomato, 239–250
  bacterial speck, 245
  bacterial spot, 245
  carotenoid, 239, 244
  clavibacter (*Clavibacter michiganense*), 245
  early blight (*Alternaria solani*), 244
  fusarium (*Fusarium oxysporum f.s. lycopersici*), 244
  flavonoid, 244
  flavor, 244
  heirloom varieties, 149, 243, 244
  hybridization, manual, 242*f*
  verticilium (*Verticillium dahliae*), 244
  late blight (*Phytophthora infestans*), 243–247
  leaf mold (*Cladosporium fulvum*), 244
  mycorrhizae, 243, 244, 253
  nematodes, 243, 244
  outdoor performance, 246, 246*f*
  outcrossing, 241*f*
  tomato mosaic virus (ToMV), 244
  tomato spotted wilt virus (TSMV), 244
  crop-weed interaction, 61
Topcross breeding, 217*f*, 218

Traceability, 92, 143
Traditional varieties, 10, 94, 105, 149
Transgenic contamination, 175
Triticale, 62, 68

UPOV, 86, 92, 126, 151, 153

Values, intrinsic, 127–129
Values, intrinsic, definition, 128b
Value and cultivation use (VCU), 7t, 92, 118, 143, 144, 146–149
Variety (*see also* Cultivar) traits, testing, 5, 6, 69, 70, 84, 87, 92, 110, 139–160, 233, 264
Variety, definition, 78b
Vigor, *see* under Crop Growth
Volatiles, 45, 50

Weed species
  Abutilon theophrasti, 205
  Agropyron repens, 71
  Alopecurus myosuroides, 62, 63, 205
  Allelopathy, 64t, 64, 68, 205
  Anagallis arvensis, 71
  Apera spica-venti, 68, 71
  Avena fatua, 63, 68, 70
  Aphanes arvensis, 70
  Brassica juncea, 63
  Bromus tectorum, 62
  Chenopodium album, 70
  Cirsium arvense, 71
  Galium aparine, 71
  Helminthia echioides, 205
  Lolium multiflorum, 63
  Lolium rigidum, 62, 63
  Matricaria recutita, 71
  Polygonum convolvulus, 70, 71
  Ranunculus sardous, 71
  Setaria italica, 205
  Sorghum halepense, 204
  Stellaria media, 71
  Vicia hirsuta, 71
  Xanthium pensylvanicum, 204
Weed
  tolerance to shading 70, 71
  competitiveness, suppression, 6, 61–76, 64t, 83, 117, 125, 163, 168, 176, 180, 204, 205, 220, 222

Wild plants/relatives/species, 26, 44, 46, 81, 108, 109t, 126, 128b, 132, 170, 216, 230–233, 233t, 239, 240
Wheat, 10, 28, 80, 90, 163–174
  Allelopathy, 68
  brown rust (*Puccinia triticina*), 42
  bulk selection, 164, 166f, 168
  common bunt (*Tilletia tritici* syn. *T. caries*), 169
  composite cross populations, *see* Composite cross populations
  durum, 117–119, 154
  ear to row progeny testing, 165f
  dwarfing genes, 19, 62
  end-use quality, 169, 170
    bread-making quality, 170
    gluten strength, 170
    grain protein content, 170
    nutritional value, 170
  evolutionary breeding, *see* Evolutionary breeding
  eye-spot (*Oculimacula spp.*), 42
  heirloom varieties, 81, 82f, 141
  leaf rust, 51
  mass selection, 163, 164
  mixtures, 83, 85, 86, 88f
  multilines, 83
  mycorrhizae, 25, 169
  nitrogen (nutrient)-efficiency, 22, 23, 29, 31, 168, 169, 171
  nutritional value, 163, 170
  phosphorus-efficient, 19, 23, 24
  seed-borne diseases, 169
  selection in randomized complete block design, 164, 164f
  take-all (*Gaeumannomyces graminis*), 42
  tillering capacity, 167
  tolerance to mechanical weed control, 168
  weed competitiveness, 62, 63, 64t, 67, 68, 70, 163, 167–169, 171
  yellow (stripe) rust (*Puccinia striiformis*), 41, 42, 52
  Zn-efficiency, 24

Yield potential, 17, 51, 64t, 144, 207, 257
Yield stability, 6, 31, 89, 120, 125, 167, 207, 208, 219, 220, 221, 228, 234, 244, 267, 268

Zinc (Zn), 22, 24–27, 170, 270
Zucchini, *see* Squash (*Cucurbita pepo*)

Keep up with critical fields

Would you like to receive up-to-date information on our books, journals and databases in the areas that interest you, direct to your mailbox?

Join the **Wiley e-mail service** - a convenient way to receive updates and exclusive discount offers on products from us.

Simply visit **www.wiley.com/email** and register online

We won't bombard you with emails and we'll only email you with information that's relevant to you. We will ALWAYS respect your e-mail privacy and NEVER sell, rent, or exchange your e-mail address to any outside company. Full details on our privacy policy can be found online.

www.wiley.com/email